Footprints in Micrometeorology and Ecology

Monique Y. Leclerc · Thomas Foken

Footprints in Micrometeorology and Ecology

With Contributions by M. J. Savage and M. Göckede

Springer

Authors
Monique Y. Leclerc
Laboratory of Environmental Physics
University of Georgia
Griffin, GA
USA

Thomas Foken
Abteilung Mikrometeorologie
Universität Bayreuth
Bayreuth
Germany

With Contributions by
Michael J. Savage
School of Agricultural, Earth and Environmental Sciences
University of KwaZulu-Natal
Pietermaritzburg
Republic of South Africa

Mathias Göckede
Department for Biogeochemical Systems
Max-Planck-Institute for Biogeochemistry
Jena
Germany

ISBN 978-3-642-54544-3 ISBN 978-3-642-54545-0 (eBook)
DOI 10.1007/978-3-642-54545-0
Springer Heidelberg New York Dordrecht London

Library of Congress Control Number: 2014939401

"Footprints" drawn by Gesa Foken, Leipzig, Germany
www.foken-gesa.de

Printed on acid-free paper

Springer is part of Springer Science+Business Media (www.springer.com)

Preface

Ever since the humble beginnings of micrometeorology over 50 years ago, micrometeorologists have pondered over ways in which to best measure surface-atmosphere exchange at non-ideal sites. In setting up their instrumentation to ensure the highest integrity of data quality, micrometeorologists went to great lengths seeking to eliminate upwind obstacles suspected to adversely degrade the quality of their dataset. Constantly present in the mind of these early pioneers, the problematic determination of the range of the upwind coverage covered by an atmospheric measurement was an ever present concern on their mind. Pasquill however, in his groundbreaking work of 1961 developed a series of empirical guidelines aimed at identifying the source area.

While *a priori* this may appear to be a moot point for non-micrometeorologists, a sensor in the atmosphere does not measure the properties at the point where the sensor is located. Indeed, the sensor measurement reflects the scalar and dynamic properties of eddies embedded in the flow advected past an atmospheric sensor, while an atmospheric flux represents the correlation of the properties of eddies going past the flux system and their vertical wind velocity. Both concentration and flux measurements are the product of a spatial average over the path length of the sensor/flux system and a temporal average dictated by the measurement period (typically 30-min period).

Since the inception of micrometeorological research up until the 1980s, experimentalists limited the scope of their measurements to smooth, flat terrain covering extending homogeneous areas. This state-of-affairs was then to undergo a profound transformation in the mid-1980s with the arrival of a fortuitous combination of cheap computers, the production of affordable data acquisition systems and data loggers and, above all, with the arrival of affordable, fast response sonic anemometers/thermometers that the common use as we know it today surfaced. These modern measurement systems then opened the door to a vast and rapid expansion of the field of micrometeorology, leading experimentalists to move into forays of considerable challenge: the scientific community relaxed their restrictions of limiting their efforts to quasi-idealized terrain to then shift their focus to frequently encountered terrain or over surfaces presenting much need in assessing atmosphere-exchange. It is then that, for the first time, measurements over tall forested canopies and over mosaic-like terrain grew to become the norm rather than the exception.

Furthermore, these techniques were soon adopted by scientists outside the meteorological community: the deceptive ease of use of the eddy-covariance technique opened the door to a myriad of experiments in the field of ecology and became extensively used at difficult sites. The footprint concept was developed in an attempt to provide leadership in this rapidly expanding field.

Why are we writing this book, the reader might well ask: With the recent and rapid proliferation of papers in the field of eddy-covariance essential, either pertaining to or resorting to the use of eddy-covariance, there has yet to be a full comprehensive 'manual' summarizing from the ground up the plethora of ways estimating footprints. We have wanted to provide a comprehensive yet easy-to-use guide to those unfamiliar with the concept. We have thus included the rudiments of micrometeorology along with measurement methods. Furthermore, the present book also offers a fresh insight into practical applications like tall tower measurements, wind power investigations, and air pollution issues.

The idea of writing this 'field manual' was spurred by the preparation of the special issue of *Agricultural and Forest Meteorology* in 2004 edited by Timo Vesala, Ullar Rannik, and colleagues including but not limited to John Finnigan, Dennis Baldocchi, Xuhui Lee, and many others; this special issue, along with the recent productions of three overviews by Vesala et al. (2008, 2010) and Rannik et al. (2012) into non-traditional readership further demonstrated the relevance of the present endeavor. This manual on footprints should provide a solid well-rounded foundation establishing the basis for robust flux experiments (tower positioning, height of measurements, difficulties with upstream inhomogeneous surfaces, and related errors) and their subsequent interpretation especially when used with the *Handbook on Micrometeorology* (Lee et al. 2004) and the recently published book on *Eddy-Covariance* (Aubinet et al. 2012).

The reader should forgive a personal note of Chap. 1. These views have formed after more than 25 years in the field. Despite this, one point should be emphasized: Writing this book was only possible thanks to the wonderful cooperation of many scientists in common projects and in the preparation of joint papers, overview papers, and book chapters. We want to thank them all; the list is extremely long as the references sections will attest.

We are particularly grateful to M. J. Savage and M. Göckede for their unwavering support, mainly for Chaps. 1, 3, 6 and 8. One of the authors (M. Y. Leclerc) wishes to sincerely express her appreciation and gratitude to Prof. Joon Kim of Seoul National University for his hospitality during a portion of the book writing. Professor Kim provided the conditions needed to foster useful discussions, concentration, and solitude. The preparation of the book was supported by the states of Georgia and Bavaria mainly by the funding of technological cooperation (BayCaTEC-Georgia) to whom we are most indebted.

Griffin, Bayreuth, January 2014

Monique Y. Leclerc
Thomas Foken

References

Aubinet M, Vesala T, Papale D (2012) Eddy covariance: apractical guide to measurement and data analysis. Springer, Dordrecht, p 438

Lee X, Massman WJ, Law B (eds) (2004) Handbook of micrometeorology: a guide for surface flux measurement and analysis. Kluwer, Dordrecht, p 250

Pasquill F (1961) Estimation of dispersion of windborne material. Meteorol Mag 90:33–49

Rannik Ü, Sogachev A, Foken T, Göckede M, Kljun N, Leclerc MY, Vesala T (2012) Footprint analysis. In: Aubinet M, et al (eds) Eddy covariance: A practical guide to measurement and data analysis. Springer, Berlin, pp 211–261

Vesala T, Kljun N, Rannik U, Rinne J, Sogachev A, Markkanen T, Sabelfeld K, Foken T, Leclerc MY (2008) Flux and concentration footprint modelling: state of the art. Environ Pollut 152:653–666

Vesala T, Kljun N, Rannik Ü, Rinne J, et al (2010) Flux and concentration footprint modelling. In: Hanrahan G (ed) Modelling of pollutants in complex environmental systems, vol II. ILM Publications, St. Albans, Glendale, pp 339–355

Contents

1 History and Definition 1
1.1 Micrometeorological Measurements 1
1.2 Towards the Footprint Definition 4
1.3 Footprint Modeling 7
1.4 Validation of Footprint Models 12
References 15

2 Surface-Layer Properties and Parameterizations 21
2.1 Atmospheric Boundary Layer and Scales 21
2.2 Turbulence Parameterization 25
2.2.1 Flux-Gradient Similarity 25
2.2.2 Profile Functions Above the Canopy 34
2.2.3 Profile Functions in the Canopy 35
2.2.4 Roughness Sublayer 37
2.2.5 Power Laws 39
2.2.6 Dispersion Profiles 40
2.2.7 Relevance of Profile Parameterizations in Footprint Models 43
2.3 Internal Boundary Layers 44
2.3.1 Mechanical Internal Boundary Layer 46
2.3.2 Thermal Internal Boundary Layer 48
2.3.3 Blending Height Concept 49
2.4 Modeling Concepts 50
2.4.1 Diffusion Model 50
2.4.2 Lagrangian Model 52
2.4.3 Higher-Order Closure Model 55
2.4.4 Large-Eddy Simulation Model 56
2.5 Averaging Surface Characteristics 57
2.5.1 Averaging Using Effective Parameters 58
2.5.2 Flux-Averaging Models in Inhomogeneous Terrain 59
References 62

3 Classification of Footprint Models 71
3.1 Analytical Footprint Models 72
3.1.1 The Schuepp et al. (1990) Approach 73
3.1.2 The Schmid and Oke (1990) approach. 74
3.1.3 The Family of Horst and Weil's (1992) Analytical Solution 74
3.1.4 Analytical Solutions Based on Lagrangian Models 81
3.2 Lagrangian Simulations 84
3.2.1 The Leclerc and Thurtell (1990) Approach 88
3.2.2 The Sabelfeld-Rannik Approach 89
3.2.3 The Kljun et al. (2002) 3D Backward Lagrangian Footprint Model 90
3.3 Higher-Order Closure Footprint Models 91
3.4 Large-Eddy Simulation Models. 92
3.5 Hybrid Footprint Models 93
3.5.1 LES-Driven Lagrangian Stochastic Models 94
3.5.2 LES-Embedded Lagrangian Stochastic Models: The Steinfeld et al. (2008) Approach. 95
3.5.3 Higher-Order Closure-Driven Lagrangian Simulation . . . 96
References 98

4 Footprint Studies 103
4.1 Footprint in the Atmospheric Boundary Layer 103
4.1.1 Tall Tower Footprints 104
4.1.2 The Influence of Coriolis Forces on Footprint 110
4.1.3 Flux Footprints in the Convective Boundary Layer 112
4.1.4 Footprint in the Roughness Sub-Layer of Plant Canopies 114
4.2 In-Canopy Footprints 115
4.3 Flux Footprint in Canopy Over Hills 120
4.4 Influence of Contrasting Adjoining Surfaces on Footprints 120
4.4.1 Role of Contrasting Thermal Land Surfaces on Fluxes and Footprints 120
4.4.2 Role of Clearcuts on Forest Fluxes/Footprints 121
4.4.3 Footprints in the Presence of a Transition from the Forest Leading Edge 126
4.5 Flux Footprints Over Complex Topography in Forests 131
4.6 Emissions of Odor and Reactive Trace Gas Fluxes Using the Flux Footprint Method 135
4.7 Footprints in Urban Areas 137
References 140

5 Model Validation 145
5.1 Model Validation Against Other Models 147
5.2 Model Validation and Comparison Against Experimental Data 151
5.3 Model Validation with Natural Tracers 153
5.4 Classification of the Comparison Results 153
References 157

6 Land Surface: Coupled Footprints 159
6.1 Grid Schema of Surface Characteristics 159
6.2 Determination of Surface Characteristics 162
6.2.1 Roughness Length 162
6.2.2 Remote-Sensing Data 164
6.3 Coupling Footprint Results with Surface Information 169
References 169

7 Application of Footprint Models to Different Measurement Techniques 171
7.1 Profile Technique 171
7.1.1 Profile Technique with Three and More Measuring Levels 172
7.1.2 Profile Technique with Two Measuring Levels 174
7.1.3 Accuracy and Footprint Issues for Profile Technique 176
7.2 Eddy-Covariance Technique 181
7.2.1 Basics of the Eddy-Covariance Method 181
7.2.2 1D Eddy-Covariance Method 183
7.2.3 Generalized Eddy-Covariance Method (3D) 184
7.2.4 Quality Control of Eddy-Covariance Data 186
7.3 Scintillometer Technique 188
7.4 Airborne Measurement Technique 190
References 194

8 Practical Applications of Footprint Techniques 199
8.1 Selection of Flux Measurement Sites 199
8.2 Interpretation of Flux Data 205
8.2.1 Footprint Climatology 206
8.2.2 Covering the Area of Interest 207
8.2.3 Footprint-Dependent Data Quality Control 209
8.3 Upscaling Point Measurements Using Footprint Models 213
8.4 Additional Practical Application 214
8.4.1 Air Pollution Application and Trace Gas Fluxes 214
8.4.2 Wind-Energy Application 215

8.5 Easily Applicable Footprint Models . . . 217
8.6 Limits of Footprint Application . . . 218
References . . . 220

9 Looking Forward to the Next Generation of Footprint Models . . . 225
References . . . 228

Glossary . . . 231

About the Authors . . . 235

Index . . . 237

Abbreviations

AmeriFlux	Research Network of the Americas
ASTER	Advanced Spaceborne Thermal Emission and Reflection radiometer
BOREAS	Experiment in the boreal forests of Northern America
CCGD	Carbon Cycle Greenhouse Gases Group
CCSM	Community Climate System Model
CHIOTTO	Continuous high-precision tall tower observations of greenhouse gases
CPU	Central processing unit
DBSAS	Displaced beam small aperture scintillometer
ERF	Environmental response function
ESDU	Engineering Sciences Data Unit
ESRL	NOAA Earth System Research Laboratory
EUROFLUX	European part of FLUXNET
EVI	Enhanced Vegetation Index
FACE	Free-Air Carbon Dioxide Enrichment
FIFE	First ISLSCP Field Experiment
FLUXNET	Flux network
FSAM	Flux footprint model by Schmid (1994, 1997)
GEWEX	Global Energy and Water Cycle Experiment
GIS	Geographical Information System
GMD	Geoscientific Model Development
HAPEX-MOBILHY	Hydrological Atmospheric Pilot Experiment—Modelisation du BiLan Hydrique
HERC	Helsinki Environment Research Centre
HYSPLIT4	Hybrid Single Particle Lagrangian Integrated Trajectory model by Draxler (1997)
IBL	Internal boundary layer
ID	Identification document
IKONOS	Commercial earth observation satellite
INSTAAR	Institute of Arctic and Alpine Research
INTAS	International association for the promotion of cooperation with scientists from the independent states of the former Soviet Union

IR	Infrared
ISLSCP	International Satellite Land Surface Climatology Project
ITCE	International Turbulence Comparison Experiment
KUREX	Kursk experiment
Landsat	Programme for acquisition of satellite imagery of Earth
LAS	Large aperture scintillometer
LES	Large Eddy Simulation
LITFASS	Lindenberg Inhomogeneous Terrain Fluxes between Atmosphere and Surface: a long-term Study
LNF	Localized near-field theory by Raupach (1989)
LPDM-B	Lagrangian footprint model by Kljun et al. (2002)
LS	Lagrangian simulation
LST	Land surface temperature
MAGS	Mackenzie Area GEWEX Study
MODIS	Moderate-resolution Imaging Spectroradiometer
NASA	National Aeronautics and Space Administration
NCAR	National Center for Atmospheric Research
NCEP	National Centers for Environmental Prediction
NDVI	Normalized Difference Vegetation Index
NEL	New equilibrium layer
NIR	Near IR
NOAA	National Oceanic and Atmospheric Administration
NOPEX	Northern Hemisphere Climate Processes *Experiment*
PALM	LES model according to Raasch and Schröter (2001)
PFT	Perfluorocarbon tracer
SAM	Footprint model according to Schmid and Oke (1990)
SCADIS	1.5-order closure model by Sogachev and Lloyd (2004)
SGS	Sub-grid scale
SRNL	Savannah River National Laboratory
STILT	Model for tall tower data according to Lin et al. (2004)
SVAT	Surface-Vegetation-Atmosphere-Transfer
TK	Turbulence knight
TKE	Turbulent kinetic energy
VOC	Volatile organic compound

Symbols

a	Footprint fraction related to the target area
a_i	Normalized footprint
Bo	Bowen ratio
C_n^2	Refraction structure-function parameter ($m^{-2/3}$)
C_T^2	Temperature structure-function parameter ($K\ m^{-2/3}$)
c	Concentration (general) (*)
c_*	Concentration scale (*)
c'	Fluctuation of the concentration (general) (*)
c_p	Specific heat at constant pressure ($J\ kg^{-1}\ K^{-1}$)
C	Weighting factor for aircraft data
C_0	Kolmogorov constant
$C_{k,j}$	Weighting function
d	Displacement height (m)
D_{ext}	Extended fetch (m)
D_{min}	Minimal fetch (m)
e	Water vapor pressure (hPa)
E	Power spectra (general) (*)
E	Turbulent kinetic energy (*)
EVI	Enhanced vegetation index
f	Frequency (s^{-1})
f	Coriolis parameter (s^{-1})
f	Footprint function*
$\bar{f}^y$	Crosswind-integrated flux footprint (*)
f^P	Footprint function of the effect level (*)
$F(x)$	Distribution density function
F	Flux (general) (*)
F_T	Total flux (*)
F_e	Flux in an elevated level (*)
F_{obs}	Observed flux (*)
F_{tar}	Flux of the target area (*)

F_{surf}	Flux of a surface (not target area) (*)
g	Acceleration due to gravity (m s^{-2})
$G(y)$	Gaussian distribution function
h	Canopy height (m)
h_c	Canopy height (m)
$H(z)$	Gaussian distribution function
H/L	Hill slope (H: hill height, L: length scale)
K	Turbulent diffusion coefficient (general) (m^2 s^{-1})
K_c	Turbulent diffusion coefficient of mass transfer (m^2 s^{-1})
K_H	Turbulent diffusion coefficient of sensible heat (m^2 s^{-1})
K_m	Turbulent diffusion coefficient of momentum (m^2 s^{-1})
$K_{x,y,z}$	Component of the turbulent diffusion Coefficient (m^2 s^{-1})
K_χ	Turbulent diffusion coefficient of mass transfer (m^2 s^{-1})
l_b	Blending height (m)
L	Obukhov length (m)
L_c	Canopy drag scale (m)
L_R	Resolution of a matrix size (m)
L_v	Obukhov length with buoyancy flux (m)
L_s	Shear scale (m)
LAI	Leaf area index (m^2 m^{-2})
$NDVI$	Normalized difference vegetation index
O_p	Photosynthetically active radiation (μmol m^{-2} s^{-1})
p	Air pressure (hPa)
p	Exponent of the power law
P	Portion of the total integrated footprint effect (*)
P	Path length of a scintillometer (m)
Pr_t	Turbulent Prandtl number
q	Specific humidity (kg kg^{-1})
q'	Fluctuation of specific humidity (kg kg^{-1})
q_*	Scale of the specific humidity (kg kg^{-1})
Q	Source density (general) (*)
Q_χ	Dry deposition (kg m^{-2} s^{-1})
Q_E	Latent heat flux (W m^{-2})
Q_G	Ground heat flux (W m^{-2})
Q_H	Sensible heat flux (W m^{-2})
Q_{Hv}	Buoyancy flux (W m^{-2})
Q_N	Normalized flux (m s^{-1})
Q_s^*	Net radiation (W m^{-2})
Q_η	Source density of the η parameter (*)
Q_0	Source density at the surface (*)
Q_χ	Concentration or mixing ratio flux (*)
r	Shape parameter
R	Number of pixel in a fly segment

R_L	Gas constant of dry air (J kg^{-1} K^{-1})
RN	Parameter of relative non-stationarity
Ri_f	Flux Richardson number
Ri_g	Gradient richardson number
Ri_b	Bulk Richardson number
Ri_c	Critical Richardson number
Sc_t	Turbulent Schmidt number
S_c	Concentration sources and sinks (*)
t	Time (s)
T	Temperature (K)
T'	Fluctuation of the temperature (K)
T_*	Temperature scale (K)
U	Mean wind speed (m s^{-1})
u	Longitudinal component of the wind velocity (m s^{-1})
u'	Fluctuation of the longitudinal component of the wind velocity (m s^{-1})
u_*	Friction velocity (m s^{-1})
v	Lateral component of the wind velocity (m s^{-1})
v'	Fluctuation of the lateral component of the wind velocity (m s^{-1})
w	Vertical component of the wind velocity (m s^{-1})
w'	Fluctuation of the vertical component of the wind velocity (m s^{-1})
w_*	Convective (Deardorff) velocity (m s^{-1})
W	Weighting factor for scintillometers
X	Dimensionless distance $X = w_* x/Uh$
x	Fetch (m)
x	Horizontal direction (length) (m)
x,	Measuring variable (general) (*)
x'	Fluctuation of a measuring variable (general) (*)
y	Horizontal direction (length, perpendicular to x) (m)
z	Height (general, geometric) (m)
z_i	Mixed-layer height (m)
z_m	Measuring height (m)
z_0	Roughness parameter, roughness height (m)
z_{0q}	Roughness height for water vapor pressure (m)
z_{0T}	Roughness height for temperature (m)
z'	Height (aerodynamic) (m)
$\overline{Z_H}$	Averaged building height (m)
δ	Depth of the internal boundary layer (m)
δ_T	Thickness of the thermal internal boundary layer (m)
$\delta\varphi$	Accuracy of the universal function
Δe	Water vapor pressure difference (hPa)
ΔT	Temperature difference (K)
Δu	Wind velocity difference (m s^{-1})
Δz	Height difference (m)

$\Delta\chi$	Concentration or mixing ratio difference (*)
$\Delta_{\chi,min}$	Minimal concentration or mixing ratio difference (*)
ε	Energy dissipation (m^2 s^{-3})
ζ	Dimensionless height z/L
η	Quantity being measured at a location (footprint) (*)
η	Kolmogorov's micro scale (m)
θ	Potential temperature (K)
θ_v	Virtual potential temperature (K)
κ	Von-Kármán constant
κ	Wave number (m^{-1})
λ	Heat of evaporation for water (J kg^{-1})
λ_F	Frontal areal index
λ_P	Plain area fraction
Λ_u	Eulerian turbulent length scale for the horizontal wind (m)
ν	Kinematic viscosity (m^2 s^{-1})
ρ	Air density (kg m^{-3})
ρ	Autocorrelation function (*)
ρ_d	Density of dry air (kg m^{-3})
σ_u	Standard deviation of the longitudinal wind component (m s^{-1})
σ_v	Standard deviation of the lateral wind component (m s^{-1})
σ_w	Standard deviation of the vertical wind component (m s^{-1})
σ_T	Standard deviation of the temperature (K)
τ	Shear stress (kg m^{-1} s^{-2})
τ	Lagrangian integral time scale (s)
τ_c	Scalar eddy diffusivity (*)
Π	Coefficient of Buckingham's Π-Theorem
φ	Geographical latitude (°)
φ_m	Universal function for momentum exchange
φ_H	Universal function for sensible heat flux
φ_E	Universal function for latent heat flux
φ_*	Correction function for the roughness sublayer
χ	Concentration or mixing ratio (general)
$\chi_{\min}$	Accuracy of the measurement system (*)
χ'	Fluctuation of the concentration or mixing ratio (general) (*)
χ_y	Crosswind integrated concentration or mixing ratio (*)
ψ	Weighting function for the land cover type
ψ	Concentration function of reactive particles (*)
ψ_H	Integral of the universal function for sensible heat
ψ_m	Integral of the universal function for momentum
ψ_{ijk}	Weighting function in Eq. 7.28
ω	Dissipation of the turbulent kinetic energy (*)
ω	Weight distribution function, isopleths
ω_P	Weight distribution function, isopleths for level P

Ω_p Footprint effect level (*)
Ω Angular velocity of the rotation of the Earth (s^{-1})

Indices

m Measurement
r Special position in coordinate system

Remark

* Dimension according to the use of the parameter

Chapter 1
History and Definition

This chapter describes the challenges and the history of micrometeorology. For sake of comprehensiveness, it also provides an overview of essential definitions that the reader might consider becoming familiar with before delving deeper into the present volume. Furthermore, this chapter provides the historical perspective of the evolution of a rapidly maturing field right up to the development of recent footprint tools used in research as in applications. It should be apparent to all that such an overview can only scratch the surface while some of the details will be described in the following chapters. It goes without saying that this overview is tinted by the experiences of the authors.

1.1 Micrometeorological Measurements

At the beginning of the last century, much progress was made in hydrodynamics beginning with the fundamental papers by Taylor (1915), Richardson (1920), and Prandtl (1925). The transition to micrometeorology was done by Schmidt (1925) in Vienna, who formulated the 'austausch coefficient' while in Munich, Geiger (1927) summarized microclimatological works in his famous book (still in print) 'The climate near the ground' (Geiger et al. 2009). A few years later in Leipzig, Lettau (1939) pioneered atmospheric turbulence investigations. Most experimental studies of that time were influenced by Albrecht, who wrote the first paper about the energy balance of the earth (Albrecht 1940). Those marked the beginning of micrometeorological studies seeking to measure and understand the energy exchange between the atmosphere and the earth surface, a field that flourished after the Second World War.

M. Y. Leclerc and T. Foken, *Footprints in Micrometeorology and Ecology*,
DOI: 10.1007/978-3-642-54545-0_1, © Springer-Verlag Berlin Heidelberg 2014

Therefore, first large field experiments were planned in quasi-ideal site conditions without large heterogeneities or obstacles. Examples include the famous O'Neill experiment in 1953 (Lettau and Davidson 1957) and several experiments at the Australian field sites like Kerang (Garratt and Hicks 1990), and the Tsimlijansk site in Russia. While the first experiments used mainly the profile approach in later experiments in the 60s, the eddy-covariance method rapidly grew in popularity. Above and beyond providing a means to provide a direct mass balance of scalar exchanged to/from a surface, it also enabled to determine universal functions of the Monin and Obukhov (1954) similarity theory and the turbulent Prandtl and Schmidt numbers.

This direct measurement method for turbulent fluxes, now known as the eddy–covariance method, was developed probably independently by Montgomery (1948), Swinbank (1951), and Obukhov (1951). This method only emerged after the development of the sonic anemometer for which the basic equations are given by Schotland (1955). After the development of a sonic thermometer (Barrett and Suomi 1949) during the O'Neill experiment in 1953 (Lettau and Davidson 1957), a vertical sonic anemometer with a 1-m path length (Suomi 1957) was already used. The design of today's anemometers was developed by Bovscheverov and Voronov (1960), and later by Kaimal and Businger (1963) and Mitsuta (1966). The phase-shift anemometers have now been replaced by running time anemometers with time measurements (Hanafusa et al. 1982). These anemometers produced by the Japanese company Kaijo-Denki were the first commercially available sonic anemometers. This history is discussed in greater detail by Moncrieff (2004).

These findings were the basis for many famous experiments (Table 1.1), including turbulence sensors intercomparison experiments along with experiments delving into the study of turbulent exchange processes (i.e. KANSAS 1968 experiment (Izumi 1971) which was the basis for the widely used universal function by Businger et al. (1971). The Minnesota experiment followed in 1973 to investigate the validity of the function (Kaimal and Wyngaard 1990). An important summary about the status of the knowledge of turbulent exchange processes between the atmosphere and the surface was given in 1973 at the Workshop on Micrometeorology (Haugen 1973). Following the workshop and inspired by a seminal paper by Elliott (1958), the transition of investigations away from homogeneous to heterogeneous surfaces was made: The arrival of studies demonstrating a step change in surface roughness and its related internal boundary layer concept marked an important development in modern micrometeorology (Busch and Panofsky 1968; Peterson 1969; Taylor 1969; Shir 1972).

Rare are measurements inside low vegetation. Most of our knowledge (Cionco 1978; Wilson et al. 1982), also applied to footprint analysis, is based on measurements made by Silversides (1974) using a split-film anemometer. Inside tall vegetation, such profiles were more often measured (see Chap. 2).

The extension to more complex surfaces first came through the FIFE experiment in the USA (Sellers et al. 1988) followed by similar experiments in France (HAPEX-MOBILHY, André et al. 1990) and in Russia KUREX-88 (Tsvang et al.

Table 1.1 Important micrometeorological experiments up to the beginning of the 80s according to Foken (2006) based on McBean et al. (1979), Garratt and Hicks (1990), and Foken (1990)

Year	Place	Surface	Type, name	References
1953	O'Neill, USA	Step	Boundary-layer experiment	Lettau and Davidson (1957)
1962	Kerang, Australia	Step	Surface-layer experiment	Swinbank and Dyer (1968)
1964	Hay, Australia	Step	Surface-layer experiment	
1965	Hanford, USA	Sage	Anemometer comparison	Businger et al. (1969)
1968	Kansas, USA	Step	Micrometeorological experiment, KANSAS 1968	Izumi (1971)
1968	Vancouver, Canada	Water	ITCE-1968	Miyake et al. (1971)
1970	Tsimlyansk, Russia	Step	ITCE-1970	Tsvang et al. (1973)
1973	Minnesota, USA	Harvested crop	Boundary-layer experiment Minnesota 1973	Readings et al. (1974)
1976	Conargo, Australia	Step	ITCE-1976	Dyer (1981); Dyer and Bradley (1982)
1981	Tsimlyansk, Russia	Step	ITCE-1961	Tsvang et al. (1985)

For experiments after 1980, see Foken (2008). ITCE: International Turbulence Comparison Experiment

1991). During these experiments, aircraft overpass were also included in these experiments raising further questions regarding the interpretation and incorporation of fluxes over different (adjoining) surfaces together to a common picture.

T. F. remembers that time: When P. Sellers visited in the KUREX-88 about 500 km South of Moscow we discussed together with L.R. Tsvang, J. Ross, J. Fazu, J. Zelený and others the problems of the heterogeneous surfaces and the limitations of the eddy-covariance method for these conditions, later on used as a data quality test method (Foken and Wichura 1996). We decided that many gaps must be filled to fully understand the processes. Zubkovskij and Sushko (1987) investigated the limits of the frozen turbulence hypothesis as a measure of how long a surface can influence the turbulence structure. Ross (1981) underlined the importance of the plant structure and the radiation distribution. Finally we decided to repeat an internal boundary layer experiment over typical agricultural fields in 1990 in Estonia (TARTEX-90, Foken et al. 1993) at the time when the former Soviet Union was dismantled and Germany was unified. This was unfortunately also the end of a successful cooperation spanning more than a ten-year period between East European groups (Foken and Bernhardt 1994).

At the end of the 80s, analytical and numerical solutions to diffusion equations proliferate in the literature for many source configurations, initial and boundary conditions and levels of idealization of diffusivity and velocity profiles (Calder 1952; Sutton 1953; Rao et al. 1974; Wilson et al. 1982; Gash 1986; Arya 1999). From these solutions, vertical scalar profiles obtained as a function of downwind distance became the basis used in footprint modeling.

1.2 Towards the Footprint Definition

The 80s marked a period in which tools aiming at improving the development of the interpretation of micrometeorological measurements. Before the advent of footprint models, other tools were used which approximated in some way the concept of the footprint. As already mentioned above, the internal boundary-layer concept was also used to define a necessary fetch for micrometeorological measurements. For more details, the reader is referred to Sect. 2.3.

In the 80s, Czech scientists made measurements on an 80-m-tower in the very complex mine area of Northern Bohemia. To assist with the interpretation of the dataset, they developed a so-called macro roughness (Zelený and Pretel 1986), which was something akin to a weighted standard deviation of the heterogeneities of the underlying surface. The number of grids was chosen using logarithmical distances. Foken and Zelený (1988) investigated different definitions of such a macro roughness and found that they are significantly correlated to different turbulence characteristics like normalized standard deviations of the wind components at different heights. This was similar to the dependence of turbulence characteristics on the footprint area presented by Foken and Leclerc (2004).

M.Y.L. remembers that time: The history of 'footprints' studies goes back to the late eighties when Peter Schuepp of McGill University visited M.Y, Leclerc at Utah State Univ. in February 1988 to see whether she could not model, using the Lagrangian stochastic simulation something both interesting and, at the time, something considered rather puzzling: The CO_2 flux uptake seen by the Canadian National Aeronautical Establishment's Twin-Otter aircraft as it passed over Ile Royale, an island located in Lake Superior, gave fluxes which peaked, not above the forested island itself, but rather downwind from it. That explicit connection of a source/sink to a point flux measurement was then coined 'footprint' in the first paper by Leclerc and Thurtell (1989). That paper was entitled 'Footprint Predictions of Scalar Fluxes and Concentration Profiles using a Markovian Analysis' presented at the American Meteorological Society at the 19th Conference of Agricultural and Forest Meteorology Conference in Charleston, South Carolina (March 7th–10th, 1989). Shortly after, in the refereed articles by Schuepp et al. (1990) and Leclerc and Thurtell (1990).

The two original companion papers, by Schuepp et al. (1990) and Leclerc and Thurtell (1990) respectively, were simultaneously written and meant to be presented as a paper series. Because of small delays in the figure preparation of the final draft of one of the papers, it was decided that the Schuepp et al. (1990) paper would be incorporated in the memory of Hans Panofsky's special issue of Boundary-Layer Meteorology, while the Leclerc and Thurtell (1990) would follow a couple of months later. The Schuepp et al. (1990) article, based on the compact analytical solution by Gash (1986), provided a quick and effective way to model footprints since the latter presented a simple method to provide a rough estimate of the sampling error which would result from an upwind step-change in evaporation rate in limited fetch conditions. It used Calder's (1952) approximation of a uniform wind field and neutral atmospheric stability. The Schuepp et al. (1990) and Leclerc and Thurtell (1990) papers explicitly provided a method to identify the portion of the flux contributed by different sources upwind, with the Schuepp et al. (1990) contribution allowing experimentalists to incorporate into signal processing routines the nearly instantaneous 'field-of-view' assessment of their measurements while the Leclerc and Thurtell (1990) study incorporated real wind profiles, the effect of atmospheric stability, and different surface roughnesses.

On the basis of these two original papers alone, the NASA FIFE field campaign (Sellers et al. 1988) was entirely redesigned using footprint predictions from these models as a tool to reconcile observations and measurements at different scales and across different towers and locations (Kanemasu et al. 1992). For the first time in micrometeorology, experimentalists could now plan upcoming experiments and intercompare measurements from different platforms: flux measurements from aircrafts flying at different altitudes could be intercompared with their respective fluxes over the *Konza prairie* (FIFE) while tower fluxes could be intercompared using a quantitative tool assessing the amount of upwind fetch contributed to the measured flux. Measurements taken at different scales, became, almost overnight, more easily discussed during their daily intercomparison sessions. The 'footprint' concept had then received its baptism by the micrometeorologists and had become well entrenched into micrometeorology. The Schuepp et al. (1990) paper provided a quick, effective, if crude, idea of the surface sensed by a flux platform while the Leclerc and Thurtell (1990) paper, laying out the Lagrangian simulation of particle trajectories in inhomogeneous turbulence, lent sophistication to the footprint concept, by expressing explicitly a more realistic wind profile, the atmospheric stability, and expanded this work to a wide range of surface roughnesses. Furthermore, it depicted the behavior of the footprint peak as a function of both surface roughness and stability and then showed the cumulative effect of adding upwind surface elements to the modeled fetch on flux results.

Nearly in parallel with the Schuepp-Leclerc-Thurtell's efforts, Tim Oke with graduate student Hans Peter Schmid had begun working on a related concept, that of the source area influencing measurements, an adaptation from Pasquill's early efforts (1972). They presented their results at the 8th Symposium on Turbulence and Diffusion, San Diego, CA. in 1988 (Schmid and Oke 1988) which led to Hans Peter Schmid's doctoral dissertation that year. Oke and Schmid defined the 'source area of an eddy-covariance measurement as the surface area containing heat sources and/or sinks influencing those air parcels carried past the sensor under given external conditions'. Schmid later changed the Oke and Schmid's source area term to the use of the term 'footprint', more in line with the original footprint papers. Schmid and Oke (1990) discussed the concept of a source area model (SAM) using a plume-diffusion model to estimate the source region. This concept, borrowed from Pasquill's work (1972) which traditionally applied to air pollution purposes (Taylor 1915; Schmid 1994). The subsequent paper by Schmid (1997) explores the matching of scales of observations and fluxes and defines criteria of representativeness of several distinct measurement methods (Schmid 1997, 2002; Schmid and Lloyd 1999).

Two years later, Horst and Weil (1992) published analytical solutions to the diffusion equation presented in a form describing the footprint. The original solution to the diffusion equation had been presented earlier by van Ulden (1978) and by Horst (1979). The Horst and Weil (1992) solution had the advantage that it provided more realism to existing analytical solutions to the advection-diffusion equation by providing a realistic wind profile and the effect of atmospheric stability in the solution. This constituted a significant step in the evolution of analytical footprint models. The following paper by the same authors (Horst and Weil 1992) brought subsequent refinement to their original paper. That article was based on the work of Horst and Weil (1992) with the concentration-source area model by Schmid and Oke (1990) extended to include conditions of stable thermal stratification and the model's solution improved.

Footprint definition: The early papers by Schuepp et al. (1990) and Leclerc and Thurtell (1990) coined the word 'footprint' to 'the effective upwind source area sensed by the observation', with 'source' understood to include negative flux densities. Formally, Horst and Weil (1992) describe the flux footprint in a mathematical form: **The footprint encompassed by a point flux measurement is the influence of the properties of the upwind source area weighted with the footprint function.** That definition, however, has been evolving more toward 'not so much an effective upwind source area' than the original definition warrants it and which implies a two-dimensional source but rather an effective upwind source volume to reflect measurements over complex tall canopies characterized with vertical distribution of sources and sinks. This has become more apparent when the footprints are examined in the light of flux measurement above a tall canopy with say, an understory and soil emissions.

Based on this definition Horst and Weil (1992) made also the mathematical formulation for the footprint: The footprint function *f* combines the source area Q_η of a measuring signal η (scalar, flux) in relation to its spatial extent and its distribution of intensity, as illustrated in Fig. 1.1, and is given by:

$$\eta(x_m, y_m, z_m) = \int_{-\infty}^{\infty}\int_{-\infty}^{\infty} Q_\eta(x', y', z' = z_0) \cdot f(x_m - x', y_m - y', z_m - z_0)dx'dy' \quad (1.1)$$

Hereby the source area is in the height $z' = z_0$ (z_0: roughness height) and the footprint is calculated for the sensor height z_m. From this follows two further definitions: one about concentration and flux footprint and one about the dimension of the footprint.

Schmid (1994) defined different source area functions Q_η for scalar or concentration footprints and for flux footprints. For scalar footprints, the source function is simply the concentration distribution

$$Q_\eta(x, y, z = z_0) = \chi(x, y, z = z_0), \quad (1.2)$$

while for flux footprints, the source function must be replaced by a flux distribution

$$Q_\eta(x, y, z = z_0) = K(z)\frac{\partial\chi(x, y)}{\partial z}, \quad (1.3)$$

where *K*(*z*) is the turbulent diffusion coefficient. He found that the extension of the flux footprint is much shorter than for the concentration footprint. This separation is not always carefully done in all models. In the case of concentration footprints, the footprint function is always between 0 and 1 while the flux footprint may also be negative in complex terrain (Finnigan 2004).

Furthermore, footprint models can be separated according to their dimension (Table 1.2). To preclude any misunderstanding, we make a distinction between the definition of the source area and that of the footprint for various dimensions.

1.3 Footprint Modeling

This chapter expands on the description of modeling concepts after the basic definitions about footprints were developed at the beginning of the 90s.

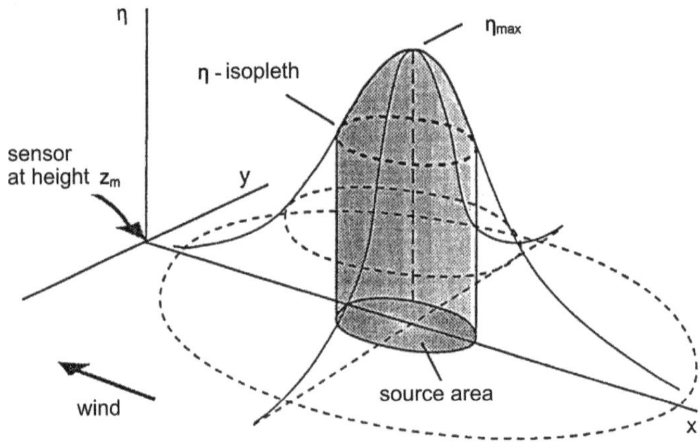

Fig. 1.1 Schematic picture of the footprint function according to Schmid (1994)

Table 1.2 Definition of dimensions of source area and footprint

Dimension	1-dimensional (1D)	2-dimensional (2D)	3-dimensional (3D)
Source area	Line source $Q_\eta(x)$	Two dimensional source in x and y, while z is constant, $Q_\eta(x,y)$	Three dimensional source, $Q_\eta(x,y,z)$
Footprint	Distribution of the concentration or flux density along a horizontal line, $\eta(x)$	Distribution of the concentration or flux density along a horizontal plane, $\eta(x,y)$	Distribution of the concentration or flux density in a non-horizontal plane like in a hilly region, $\eta(x,y,z)$

The footprint idea was extended from the surface layer to the lower convective boundary layer by Leclerc et al. (1997) with the use of Large Eddy Simulation (LES). This study quantified the degree of connection between the surface and an airborne flux platform in the lower convective boundary layer.

Footprint climatologies added to the body of work on footprints (Amiro 1998). The Amiro study was the starting point to estimate the footprint climatology in the FACE (Free-Air Carbon Dioxide Enrichment) experiment at the Duke forest (Stoughton et al. 2000). Footprint climatologies were also the basis used to screen the eddy-covariance data of about twenty European FLUXNET stations by Rebmann et al. (2005). This was subsequently broadened to most European FLUXNET stations by Göckede et al. (2008).

Wilson and Swaters (1991) derived analytical solutions to derive the footprint functions using one and two layers within which the dispersion was parameterized using *K*-theory. They calculated both the 'footprint' and the contact distance of a particle since it last touched the surface. The solutions, simple in nature, rely on the fact that travel times of the particles are large compared with the characteristic turbulence timescale, so that the error in this simplification is small. The method, which uses Monin-Obukhov similarity, has, as of today, not yet been tested. Related to this approach, backward Lagrangian stochastic models came about in the nineties (Flesch 1996). That study used backward Lagrangian stochastic models to provide a measure of the footprint given a measured atmospheric flux. In a manner analogous to that of Flesch (1996), Kljun et al. (2002) used a three-dimensional backward Lagrangian footprint for a wide range of atmospheric stabilities to determine the source area, which was presented also as an analytical approximation for homogeneous surfaces (Kljun et al. 2004b). Kljun et al. (2003) subsequently compared the three-dimensional Lagrangian footprint model by Kljun et al. (2002) against an analytical model of Kormann and Meixner (2001). Shortly after, Kljun et al. (2004a) introduced a scaling procedure for flux footprint functions over a wide range of stabilities with receptor heights ranging from the surface to the middle of the boundary layer. Kljun et al. (2004a), using SF_6 tracer release experiments in a wind tunnel, tested the three-dimensional Lagrangian stochastic footprint model and obtained a general agreement of both the peak location and shape of the resulting footprint functions between modeled and measured fluxes.

Further refinements in our understanding of footprints occur with Luhar and Rao (1994). That study integrated both approaches, analytical and stochastic, into a study involving a step-change in surface roughness and scalar fluxes and its influence on footprint fluxes.

Countless special sessions at meetings, workshops, and scholarly articles have appeared on the subject of footprints, including a recent special issue edited by Vesala, Rannik, Leclerc, Foken and Sabelfeld in Agricultural and Forest Meteorology (Vesala et al. 2004). Testing of these models (Rannik et al. 2000) have been taking place in parallel with the refinement or development of several models (Kurbanmuradov et al. 1999). Several scientific workshops based on footprints have been the subject of an INTAS (International association for the promotion of cooperation with scientists from the independent states of the former Soviet Union) project (2000, 2001, 2003), a European effort aimed at bringing together Eastern and Western scientists in the pursuit of advanced research.

It is with the rise of the Vesala group in Helsinki that the explicit effect of leaf area distributions and canopy density on footprint have been examined (Markkanen et al. 2003). Furthermore, this group further studied the influence of turbulence statistics inside and above a forest canopy as a source of variability on the footprint behavior in a Scots pine forest canopy in Finland (Rannik et al. 2000, 2003). To this end, the Vesala group used a 3D Lagrangian stochastic simulation based on the Thomson (1987) approach. They also, most interestingly, studied flux footprints over simple and complex terrain covered by heterogeneous forests using a canopy—atmospheric boundary layer and scalar transport one-and-half order closure model (Sogachev and Lloyd 2004).

A significant extension to the footprint work came with the application of flux footprints within and over forest canopies by Baldocchi (1998). In this study, the Lagrangian simulation technique was used to calculate footprints at different levels within the canopy. These in-canopy footprint results have been subsequently tested by Leclerc and her group in 2002 for short diffusion times, and in 2004 for longer diffusion distances and a wider range of atmospheric stabilities in the pine canopy of the Florida AmeriFlux site (US-Akn).

Lee (2003) used a combination of both the Raupach's (1989) localized near-field (LNF) theory and parameterization of the turbulence inside a canopy to investigate how atmospheric stability and source configuration influence the flux footprint over the canopy. Lee (2004) extended the above model to examine scalar advection from elevated sources inside plant canopies and used it to describe the behavior of footprints inside plant canopies.

If footprint modeling had been deemed a success, the need for validating these models, in particular those using the analytical solutions to the diffusion equation, would stand the 'litmus test' of flux footprints evaluated when flux sensors are placed both above rough surfaces close to sources and sinks or right amongst sources and sinks as inside a canopy layer.

Soegaard et al. (2003) applied the Schuepp et al. (1990) analytical solution as part of a large experimental campaign in Denmark to quantify the carbon dioxide budget within an intensive, highly heterogeneous agricultural area of Denmark. That footprint study was embedded in a large project involving carbon dioxide exchange measurements throughout the landscape, a scaling up to the landscape effort using satellite land-use maps, a validation of an aerial integration technique, and a quantification of the annual carbon budget from an agricultural landscape.

Falk and Gryning (2000) did a footprint analysis using a stochastic 1D model and validated their results against convective water tank experiments for atmospheric dispersion for the planetary boundary layer. Furthermore they investigated the sensitivity of the footprints to the model boundary conditions. They concluded that, if the turbulence is skewed at the ground, the footprints calculated from backward trajectories are very sensitive to the surface reflection scheme.

Kaharabata et al. (1999) applied the footprint concept to the interpretation of above-canopy sampling of trace gases to interpret VOC emissions data. Strong et al. (2004) incorporated active chemistry into a footprint Lagrangian stochastic model to which prescribed vertical profiles of required turbulence statistics were obtained using a 1D atmospheric turbulence model. It was concluded that active scalar flux estimates can be substantially improved by incorporating an active chemistry term to footprint modeling of a canopy in active defoliation.

Hsieh et al. (2000) developed an analytical model based on Ley and Thomson (1983), Gash (1986), and Horst and Weil (1992) models and tested their data against footprint data and latent heat fluxes downwind from a desert into an irrigated potato site.

Kormann and Meixner (2001) proposed a generalization of the Schuepp et al. (1990) model. They used power-law profiles of the mean velocity and eddy diffusivity based on the model by Huang (1979). Their method is based on profiles described by the Monin-Obukhov similarity theory. It has the advantage of predicting flux footprints for a wide range of atmospheric stabilities while preserving the property of remaining simple and thus well suited to online analysis of flux data. This model was made available in a simplified way for users (Neftel et al. 2008). The other advantage of their analytical solution is that it bypasses the use of the shape factor taking into consideration the related remarks of Haenel and Grünhage (1999) on the shape parameter used in Horst and Weil (1994). Their study examined the possible departure from Monin-Obukhov similarity profiles made using power-law profiles and found the deviations from Monin-Obukhov profiles to be less than 15 % in most conditions.

A novel and creative approach was more recently proposed by Kim et al. (2005) to address the issue of spatial and temporal variability in the scaling-up of tower flux measurements to the landscape. That study used high-resolution satellite maps of surface cover to which flux footprint model outputs were superimposed. The footprint model calculations were based on the Horst and Weil (1994) model. Their study is of importance since they showed that, using semi-variograms and window size techniques, this approach can be a useful tool to select the best tower location for a particular site and to analyze spatial heterogeneousness without a detailed knowledge of site meteorological information.

Finnigan (2004) also examined the footprint concept in complex terrain. In discussing footprint functions, he showed that, using Eulerian and Lagrangian arguments, the concentration footprint can be viewed as the Green function of the Eulerian mass conservation equation or as a Lagrangian transition probability but that the flux footprint cannot be described by the Green function of the flux-transport equation. Finnigan (2004) further argued that the flux footprint is a construction from both the scalar conservation equation and the concentration footprint. He also showed that, in complex flows, such as those encountered in vegetated covers on hilly terrain, an anomalous behavior is expected of the flux footprint so that it is an unreliable guide to the source area affecting tower flux measurements.

The versatility of the Large Eddy Simulation (LES) has been recognized as a potential tool to describe the flow over (Chandrasekar et al. 2003), near (Shen and Leclerc 1997) or inside very strongly sheared atmospheric flows such as within plant canopies (Shen and Leclerc 1997; Su et al. 1998; Watanabe 2009) and urban canopies (Letzel et al. 2008). Prabha et al. (2008) compared the in-canopy footprints obtained using a Lagrangian simulation with those obtained against a LES. In that model, the Lagrangian stochastic model was driven by flow statistics derived from the LES. Recently Steinfeld et al. (2008) embedded a Lagrangian footprint model into a LES model and compared the results with the calculations by Leclerc et al. (1997).

More recently, several overview papers were written (Schmid 2002; Vesala et al. 2004, 2008, 2010; Rannik et al. 2012) to complete this overview. The most important models are shown in Table 1.3.

Table 1.3 Overview about some of the most important footprint models with their dimension (if no remark: analytical model), adapted from Foken (2008) and Vesala et al. (2010) and updated

Author	Remarks
Schuepp et al. (1990)	Analytical footprint model; use of source areas, but neutral stratification and averaged wind velocity (1D)
Leclerc and Thurtell (1990)	Lagrangian footprint model (1D)
Horst and Weil (1992)	Analytical footprint model (1D)
Schmid (1994)	Separation of footprints for scalars and fluxes (1D)
Schmid (1997)	2D version of Horst and Weil (1992)
Kaharabata et al. (1997)	Analytical footprint model (2D)
Leclerc et al. (1997)	LES model for footprints (1D)
Baldocchi (1997)	Lagrangian footprint model within forests (1D)
Rannik et al. (2000; 2003)	Lagrangian model for forests (2D)
Hsieh et al. (2000)	Analytical footprint model (1D)
Kormann and Meixner (2001)	Analytical model with exponential wind profile (1D)
Kljun et al. (2002)	Back trajectories Lagrangian model for varying stratifications and heterogeneous surfaces (3D), 1D analytical version by Kljun et al. (2004b)
Sogachev and Lloyd (2004)	Boundary-layer model with 1.5 order closure (2 and 3D)
Cai and Leclerc (2007)	Concentration footprints from backward and forward in-time particle simulations driven with LES data (3D)
Prabha et al. (2008)	Footprint inside a canopy using LES (3D)
Steinfeld et al. (2008)	Footprint model with LES embedded particles (3D)
Hsieh and Katul (2009)	Second order closure model for heterogeneous surfaces (2D)

1.4 Validation of Footprint Models

Despite the body of works on predictions quantifying source-receptor relations, footprint models and their effectiveness, realism and applicability had not been tested. B. Lamb and M. Y. Leclerc, in collaboration with J. Businger, performed a SF_6 flux experiment with the logistical support of the National Center for Atmospheric Research (NCAR). In 1992, this experiment took place on the premises of the Battelle National Laboratory at the Hanford facility with the help of J. Allwine, a senior scientist at Battelle. D. Finn, a Washington State University PhD student supervised jointly by B. Lamb and by M. Y. Leclerc, not only participated in that experiment but also took a key role in the experimental data analysis that ensued and the subsequent use of the various models available. He visited M. Y. Leclerc several times in Montreal to discuss the 'insides' of the different models and their respective formulations. After several years of painstaking data analysis of the 'rambunctious' fast response continuous tracer analyzers, the group published the Finn et al. (1996) paper. With T. Horst on board, that team took advantage of their dataset to validate the shape-function (see Sect. 2.4.1) derived from Gryning et al. (1983), a necessary shape parameter used in the Horst and Weil (1992, 1994) papers, and whose formulation was found to be in general agreement with the experimental results. The results from the experiment were found to be in very good agreement with predictions from those models.

Table 1.4 Important footprint validation experiments

Year	Place	Surface	Type, name	Reference
1992	Hanford Diffusion Grid	Sagebrush	Artificial tracer (SF_6)	Finn et al. (1996)
1997	Boreal forest, Canada	Mixed sparse forest canopy	Artificial tracer (SF_6)	Kaharabata et al. (1997)
1998	Hollonville, Georgia	Peach orchard	Artificial tracer (SF_6)	Leclerc et al. (2003a)
1998	Waldstein Weidenbrunnen, Germany	Spruce	EUROFLUX site measurements 'natural tracers'	Foken et al. (1999), Foken and Leclerc (2004)
2000	Gainesville, Florida	Pine forest canopy	AmeriFlux site measurements, artificial tracer (SF_6)	Leclerc et al. (2003b)
2000	Socorro, New Mexico	Salt cedar canopy	'Natural tracer'	Cooper et al. (2003)
2000	Vielsam, Belgium	Mixed forest canopy	EUROFLUX site measurements	Rannik et al. (2000)
2002	Gainesville, Florida	Within pine canopy	AmeriFlux site measurements, six different Perfluorocarbon tracers (PFT)	Leclerc et al., unpublished
2003	Lindenberg, Germany	Grass, bare soil	LITFASS-2003, 'natural tracers'	Göckede et al. (2005)
2004	Gainesville, Florida	Within pine canopy	AmeriFlux site measurements, PFT	Leclerc et al., unpublished
2004	Karlshruhe, Germany	Wind tunnel experiment	Artificial tracer (SF_6)	Kljun et al. (2004a)

In a manner similar to that used earlier in the FIFE experiment, the BOREAS study used footprint predictions using not only analytical solutions to the diffusion equation and tracer experiments but also outputs to the LES to help improve the assessment, understanding and intercomparison of fluxes between the different tower sites and the different locations within the boreal forest (Kaharabata et al. 1997), see Table 1.4.

M. Y. Leclerc, with graduate student N. Meskidze, performed a tracer experiment, this time over a surface of intermediate roughness as found in a peach orchard. Initially, the data were collected to examine the robustness of these models when the flux system was outside the roughness sub-layer of that rough canopy. Using to their advantage the fact that peach orchard canopies are not only of intermediate roughness but that they also grow rather quickly throughout the summer, they collected additional data, this time with the same flux system transitioning into that rough sublayer close to sources and sinks. They placed two line sources perpendicular to a horizontal array of flux towers and collected data in such a way that experiments could be carried out when the wind came from one of two directions. That data formed the basis for the Leclerc et al. (2003a) paper. If they had found thus far that these models had demonstrated robustness within the range of applicability prescribed, it was not at all clear that these models, and in particular those using analytical solutions, would withstand one of the regimes where they were most needed: over extremely rough surfaces and within the roughness sublayer. There, the team built no less than a dozen of prototypes of line sources which had to be strong enough to sustain the strong winds, storms, and tail ends of hurricanes and the twisting and turning of the line sources and the battering of the latter in those conditions. The first prototype of line sources was the one that had been used previously both in the Hanford Diffusion Grid (sagebrush) experiment and in the peach orchard experiment and putting a 400 m long line source, tied at treetops onto the trees themselves. By the time the flux towers were built into the forest, mobile laboratories and gas cylinders had been brought to each site, while instrumentation shelters were built and installed on each tower, gas chromatographs working in good order, the sonic and the fast response continuous analyzers set up and data loggers programmed and in operating condition, along with other supporting instrumentation, that line source had experienced fatigue. Leclerc, accompanied by several undergraduate students, build several line sources prototypes. Either the ultra-violet radiation would weaken the lines quickly enough before the experiment could unfold or strong winds would tear the line source down. The building and withstanding of a sturdy line capable of withstanding the harsh sunny and stormy conditions of the Florida climate and weather was a daunting challenge. The basic idea of the new fully functioning line source was born. In January 2000, A. Karipot and T. Prabha, both former PhD students of Inge Dirmhirn and Erich Mursch-Radlgruber in Vienna, who some years ago had also been in contact with Foken's group, joined Leclerc's team and participated in the experiment. The new prototype of the line source was then fully built, orifices mounted on old ports from the original copper line source of the Hanford Diffusion Grid and peach orchard experiments, and flow rates checked across the line. The data collection, high quality data, had begun in earnest and now, the Leclerc group was waiting and praying for the wind direction to be favorable. Finally, with much persistence, the data was collected which formed the basis for the paper by Leclerc et al. (2003b).

If, for a short time, the Leclerc group relished in their accomplishments, at the time of data analysis and proposal writing, Leclerc, after hours of examining what appeared to be noisy sodar data burning the midnight oil in the wee hours of the night preparing for a proposal, noticed that some of the apparent scatter in the flux footprint might be connected in some form, to patterns in the sodar data. Curious, she then examined the many plots from the sodar placed above the group's mobile laboratory and started to reconstruct the puzzle. What was happening almost every morning around 10:00 at that site? Recreating the experimental scene at the site, she could hear in her ears the sudden shift in wind direction for the better part of the day. With a vector analysis of the synoptic wind and the observed wind, there must have been some forcing that skewed the flow when the winds came from the west. She then remembered having seen logging trucks in the fall of 1999. Intrigued, she investigated and found a large, very large freshly logged area about

300 m west of the tower. From then on, the data was re-examined in the light of wind direction, and the footprint predictions were found to be in very good agreement when the wind came from the east and departed by several orders of magnitude when the wind came from the west (clearing). The sodar data was double checked to make sure that the unusual signal could be trusted. That was then demonstrated in a guest paper written in the honor of G. W. Thurtell in a special issue of Agricultural and Forest Meteorology (Leclerc et al. 2003b).

References

Albrecht F (1940) Untersuchungen über den Wärmehaushalt der Erdoberfläche in verschiedenen Klimagebieten. Reichsamt Wetterdienst, Wiss Abh. Bd. VIII, Nr. 2:1–82

Amiro BD (1998) Footprint climatologies for evapotranspiration in a boreal catchment. Agric Forest Meteorol 90:195–201

André J-C, Bougeault P, Goutorbe J-P (1990) Regional estimates of heat and evaporation fluxes over non-homogeneous terrain, Examples from the HAPEX-MOBILHY programme. Boundary-Layer Meteorol 50:77–108

Arya SP (1999) Air pollution meteorology and dispersion. Oxford University Press, New York, Oxford, p 310

Baldocchi D (1997) Flux footprints within and over forest canopies. Boundary-Layer Meteorol 85:273–292

Baldocchi D, Meyers T (1998) On using eco-physiological, micrometeorological and biochemical theory to evaluate carbon dioxide, water vapor and trace gas fluxes over vegetation. Agric Forest Meteorol 90:1–25

Barrett EW, Suomi VE (1949) Preliminary report on temperature measurement by sonic means. J Meteorol 6:273–276

Bovscheverov VM, Voronov VP (1960) Akustitscheskii fljuger (acoustic rotor). Izv AN SSSR, ser Geofiz. 6:882–885

Busch NE, Panofsky HA (1968) Recent spectra of atmospheric turbulence. Quart J Roy Meteorol Soc 94:132–148

Businger JA, Miyake M, Inoue E, Mitsuta Y, Hanafusa T (1969) Sonic anemometer comparison and measurements in the atmospheric surface layer. J Meteor Soc Japan 47:1–12

Businger JA, Wyngaard JC, Izumi Y, Bradley EF (1971) Flux-profile relationships in the atmospheric surface layer. J Atmos Sci 28:181–189

Cai XH, Leclerc MY (2007) Forward-in-time and backward-in-time dispersion in the convective boundary layer: The concentration footprint. Boundary-Layer Meteorol 123:201–218

Calder KL (1952) Some recent British work on the problem of diffusion in the lower atmosphere. In: Proceedings of the US technology conference on air pollution. McGraw Hill, New York, pp 787–792

Chandrasekar A, Philbrick CR, Clark R, Doddridge B, Georgopoulos P (2003) A large-eddy simulation study of the convective boundary layer over Philadelphia during the 1999 summer NE-OPS campaign. Environ Fluid Mech 3:305–329

Cionco RM (1978) Analysis of canopy index values for various canopy densities. Boundary-Layer Meteorol 15:81–93

Cooper DI, Eichinger WE, Archuleta J, Hipps L, Kao J, Leclerc MY, Neale CM, Prueger JH (2003) Spatial source-area analysis of three-dimensional moisture fields from lidar, eddy covariance, and a footprint model. Agric Forest Meteorol 114:213–234

Dyer AJ (1981) Flow distortion by supporting structures. Boundary-Layer Meteorol 20:363–372

Dyer AJ, Bradley EF (1982) An alternative analysis of flux-gradient relationships at the 1976 ITCE. Boundary-Layer Meteorol 22:3–19

Elliott WP (1958) The growth of the atmospheric internal boundary layer. Trans Am Geophys Union 39:1048–1054
Falk AKV, Gryning SE (2000) Footprint analysis from random walk models—sensitivity to boundary conditions. In: Gryning SE, Batchvarova E (eds.) Air pollution modelling and its applications, pp 393–401
Finn D, Lamb B, Leclerc MY, Horst TW (1996) Experimental evaluation of analytical and Lagrangian surface-layer flux footprint models. Boundary-Layer Meteorol 80:283–308
Finnigan J (2004) The footprint concept in complex terrain. Agric Forest Meteorol 127:117–129
Flesch TK (1996) The footprint for flux measurements, from backward Lagrangian stochastic models. Boundary-Layer Meteorol 78:399–404
Foken T, Zelený J (1988) The influence of roughness on turbulence parameters. In: Pretel J (ed) Structure of the boundary layer over non-homogeneous terrain - field experiment KOPEX-861988. Ustav fiziky atmosfery, Prague, pp 81–90
Foken T (1990) Turbulenter Energieaustausch zwischen Atmosphäre und Unterlage—Methoden, meßtechnische Realisierung sowie ihre Grenzen und Anwendungsmöglichkeiten. Ber Dt Wetterdienstes 180:287
Foken T et al (1993) Study of the energy exchange processes over different types of surfaces during TARTEX-90. Dt Wetterdienst, Forsch. Entwicklung, Arbeitsergebnisse 4:34
Foken T, Bernhardt K (1994) Atmospheric boundary layer research in Central and East European countries with KAPG, 1981–1990. Geophys Rep 01:1–58
Foken T, Wichura B (1996) Tools for quality assessment of surface-based flux measurements. Agric Forest Meteorol 78:83–105
Foken T, Mangold A, Hierteis M, Wichura B, Rebmann C (1999) Characterization of the heterogeneity of the terrain by normalized turbulence characteristics. In: 13th symposium on boundary layer and turbulence, Dallas, TX, Am Meteorol Soc, pp 26–27, 10–15 Jan 1999
Foken T, Leclerc MY (2004) Methods and limitations in validation of footprint models. Agric Forest Meteorol 127:223–234
Foken T (2006) 50 years of the Monin-Obukhov similarity theory. Boundary-Layer Meteorol 119:431–447
Foken T (2008) Micrometeorology. Springer, Berlin, p 308
Garratt JR, Hicks BB (1990) Micrometeorological and PBL experiments in Australia. Boundary-Layer Meteorol 50:11–32
Gash JHC (1986) A note on estimating the effect of a limited fetch on micrometeorological evaporation measurements. Boundary-Layer Meteorol 35:409–414
Geiger R (1927) Das Klima der bodennahen Luftschicht. Friedr. Vieweg & Sohn, Braunschweig, p 246
Geiger R, Aron RH, Todhunter P (2009) The climate near the ground. Rowman & Littlefield Publishers Inc, Lanham, p 623
Göckede M, Markkaken T, Mauder M, Arnold K, Leps JP, Foken T (2005) Validation of footprint models using natural tracer measurements from a field experiment. Agric Forest Meteorol 135:314–325
Göckede M et al (2008) Quality control of CarboEurope flux data—Part 1: Coupling footprint analyses with flux data quality assessment to evaluate sites in forest ecosystems. Biogeoscience 5:433–450
Gryning S-E, van Ulden AP, Larsen S (1983) Dispersions from a ground level source investigated by a K model. Quart J Roy Meteorol Soc 109:355–364
Haenel H-D, Grünhage L (1999) Footprint analysis: a closed analytical solution based on height-dependent profiles of wind speed and eddy viscosity. Boundary-Layer Meteorol 93:395–409
Ham JM, Knapp AK (1998) Fluxes of CO_2, water vapor, and energy from a prairie ecosystem during the seasonal transition from carbon sink to carbon source. Agric Forest Meteorol 89:1–14
Hanafusa T, Fujitana T, Kobori Y, Mitsuta Y (1982) A new type sonic anemometer-thermometer for field operation. Pap Meteorol Geophys 33:1–19

Haugen DA (ed) (1973) Workshop on micrometeorology. American Meteorology Society, Boston, pp 392
Horst TW (1979) Lagrangian similarity modeling of vertical diffusion from a ground level source. J Appl Meteorol 18:733–740
Horst TW, Weil JC (1992) Footprint estimation for scalar flux measurements in the atmospheric surface layer. Boundary-Layer Meteorol 59:279–296
Horst TW, Weil JC (1994) How far is far enough?: the fetch requirements for micrometeorological measurement of surface fluxes. J Atm Oceanic Techn 11:1018–1025
Hsieh C-I, Katul G, Chi T-W (2000) An approximate analytical model for footprint estimation of scalar fluxes in thermally stratified atmospheric flows. Adv Water Res 23:765–772
Hsieh C-I, Katul G (2009) The Lagrangian stochastic model for estimating footprint and water vapor fluxes over inhomogeneous surfaces. Int J Biometeorol 53:87–100
Huang CH (1979) A theory of dispersion in turbulent shear flow. Atmos Environm 13:453–463
Izumi Y (1971) Kansas 1968 field program data report. Air Force Cambridge Research Laboratory, Bedford, p 79
Kaharabata SK, Schuepp PH, Ogunjemiyo S, Shen S, Leclerc MY, Desjardins RL, MacPherson JI (1997) Footprint considerations in BOREAS. J Geophys Res 102(D24):29113–29124
Kaharabata SK, Schuepp PH, Fuentes JD (1999) Source footprint considerations in the determination of volatile organic compound fluxes from forest canopies. J Appl Meteorol 38:878–884
Kaimal JC, Businger JA (1963) A continuous wave sonic anemometer-thermometer. J Climate Appl Meteorol 2:156–164
Kaimal JC, Wyngaard JC (1990) The Kansas and Minnesota experiments. Boundary-Layer Meteorol 50:31–47
Kanemasu ET et al. (1992) Surface flux measurements in FIFE: an overview. J Geophys Res 97:18.547–518.555
Kim SW, Moeng CH, Weil JC, Barth MC (2005) Lagrangian particle dispersion modeling of the fumigation process using large-eddy simulation. J Atmos Sci 62:1932–1946
Kljun N, Rotach MW, Schmid HP (2002) A three-dimensional backward Lagrangian footprint model for a wide range of boundary layer stratification. Boundary-Layer Meteorol 103:205–226
Kljun N, Kormann R, Rotach M, Meixner FX (2003) Comparison of the Lagrangian footprint model LPDM-B with an analytical footprint model. Boundary-Layer Meteorol 106:349–355
Kljun N, Kastner-Klein P, Federovich E, Rotach MW (2004a) Evaluation of Lagrangian footprint model using data from wind tunnel convective boundary layer. Agric Forest Meteorol 127:189–201
Kljun N, Calanca P, Rotach M, Schmid HP (2004b) A simple parameterization for flux footprint predictions. Boundary-Layer Meteorol 112:503–523
Kormann R, Meixner FX (2001) An analytical footprint model for non-neutral stratification. Boundary-Layer Meteorol 99:207–224
Kurbanmuradov O, Rannik U, Sabelfeld KK, Vesala T (1999) Direct and adjoint Monte Carlo for the footprint problem. Monte-Carlo Meth Appl 5:85–111
Leclerc MY, Thurtell GW (1989) Footprint predictions of scalar fluxes and concentration profiles using a markovian analysis. In: 19th conference of agricultural and forest meteorology, Charleston, SC, American Meteorology Society, 7–10 Mar 1989
Leclerc MY, Thurtell GW (1990) Footprint prediction of scalar fluxes using a Markovian analysis. Boundary-Layer Meteorol 52:247–258
Leclerc MY, Shen S, Lamb B (1997) Observations and large-eddy simulation modeling of footprints in the lower convective boundary layer. J Geophys Res 102(D8):9323–9334
Leclerc MY, Meskhidze N, Finn D (2003a) Comparison between measured tracer fluxes and footprint modeling predictions over a homogeneous canopy of intermediate roughness. Agric Forest Meteorol 117:145–158

Leclerc MY, Karipot A, Prabha T, Allwine G, Lamb B, Gholz HL (2003b) Impact of non-local advection on flux footprints over a tall forest canopy: a tracer flux experiment (special issue: advances in micrometeorology: tribute to G. W. Thurtell). Agric Forest Meteorol 115:19–30

Lee X (2003) Fetch and footprint of turbulent fluxes over vegetative stands with elevated sources. Boundary-Layer Meteorol 107:561–579

Lee X (2004) A model for scalar advection inside canopies and application to footprint investigation. Agric Forest Meteorol 127:131–141

Lettau H (1939) Atmosphärische Turbulenz. Akad. Verlagsges, Leipzig, p 283

Lettau HH, Davidson B (eds.) (1957) Exploring the atmosphere's first mile. Pergamon Press, London, New York, pp 376

Letzel MO, Krane M, Raasch S (2008) High resolution urban large-eddy simulation studies from street canyon to neighbourhood scale. Atmos Environ 42:8770–8784

Ley AJ, Thomson DJ (1983) A random walk model of dipersion in the diabatic surface layer. Quart J Roy Meteorol Soc 109:867–880

Luhar AK, Rao KS (1994) Source footprint analysis for scalar fluxes measured in flows over an inhomogeneous surface. In: Gryning SE, Millan MM (eds) Air pollution modeling and its applications. Plenum Press, New York, pp 315–323

Markkanen T, Rannik Ü, Marcolla B, Cescatti A, Vesala T (2003) Footprints and fetches for fluxes over forest canopies with varying structure and density. Boundary-Layer Meteorol 106:437–459

McBean GA, Bernhardt K, Bodin S, Litynska Z, van Ulden AP, Wyngaard JC (1979) The planetary boundary layer. WMO Note 530:201

Mitsuta Y (1966) Sonic anemometer-thermometer for general use. J Meteor Soc Japan. Ser II 44:12–24

Miyake M, Stewart RW, Burling RW, Tsvang LR, Kaprov BM, Kuznecov OA (1971) Comparison of acoustic instruments in an atmospheric flow over water. Boundary-Layer Meteorol 2:228–245

Moncrieff J (2004) Surface turbulent fluxes. In: Kabat P et al (eds) Vegetation, water, humans and the climate: a new perspective on an interactive system. Springer, Berlin, pp 173–182

Monin AS, Obukhov AM (1954) Osnovnye zakonomernosti turbulentnogo peremesivanija v prizemnom sloe atmosfery (Basic laws of turbulent mixing in the atmosphere near the ground). Trudy geofiz inst AN SSSR 24(151):163–187

Montgomery RB (1948) Vertical eddy flux of heat in the atmosphere. J Meteorol 5:265–274

Neftel A, Spirig C, Ammann C (2008) Application and test of a simple tool for operational footprint evaluations. Environ Pollut 152:644–652

Obukhov AM (1951) Charakteristiki mikrostruktury vetra v prizemnom sloje atmosfery (Characteristics of the micro-structure of the wind in the surface layer of the atmosphere). Izv AN SSSR, ser Geofiz 3:49–68

Pasquill F (1972) Some aspects of boundary layer description. Quart J Roy Meteorol Soc 98:469–494

Peterson EW (1969) Modification of mean flow and turbulent energy by a change in surface roughness under conditions of neutral stability. Quart J Roy Meteorol Soc 95:561–575

Prabha T, Leclerc MY, Baldocchi D (2008) Comparison of in-canopy flux footprints between Large-Eddy Simulation and the Langrangian simulation. J Appl Meteorol Climatol 47:2115–2128

Prandtl L (1925) Bericht über Untersuchungen zur ausgebildeten Turbulenz. Z Angew Math Mech 5:136–139

Rannik Ü, Aubinet M, Kurbanmuradov O, Sabelfeld KK, Markkanen T, Vesala T (2000) Footprint analysis for measurements over heterogeneous forest. Boundary-Layer Meteorol 97:137–166

Rannik Ü, Markkanen T, Raittila T, Hari P, Vesala T (2003) Turbulence statistics inside and above forest: influence on footprint prediction. Boundary-Layer Meteorol 109:163–189

Rannik Ü, Sogachev A, Foken T, Göckede M, Kljun N, Leclerc MY, Vesala T (2012) Footprint analysis. In: Aubinet M et al (eds) Eddy covariance: a practical guide to measurement and data analysis. Springer, Berlin, pp 211–261

Rao KS, Wyngaard JC, Coté OR (1974) The structure of the two-dimensional internal boundary layer over a sudden change of surface roughness. J Atmos Sci 31:738–746

Raupach MR (1989) A practical Lagrangian method for relating scalar concentrations to source distributions in vegetation canopies. Quart J Roy Meteorol Soc 115:609–632

Readings CJ, Haugen DH, Kaimal JC (1974) The 1973 Minnesota atmospheric boundary layer experiment. Weather 29:309–312

Rebmann C et al (2005) Quality analysis applied on eddy covariance measurements at complex forest sites using footprint modelling. Theor Appl Climat 80:121–141

Richardson LF (1920) The supply of energy from and to atmospheric eddies. Proc Roy Soc A 97:354–373

Ross J (1981) The radiation regime and architecture of plant stands. Dr. W. Junk Publishers, The Hague, p 391

Schmid HP, Oke TR (1988) Estimating the source area of a turbulent flux measurement over a patchy surface. In: 8th symposium on turbulence and diffusion, San Diego, CA, American Meteorology Society, 26–29 April 1988, pp 123–126

Schmid HP, Oke TR (1990) A model to estimate the source area contributing to turbulent exchange in the surface layer over patchy terrain. Quart J Roy Meteorol Soc 116:965–988

Schmid HP (1994) Source areas for scalars and scalar fluxes. Boundary-Layer Meteorol 67:293–318

Schmid HP (1997) Experimental design for flux measurements: matching scales of observations and fluxes. Agric Forest Meteorol 87:179–200

Schmid HP, Lloyd CR (1999) Spatial representativeness and the location bias of flux footprints over inhomogeneous areas. Agric Forest Meteorol 93:195–209

Schmid HP (2002) Footprint modeling for vegetation atmosphere exchange studies: a review and perspective. Agric Forest Meteorol 113:159–184

Schmidt W (1925) Der Massenaustausch in freier Luft und verwandte Erscheinungen. Henri Grand Verlag, Hamburg, p 118

Schotland RM (1955) The measurement of wind velocity by sonic waves. J Meteorol 12:386–390

Schuepp PH, Leclerc MY, MacPherson JI, Desjardins RL (1990) Footprint prediction of scalar fluxes from analytical solutions of the diffusion equation. Boundary-Layer Meteorol 50:355–373

Sellers PJ, Hall FG, Asrar G, Strebel DE, Murphy RE (1988) The first ISLSCP field experiment (FIFE). Bull Amer Meteorol Soc 69:22–27

Shen S, Leclerc MY (1997) Modelling the turbulence structure in the canopy layer. Agric Forest Meteorol 87:3–25

Shir CC (1972) A numerical computation of the air flow over a sudden change of surface roughness. J Atmos Sci 29:304–310

Silversides RH (1974) On scaling parameters for turbulence spectra within plant canopies. Agric Meteorol 13:203–211

Soegaard H, Jensen NO, Boegh E, Hasager CB, Schelde K, Thomsen A (2003) Carbon dioxide exchange over agricultural landscape using eddy correlation and footprint modelling. Agric Forest Meteorol 114:153–173

Sogachev A, Lloyd J (2004) Using a one-and-a-half order closure model of atmospheric boundary layer for surface flux footprint estimation. Boundary-Layer Meteorol 112:467–502

Steinfeld G, Raasch S, Markkanen T (2008) Footprints in homogeneously and heterogeneously driven boundary layers derived from a Langrangian stochastic particle model embedded into large-eddy simulation. Boundary-Layer Meteorol 129:225–248

Stoughton TE, Miller DR, Yang X, Hendrey GM (2000) Footprint climatology estimation of potential control ring contamination at the Duke Forest FACTS-1 experiment site. Agric Forest Meteorol 100:73–82

Strong C, Fuentes JD, Baldocchi D (2004) Reactive hydrocrbon flux footprints during canopy senescence. Agric Forest Meteorol 127:159–173
Su H-B, Shaw RH, Paw UKT, Moeng C-H, Sullivan PP (1998) Turbulent statistics of neutrally stratified flow within and above sparse forest from large-eddy simulation and field observations. Boundary-Layer Meteorol 88:363–397
Suomi VE (1957) Sonic anemometer—University of Wisconsin. In: Lettau HH, Davidson B (eds) Exploring the atmosphere's first mile, vol 1., Pergamon PressLondon, New York, pp 256–266
Sutton OG (1953) Micrometeorology. McGraw Hill, New York, p 333
Swinbank WC (1951) The measurement of vertical transfer of heat and water vapor by eddies in the lower atmosphere. J Meteorol 8:135–145
Swinbank WC, Dyer AJ (1968) An experimental study on mircrometeorology. Quart J Roy Meteorol Soc 93:494–500
Taylor GI (1915) Eddy motion in the atmosphere. Phil Trans R Soc London A 215:1–26
Taylor PA (1969) On wind and shear stress profiles above a change in surface roughness. Quart J Roy Meteorol Soc 95:77–91
Thomson DJ (1987) Criteria for the selection of stochastic models of particle trajectories in turbulent flows. J Fluid Mech 189:529–556
Tsvang LR, Kaprov BM, Zubkovskij SL, Dyer AJ, Hicks BB, Miyake M, Stewart RW, McDonald JW (1973) Comparison of turbulence measurements by different instruments; Tsimlyansk field experiment 1970. Boundary-Layer Meteorol 3:499–521
Tsvang LR et al (1985) International turbulence comparison experiment (ITCE-81). Boundary-Layer Meteorol 31:325–348
Tsvang LR, Fedorov MM, Kader BA, Zubkovskii SL, Foken T, Richter SH, Zelený J (1991) Turbulent exchange over a surface with chessboard-type inhomogeneities. Boundary-Layer Meteorol 55:141–160
van Ulden AP (1978) Simple estimates for vertical diffusion from sources near the ground. Atmos Environ 12:2125–2129
Vesala T, Rannik U, Leclerc MY, Foken T, Sabelfeld KK (2004) Foreword: flux and concentration footprints. Agric Forest Meteorol 127:111–116
Vesala T, Kljun N, Rannik U, Rinne J, Sogachev A, Markkanen T, Sabelfeld K, Foken T, Leclerc MY (2008) Flux and concentration footprint modelling: state of the art. Environ Pollut 152:653–666
Vesala T, Kljun N, Rannik Ü, Rinne J, Sogatchev A, Markkanen T, Sabelfeld K, Foken T, Leclerc MY (2010) Flux and concentration footprint modelling. In: Hanrahan G (ed.) Modelling of pollutants in complex environmental systems, vol II, pp 339–355
Watanabe T (2009) LES study on the structure of coherent eddies including predominant perturbations in velocities in the roughness sublayer over plant canopies. J Meteor Soc Japan 87:39–56
Wilson JD, Ward DP, Thurtell GW, Kidd GE (1982) Statistics of atmospheric turbulence within and above a corn canopy. Boundary-Layer Meteorol 24:495–519
Wilson JD, Swaters GE (1991) The source area influencing a measurement in the planetary boundary layer: the footprint and the distribution of contact distance. Boundary-Layer Meteorol 55:25–46
Zelený J, Pretel J (1986) Zur Problematik der Bestimmung der aerodynamischen Rauhigkeit der Erdoberfläche. Z Meteorol 36:325
Zubkovskij SL, Sushko AA (1987) Eksperimentalnoe issledovanie prostranstvennoj struktury temperaturnogo polja v prizemnom sloe atmosfery (Experimental investigations spatial structure of the temperature field in the near surface layer of the atmosphere). Meteorol Issledovanija 28:36–41

Chapter 2
Surface-Layer Properties and Parameterizations

Footprint models are generally based both on parameterizations and simplified assumptions typically for the lower atmospheric boundary layer. For the experimentalist in need of footprint models, it is important to know the spatial extent and the range of atmospheric conditions of each footprint model so that the most appropriate one can be selected for the purpose at hand. This chapter therefore introduces the reader to the concept of atmospheric boundary layer and parameterizations, linking those to footprint models where these parameterizations are used. More details can be found in textbooks and in relevant papers (Stull 1988; Garratt 1992; Kaimal and Finnigan 1994; Arya 1999, 2001; Hatfield and Baker 2005; Foken 2008; Monteith and Unsworth 2008; Wyngaard 2010; Moene and van Dam 2014).

2.1 Atmospheric Boundary Layer and Scales

The atmospheric boundary layer is the lowest layer of the troposphere near the ground where the friction decreases with height. In that layer, the wind velocity decreases significantly from the geostrophic wind above the boundary layer to the wind near the surface and the wind direction changes counter-clockwise on the Northern hemisphere by up to 30°–45°. The upper boundary is a mostly static stable layer (inversion) characterized by intermittent turbulence. The exchange processes between the atmospheric boundary layer and the free troposphere take place in the entrainment zone (Fig. 2.1). The thickness of this layer is approximately 10 % of the atmospheric boundary layer, which has a thickness of about 1–2 km over land and 0.5 km over the oceans. In strong stable stratification, its thickness can be as little as 10 m.

In addition, the diurnal cycles of solar radiation, temperature, humidity, and wind are also highly variable (Stull 1988), see Fig. 2.2. After sunrise, the atmosphere is warmed up by the heat transported from the surface upward and the inversion layer created during the night breaks up. The new layer is very turbulent, well mixed (mixed layer) and topped by the entrainment zone. Shortly before sunset, the stable (nightly) boundary layer develops near the ground. It has the

M. Y. Leclerc and T. Foken, *Footprints in Micrometeorology and Ecology*,
DOI: 10.1007/978-3-642-54545-0_2, © Springer-Verlag Berlin Heidelberg 2014

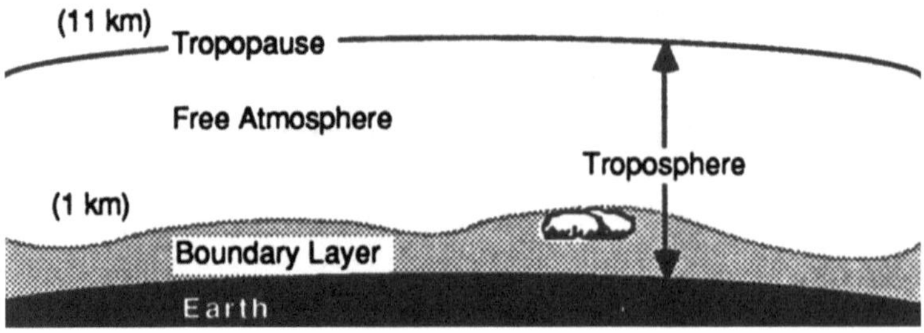

Fig. 2.1 The troposphere and its two parts: the atmospheric boundary layer and the free atmosphere (Stull 2000)

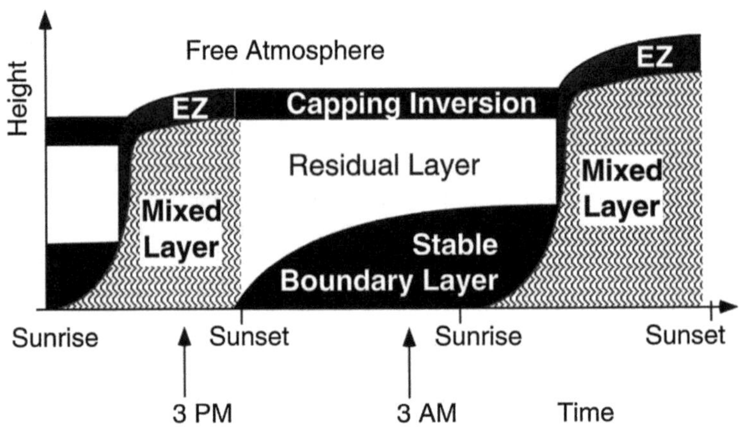

Fig. 2.2 Daily cycle of the structure of the atmospheric boundary layer (Stull 2000), *EZ* Entrainment zone

characteristics of a surface inversion and spans only approximately 100 m in depth. Above this layer, the mixed layer of the day is not very turbulent and is called the residual layer. The latter is capped by a free (capping) inversion—the upper border of the boundary layer (Seibert et al. 2000). At sunrise, the growing mixed layer rapidly erodes both the stable boundary layer and the residual layer. On overcast days, the life time of the residual layer is longer and the boundary layer is more layered than during sunny convective days.

On days with high solar irradiation, the layer structure is destroyed by convective cells. These occupy relatively small updrafts areas and develop typically over larger areas with uniform surface heating in relation to the surrounding areas like land-lake or dry-wet areas. This is according to modeled studies over areas larger than 200–500 m (Shen and Leclerc 1995).

In the upper boundary layer (*upper layer, Ekman layer*) the change of wind direction takes place in the lowest 10 %. That region is called the *surface* or the *Prandtl layer* (Fig. 2.3). Its height is approximately 20–50 m in the case of

height in m	name		exchange		stability
1000	upper layer (Ekman-layert)		turbulent	no constant flux	influence of stability
20	turbulent layer	surface layer (Prandtl-layer)		flux nearly constant with height	
1	dynamic sublayer				no influence of stability
0.01	viscous sublayer		molecular/ turbulent		
0.001	laminar/ molecular boundary layer		laminar/ molecular		

Fig. 2.3 Structure of the atmospheric boundary layer (Foken 2008)

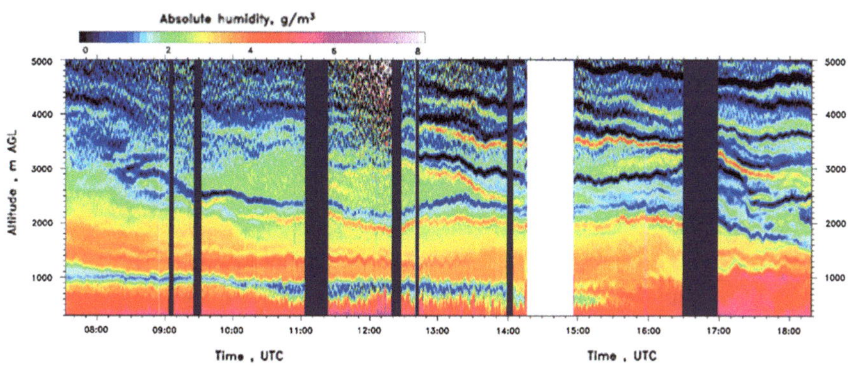

Fig. 2.4 Highly variable structure of the atmospheric boundary layer and of the boundary layer height measured with a Lidar (Behrendt et al. 2009, Published with kind permission of © International Society for Optics and Photonics, 2009. All Rights Reserved)

unstable conditions and a few meters in stable stratification. It is also called the *constant flux layer* because of the assumption of nearly constant fluxes with height. In this layer, the vertical wind profile is logarithmic (*inertial sublayer*). This assumption forms the basis of similarity theory, first attributed to Monin and Obukhov (1954). Because of this height invariance in energy and mass fluxes, these fluxes can be measured anywhere within the surface layer. The very thin *viscous* and *laminar* (term used for the flow field) or *molecular layer* (term used for scalars, is not very relevant for measurements but nevertheless used in several models (Fig. 2.3). According to the similarity theory by Monin and Obukhov (1954), a layer with a thickness of approximately 1 m (*dynamical sublayer*) is not influenced by atmospheric stability—this layer is nearly always neutral. In the real atmosphere the atmospheric boundary layer is highly variable (Fig. 2.4), which can partly described in footprint models (Kljun et al. 2002; Steinfeld et al. 2008).

Atmospheric processes are characterized by time scales extending from seconds (e.g. turbulent exchange) to several days (e.g. Rossby waves, horizontal

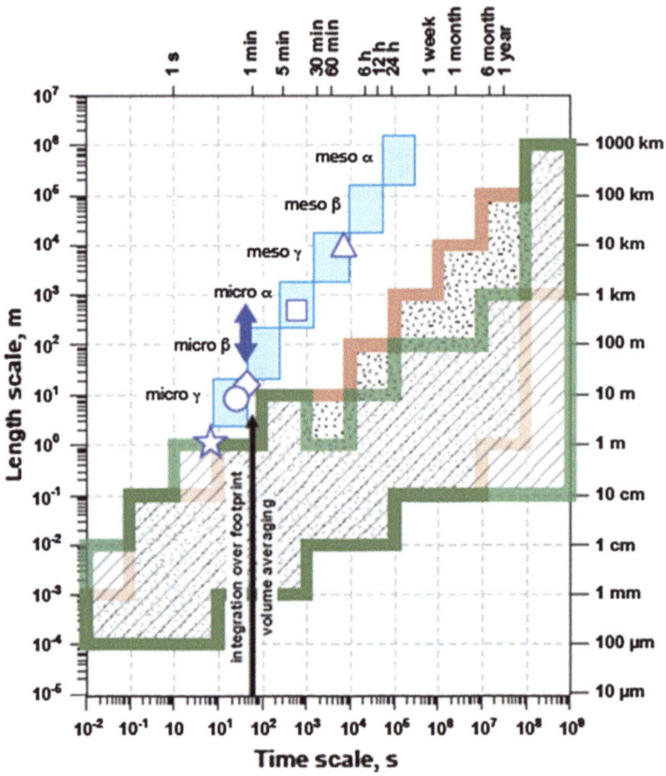

Fig. 2.5 Temporal and spatial scales of atmospheric (turbulent), plant (physiological), and soil processes. Atmospheric processes (Orlanski 1975) are given in *light blue squares* of one order of magnitude (from micro γ to meso α). Forest canopy related transport processes comprise turbulent transport in canopy (*white star*), vertical advection in canopy (*white circle*), transport above canopy (*white diamond*), coherent structures (*blue double arrow*), footprint averaged turbulent flux (*white square*), and horizontal advection at canopy top (*white triangle*). The scales of plant processes, relevant for energy and matter exchange with the atmosphere (Schoonmaker 1998), is the spotted area, those of soil processes (Blöschl and Sivapalan 1995; Vogel and Roth 2003) are shown by the brown framed (Foken et al. 2012, designed by E. Falge, modified, Published with kind permission of © Copernicus Publications, distributed under the Creative Commons Attribution 3.0 License, 2012. All Rights Reserved)

advection), and from millimetres (e.g. smallest eddies) to the size of high and low pressure areas (up to 10 × 10 km^4). For atmospheric processes, scales (defined e.g. by Orlanski 1975) range between 10^0–10^7 m and 10^0–10^6 s, respectively (see Fig. 2.5, diagonal orientated boxes). Atmospheric scales of exchange processes of energy and mass related to the issue of this book comprise both turbulent transport and coherent structures inside and above canopies, footprint-related turbulent fluxes, and horizontal advection in and at the canopy top in a range between 10^0–10^4 m and 10^0–10^4 s, respectively.

In contrast, soil and plant processes cover similar time scales but smaller length scales. While flux measuring methods (like the eddy-covariance technique) are working in the typical atmospheric scale mainly the micro-α, β, γ scales dependent on the measuring height, the footprint method is a tool to average the smaller soil and plant scales with the atmospheric scales typically on the micro-α, β scale. This is in a simplified format shown in Fig. 2.5. Because most footprint models assume a homogeneous surface, special area-averaging techniques must be used, which is a topic of Sect. 2.5.

2.2 Turbulence Parameterization

Footprint models rest on the assumptions of vertical profiles of wind, temperature and scalar as well as profiles of turbulence parameters. These depend on fluxes of momentum, sensible and latent heat or on the concentration of trace gases. Since turbulence variables are often limited, they must be parameterized using other meteorological data. The basis for this lies in the typical similarity or simplifications of the latter. The most important is the flux-profile similarity and the flux-variance similarity. The first is identical with the Monin-Obukhov similarity theory (Monin and Obukhov 1954; Foken 2006), which expresses the relationship between the turbulent flux and the vertical gradient of its state parameter and the gradient of the wind velocity under the assumption of a stratified surface layer. A more simple relation is the well-known logarithmic profile according to Prandtl (1925) in neutral conditions. Both can be simplified using the Bowen-ratio similarity (Bowen 1926), i.e. the ratio of two fluxes is proportional to the difference of its state parameters between two levels mathematically. The flux-variance similarity describes the relation between the turbulent flux and the variance of the state parameter (Obukhov 1960) which is often also a function of stability in the surface layer. Both similarity relations will be described below in addition to often used empirical functions. The similarity theory based on the assumption of horizontal homogeneity, low vegetation, and steady state conditions. These assumptions are often not fulfilled and limit the application of footprint models, as turbulence properties and variables are typically inhomogeneous in the nature.

2.2.1 *Flux-Gradient Similarity*

According to Prandtl (1925) and its mixing length theory for neutral conditions, the turbulent fluxes follow the flux-gradient similarity or the so-called K-approach. In general terms, K represents the sum of the molecular diffusion and the turbulent diffusion coefficient. Because the turbulent coefficient is up to five orders of magnitude larger than the molecular coefficient, only the latter is used. This however does not apply to the viscous sublayer. The turbulent fluxes are proportional to the gradient of the state variable with the turbulent diffusion coefficient K. Therefore, the momentum flux τ, the sensible heat flux Q_H and the mass flux Q_χ (for water vapour the latent heat flux Q_E) are

$$\tau = \rho u_*^2 = -\rho\overline{u'w'} = \rho K_m \frac{\partial \overline{u}}{\partial z} \tag{2.1}$$

$$Q_H = c_p \rho \overline{w'T'} = -c_p \rho K_H \frac{\partial \overline{T}}{\partial z} \tag{2.2}$$

$$Q_\chi = \rho \overline{w'\chi'} = -\rho K_\chi \frac{\partial \overline{c}}{\partial z} \tag{2.3}$$

where K_m, K_H and K_χ are the turbulent diffusion coefficients for momentum, sensible heat and trace gases, ρ is the air density, c_p is the specific heat for constant pressure, u is the horizontal wind velocity, T is the temperature, c is the trace gas concentration and w', u', T' and c' are the turbulent fluctuations of the vertical and horizontal wind components, the temperature and the trace gas concentration. The equation for the friction velocity u_* is only valid, if u is aligned in the direction of the mean wind velocity. This can be expressed in Cartesian coordinate as

$$u_* = \left[\left(\overline{u'w'}\right)^2 + \left(\overline{v'w'}\right)^2\right]^{1/4}, \tag{2.4}$$

where are $\overline{u'w'}$ and $\overline{v'w'}$ are the two components of the momentum tensor in the direction of the horizontal wind components u and v and $\overline{w'T'}$ and $\overline{w'c'}$ are the temperature and concentration flux with the vertical wind component w, the temperature T and the concentration of a trace gas (e.g. water vapour) c. The averaging operator obeys the Reynolds averaging of the total flux

$$\overline{xw} = \overline{x}\,\overline{w} + \overline{x'w'}, \tag{2.5}$$

where x can be replaced by the variable of the mean quantity of interest and x' the instantaneous component. Because of the assumption that $\overline{w} = 0$, the total flux can be replaced by the turbulent flux, which can be measured as a covariance $\overline{x'w'}$ (with the eddy-covariance method). This assumption is far from trivial, since it is seldom fulfilled for several reasons in the surface layer including surface heterogeneity, vegetation, topography, or instrumental reasons (Aubinet et al. 2012). To realize this, the coordinate system must be rotated into the streamlines (Kaimal and Finnigan 1994; Wilczak et al. 2001; Finnigan et al. 2003). Furthermore, there are instances in the stable boundary layer where Eq. 2.5 must see the addition of a wave component (Foken and Wichura 1996; Durden et al. 2013). The wave component is embedded in the signal and is superimposed to the turbulent flux.

The relation between the turbulent diffusion coefficient for momentum K_m and heat K_H is given by the turbulent Prandtl number

$$K_m = \Pr_t \cdot K_H \tag{2.6}$$

which is $Pr_t \sim 0.8$. The relation between the turbulent diffusion coefficients of momentum and water vapour is called the turbulent Schmidt number Sc_t. In the case of neutral stratification, K_m can be written according to the concept of the flow near the wall with the von-Kármán constant κ (Prandtl 1925):

$$K_m = \kappa \cdot z \cdot u_*. \tag{2.7}$$

Combining these relations expressing the turbulent diffusion coefficients with Eqs. (2.1)–(2.3), the equations for the friction velocity, the sensible and the latent heat flux can be expressed in kinematic units, where q is the specific humidity:

$$u_* = \sqrt{-\overline{u'w'}} = \kappa \cdot z \cdot \frac{\partial u}{\partial z} = \kappa \cdot \frac{\partial u}{\partial \ln z} \tag{2.8}$$

$$\overline{w'T'} = -\frac{1}{\Pr_t} \cdot \kappa \cdot u_* \cdot \frac{\partial T}{\partial \ln z} \tag{2.9}$$

$$\overline{w'q'} = -\frac{1}{Sc_t} \cdot \kappa \cdot u_* \cdot \frac{\partial q}{\partial \ln z}. \tag{2.10}$$

The turbulent fluxes of sensible heat can be transferred into energetic units by multiplication with the air density for dry air (pressure p in hPa and temperature in K)

$$\rho = \frac{p \cdot 100}{287.0586 \cdot T} \left[\text{kgm}^{-3}\right]. \tag{2.11}$$

For wet air, the air temperature must be replaced by the virtual temperature

$$T_v = T(1 + 0.61 \cdot q), \tag{2.12}$$

which includes the influence of moisture on air density. The heat capacity for constant pressure is

$$c_p = 1004.832 \left[\text{JK}^{-1}\text{kg}^{-1}\right]. \tag{2.13}$$

The latent heat flux in energetic units follows the multiplication of Eq. (2.10) with air density and the specific heat of evaporation

$$\lambda = 2500827 - 2360(T - 273.15)\left[\text{Jkg}^{-1}\right]. \tag{2.14}$$

If the latent heat flux in kinematic units were not determined with the specific humidity but instead with water vapour pressure, an additional multiplication with the factor $\frac{0.622}{p}$, where p is in hPa, is necessary.

Table 2.1 Roughness length in m from different sources (Reithmaier et al. 2006, updated)

surface	ESDU (1972)	Troen and Peterson (1989)	Wieringa (1992)	Fiedler according to Hasager and Jensen (1999)	Davenport et al. (2000)
Ice	10^{-5}				
Water	10^{-4}–10^{-3}				
Snow	0.002				
Bare soil		0.03	0.004	0.03	0.005
Grassland	0.005–0.02	0.03	0.06	0.08	0.03
Winter crops (winter)		0.1	0.09	0.12	0.1
Winter crops	0.05	0.1	0.18	0.09	0.25
Summer crops	0.05	0.1	0.18	0.09	0.25
Clearings		0.1	0.35	0.004	0.2
Shrubs	0.2	0.4	0.45	0.3	0.5
Conifer forest	1–2	0.4	1.6	0.9	1.0
Deciduous forest	1–2	0.4	1.7	1.2	2.0
Settlement	0.5–2	0.4	0.7	0.5	2.0

Furthermore, for heights above approximately 10 m, the temperature must be replaced in all equations given above by the potential temperature

$$\theta = T\left(\frac{1000}{p}\right)^{R_L/c_p}. \tag{2.15}$$

The integration of Eq. (2.8) is given by

$$u(z) - u(z_0) = u(z) = \frac{u_*}{\kappa}\ln\frac{z}{z_0}, \tag{2.16}$$

where z_0 is an integration constant. Because this parameter is dependent on the characteristics of the underlying surface, it is called the roughness parameter or the roughness length. It varies from 10^{-3} to 10^{-5} m for water and ice, 10^{-2} m for grassland up to 0.2 m for small trees. More data are given in Table 2.1. Additional details about the application of the different schema are given in Sect. 6.2.1.

The integration of the equations for the sensible (2.9) and the latent heat (2.10) flux is formally identical to those of the momentum flux. The integration constants are so-called roughness temperature z_{0T} and roughness humidity z_{0q}. Both are approximately 10 % of the roughness length. In this region, the temperature and the humidity have approximately the value of those near the surface. These roughness lengths are usually parameterized in models.

$$T(z) - T(z_{0T}) = -\frac{\Pr_t \cdot T_*}{\kappa} \ln \frac{z}{z_{0T}} \tag{2.17}$$

$$q(z) - q(z_{0q}) = -\frac{Sc_t \cdot q_*}{\kappa} \ln \frac{z}{z_{0q}} \tag{2.18}$$

with the dynamical temperature or temperature scale

$$T_* = -\frac{\overline{w'T'}}{u_*} \tag{2.19}$$

and the dynamical moisture scale

$$q_* = -\frac{\overline{w'q'}}{u_*}. \tag{2.20}$$

The extension of the profile equation for non-neutral conditions is given by Monin-Obukhov's similarity theory (Monin and Obukhov 1954). This theory defines a dimensionless Obukhov parameter

$$\varsigma = {}^{z}\!/_{L} \tag{2.21}$$

which describes the effects of friction, sensible heat flux and buoyancy. The parameter L is called Obukhov length (Obukhov 1946, 1971; Businger and Yaglom 1971; Foken 2006).

$$L = -\frac{u_*^3}{\kappa \frac{g}{T} \frac{Q_H}{\rho \cdot c_p}} \tag{2.22}$$

This definition is valid near the surface and provides low moisture content. In the case that air density is influenced by moisture, the use of the virtual temperature is more exact, Eq. (2.12). In addition, the temperature should be replaced by virtual temperature in the air density Eq. (2.12). The more exact definition of the Obukhov length is with the virtual potential temperature:

$$L_v = -\frac{u_*^3}{\kappa \frac{g}{\theta_v} \frac{Q_{Hv}}{\rho \cdot c_p}}. \tag{2.23}$$

In this equation, Q_{Hv} is is called the buoyancy flux because it includes also the motion due to the moisture effect on air density. The buoyancy flux can be determined with Eq. (2.9) by replacing the temperature by the virtual temperature, which is nearly equal to the sonic temperature (Kaimal and Gaynor 1991) measured with sonic anemometers (Sect 7.2).

Table 2.2 Determination of the stability the surface layer dependent on the dimensionless parameter ς and the universal function $\varphi(\varsigma)$ adopted from Foken (2008)

Stability	Remark	ς	$\varphi(\varsigma)$
Unstable	Free convection, independent from u_*	$-1 > \varsigma$	No definition
	Dependent from u_*, T_*	$-1 < \varsigma < 0$	$\varphi(\varsigma) < 1$
Neutral	Dependent from u_*	$\varsigma \sim 0$	$\varphi(\varsigma) = 1$
Stable	Dependent from u_*, T_*	$0 < \varsigma < 0.5\ldots2$	$1 < \varphi(\varsigma) < 3\ldots 5$
	Independent from z	$0.5\ldots1 < \varsigma$	$\varphi(\varsigma) \sim \text{const} \sim 3\ldots 5$

From the application of Monin-Obukhov similarity theory on profiles, Eqs. (2.8)–(2.10), follows a dependency on the dimensionless parameter ζ (Table 2.2), which is the basis of the universal functions $\varphi_m(\varsigma)$, $\varphi_H(\varsigma)$ and $\varphi_E(\varsigma)$ for the momentum, sensible and latent heat exchange:

$$u_* = \sqrt{-\overline{u'w'}} = \frac{\kappa \cdot z}{\varphi_m(\varsigma)} \cdot \frac{\partial u}{\partial z} = \frac{\kappa}{\varphi_m(\varsigma)} \cdot \frac{\partial u}{\partial \ln z} \tag{2.24}$$

$$\overline{w'T'} = -\frac{\kappa \cdot u_*}{\mathrm{Pr}_t \cdot \varphi_H(\varsigma)} \cdot \frac{\partial T}{\partial \ln z} \tag{2.25}$$

$$\overline{w'q'} = -\frac{\kappa \cdot u_*}{Sc_t \cdot \varphi_E(\varsigma)} \cdot \frac{\partial q}{\partial \ln z}. \tag{2.26}$$

The present recommendation for the use of universal functions (Fig. 2.6) is the universal functions by Businger et al. (1971) in the re-evaluated form by Högström (1988). Given that several footprint models use other functions, a selection is given in Table 2.3 for momentum flux and Table 2.4 for the sensible and latent heat fluxes. There is a paucity of universal functions for the stable stratification because of the complexity of the nocturnal stable boundary layer (Andreas 2002). The universal functions can be assumed to be constant for $\zeta > 0{,}8$ (see e.g. Handorf et al. 1999).

The accuracy of the profile method depends on those of the turbulent Prandtl or Schmidt numbers, the von-Kármán constant and the universal functions. For the turbulent Prandt number, an overview of data several authors is given in Table 2.5. The von-Kármán constant is presently accepted as $\kappa = 0.40 \pm 0{,}01$ (Högström 1996). For the universal function, the following accuracies are given by Högström (1996):

$$\begin{aligned} |z/L| \le 0.5: &\quad |\delta\varphi_H| \le 10\,\% \\ |z/L| \le 0.5: &\quad |\delta\varphi_m| \le 20\,\% \\ z/L > 0.5: &\quad \varphi_m,\ \varphi_H = const\ ? \end{aligned} \tag{2.27}$$

Table 2.3 Universal function for the momentum exchange including the re-evaluated form by Högström (1988) with a von-Kármán constant of 0.40 and marked with *, adapted from Foken (2008)

References	κ	Universal function for momentum exchange, $\varphi_m(\varsigma)$	
Webb (1970)	–	$1+4.5\,z/L$	$z/L<-0.03$
Businger et al. (1971)	0.35	$(1-15\,z/L)^{-1/4}$	$-2<z/L<0$
		$1+4.7\,z/L$	$0<z/L<1$
Businger et al. (1971), Högström (1988)	0.40*	$(1-19.3\,z/L)^{-1/4}$	$-2<z/L<0$
		$1+6\,z/L$	$0<z/L<1$
Dyer (1974)	0.41	$(1-16\,z/L)^{-1/4}$	$-1<z/L<0$
		$1+5\,z/L$	$0<z/L$
Dyer (1974), Högström (1988)	0.40*	$(1-15.2\,z/L)^{-1/4}$	$-1<z/L<0$
		$1+4.8\,z/L$	$0<z/L$

Table 2.4 Universal function for the exchange of sensible and latent heat including the re-evaluated form by Högström (1988) with a von-Kármán constant of 0.40 and marked with *, adapted from Foken (2008)

References	κ	Universal function for the exchange of sensible and latent heat	
Webb (1970)	–	$1+4.5\,z/L$	$z/L<-0.03$
Businger et al. (1971)	0.35	$0.74\,(1-9\,z/L)^{-1/2}$	$-2<z/L<0$
		$0.74+4.7\,z/L$	$0<z/L<1$
Businger et al. (1971), Högström (1988)	0.40*	$0.95\ \ (1-11.6\,z/L)^{-1/2}$	$-2<z/L<0$
		$0.95+7.8\,z/L$	$0<z/L<1$
Dyer (1974)	0.41	$(1-16\,z/L)^{-1/2}$	$-1<z/L<0$
		$1+5\,z/L$	$0<z/L$
Dyer (1974), Högström (1988)	0.40*	$0.95(1-15.2\,z/L)^{-1/2}$	$-1<z/L<0$
		$0.95+4.5\,z/L$	$0<z/L$

For the re-evaluation by Högström (1988) use $\varphi_H(\varsigma)\sim\varphi_E(\varsigma)$, $Pr_t=Sc_t=1$ because both numbers are already included into the universal function

It must be assumed that surface-layer parameterizations are influenced by boundary-layer conditions, specially by those of the mixed-layer height (Johansson et al. 2001). It should be pointed out however that the influence of the latter is still second to the influence of atmospheric stratification.

Integrating Eqs (2.24) and (2.26) and using the universal functions presented in Tables 2.3 and 2.4 was first shown by Paulson (1970). The integration from the roughness length z_0 to z in the wind profile applies the definition $u(z_0)=0$

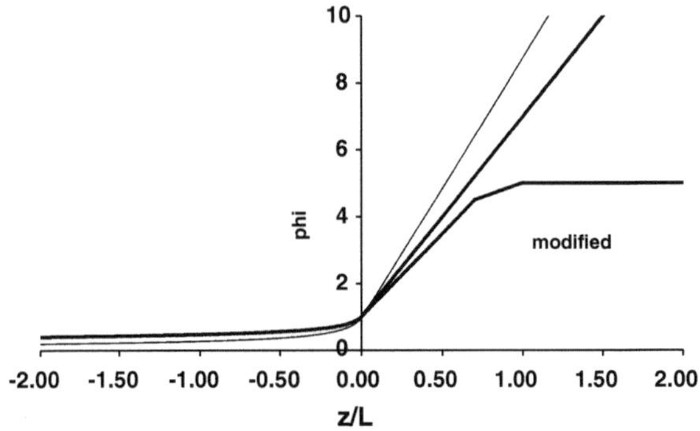

Fig. 2.6 Typical universal function for momentum (*bold line*) and the heat and mass exchange (*thin line*). The line 'modified' uses a height-independent range (Foken 2008). Well defined is the function only in the range |z/L| < 1 (Tables 2.2, 2.3, and 2.4)

Table 2.5 The reciprocal turbulent Prandtl number according to different authors (Foken 2008)

Authors	Pr_t^{-1}
Businger et al. (1971)	1.35
– correction according to Wieringa (1980)	1.00
– correction according to Högström (1988)*	1.05
Kader and Yaglom (1972)	1.15–1.39
Foken (1990)	1.25
Högström (1996)	1.09 ± 0.04

*Högström (1988) uses $Pr_t = 1$ in the profile equation, but has included $Pr_t = 1.05$ in the universal function (see Tables 2.3 and 2.4)

$$\begin{aligned} u(z) - u(z_0) = u(z) &= \frac{u_*}{\kappa}\left[\ln\frac{z}{z_0} - \int \phi_m(z/L)\ dz\right] \\ u(z) &= \frac{u_*}{\kappa}\left[\ln\frac{z}{z_0} \quad - \quad \psi_m(z/L)\right] \end{aligned} \tag{2.28}$$

with the integrated universal function:

$$\psi_m(\varsigma) = \int\limits_{z_0/L}^{z/L} [1 - \phi_m(\varsigma)]\ \frac{d\varsigma}{\varsigma}. \tag{2.29}$$

The integration of the universal function by Businger et al. (1971) and subsequently reformulated by Högström (1988) is for the momentum exchange and the flux of sensible heat in the unstable case:

$$\psi_m(\varsigma) = \ln\left[\left(\frac{1+x^2}{2}\right)\left(\frac{1+x}{2}\right)^2\right] - 2\tan^{-1}x + \frac{\pi}{2} \quad for \quad \varsigma < 0 \tag{2.30}$$

$$\psi_H(\varsigma) = 2\ln\left(\frac{1+y}{2}\right) \quad for \quad \varsigma<0 \tag{2.31}$$

with

$$x = (1 - 19.3\varsigma)^{1/4} \quad y = 0.95\ (1 - 11.6\varsigma)^{1/2}. \tag{2.32}$$

In the stable case, the integration is very simple:

$$\psi_m(\varsigma) = -\ 6\ \varsigma \quad for\ \varsigma \geq 0 \tag{2.33}$$

$$\psi_H(\varsigma) = -\ 7.8\ \varsigma \quad for\ \varsigma \geq 0. \tag{2.34}$$

As far as other universal functions are concerned, according to Tables 2.3 and 2.4, the parameters x and y in Eq. (2.32) must be defined differently.

Besides the stability parameter, ς represents also another stability parameter which can be formulated using the equation of turbulent energy (TKE). The ratio of the buoyancy production term and the mechanical production term is called the flux Richardson number (Richardson 1920; Stull 1988)

$$Ri_f = \frac{g}{\overline{T}} \cdot \frac{\overline{w'T'}}{\overline{w'u'} \cdot \left(\partial u / \partial z\right)}. \tag{2.35}$$

Because fluxes are proportional to gradients, a gradient Richardson number can also be defined:

$$Ri_g = -\ \frac{g}{\overline{T}} \cdot \frac{\partial T / \partial z}{\left(\partial u / \partial z\right)^2}. \tag{2.36}$$

A further simplification is the bulk Richardson number

$$Ri_b = -\ \frac{g}{\overline{T}} \cdot \frac{\Delta T \cdot \Delta z}{(\Delta u)^2}, \tag{2.37}$$

used in the meteorology. In analogy to the Obukhov length, the Richardson number definition can also be given using the potential and virtual temperatures. If fluxes are available, the Richardson flux number should be used. Otherwise, the gradient or bulk number will be substituted.

The critical Richardson number, which characterizes the change from turbulent to laminar or molecular conditions, is $Ri_{gc} = 0.2$ or $R_{fc} = 1.0$. The recalculation from ς into Ri_g is stability dependent according to the following relations (Businger et al. 1971; Arya 2001):

Table 2.6 Overview of different stability parameters (Foken 2008), added by the potential temperature $\theta\ (z) = T(0\ m) + 0.0098\ \mathrm{K} \cdot z$

Stability	Temperature for $z < 10$ m	Potential temperature	Ri	L	$\varsigma = z/L$
Unstable	$T(0) > T(z)$	$\Theta(0) > \Theta(z)$	< 0	< 0	< 0
Neutral	$T(0) \sim T(z)$	$\Theta(0) \sim \Theta(z)$	~ 0	$\pm\infty$	~ 0
Stable	$T(0) < T(z)$	$\Theta(0) < \Theta(z)$	$0 < Ri_g < 0.2$ $0 < Ri_f < 1.0$	> 0	$0 < \varsigma < \sim 1$

$$\begin{aligned} \varsigma &= Ri_g \quad f\ddot{u}r \quad Ri_g < 0 \\ \varsigma &= \frac{Ri_g}{1 - 5\,Ri_g} \quad for \quad 0 \leq Ri_g \leq 0.2 = Ri_c. \end{aligned} \tag{2.38}$$

An overview over different parameters is given in Table 2.6.

2.2.2 Profile Functions Above the Canopy

Over dense vegetation (forests, crops, etc.) due to the logarithmical wind profile the surface according to Eq. (2.16) is an apparent surface at height d (displacement height, zero-plane displacement height), for which the wind profile fulfil these equations. The new height is called the aerodynamic height $z'(d) = 0$. In contrast, the geometric height is measured from the ground surface, is $z = z' + d$. Because Eq. (2.16) is valid for the aerodynamic heights (Fig. 2.7), the equation with geometric heights measured from the surface is given as:

$$u(z) = \frac{u_*}{\kappa} ln \frac{z - d}{z_0}. \tag{2.39}$$

Consequently, all profile equations and equations related to integral turbulence characteristics in the following chapters must be modified for vegetation by replacing "z" with "$z + d$" or by assuming that all heights are aerodynamic heights. Usually, $d = 0.67\ h_c$ is applied with h_c as canopy height. Under these conditions, the roughness length is simply approximated by

$$z_0 = 0.1\ h_c. \tag{2.40}$$

Foken (2008) recommended to determine the canopy height using the tallest plants or trees that cover 10 % or more of the vegetation at the site.

More complicated is the determination of the displacement height in an urban surface. If profile and flux measurements are available, the displacement height can be calculated using the constant flux layer assumption within the surface layer. This is referred to as the aerodynamic approach. Thereby, the displacement height

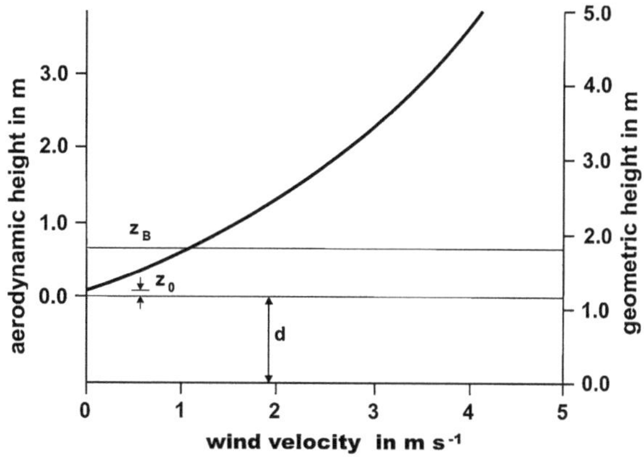

Fig. 2.7 Aerodynamic and geometric height for dense vegetation ($d = 1.2$ m) according to Foken (2008)

is the level where fluxes from both profile and flux measurements become identical. The displacement height can also be determined using two scintillometers placed at two different levels (Kanda et al. 2002).

The morphometric method (Grimmond and Oke 1999) has been more often used, in the simplest way

$$d = f_d\, \overline{z_H}, \quad z_0 = f_o\, \overline{z_H} \tag{2.41}$$

with $\overline{z_H}$ as averaged building height and using the above given values of $f_d = 0.67$ and $f_O = 0.1$ or $f_d = 0.8$ for densely built-up cities (Roth et al. 2006).

Another method was proposed by MacDonald et al. (1998) who use also the density of the buildings as well as the mean building height

$$d = \overline{z_H}\left[1 + \alpha^{-\lambda_P}(\lambda_P - 1)\right] \tag{2.42}$$

with empirical coefficient $\alpha = 4.43$ and λ_P the plain area fraction calculated as the area fraction occupied by built-up elements.

Another approach was introduced by Raupach (1994), which uses the frontal areal index λ_F

$$d = \overline{z_H}\left(1 - \frac{1 - e^{-\sqrt{c_{d1}\lambda_F}}}{\sqrt{c_{d1}\lambda_F}}\right) \tag{2.43}$$

with the empirical coefficient $c_{dl} = 7.5$.

2.2.3 Profile Functions in the Canopy

Measurements of profile functions in low vegetation are very rare because of experimental problems. The sensors are often large in relation to the canopy height

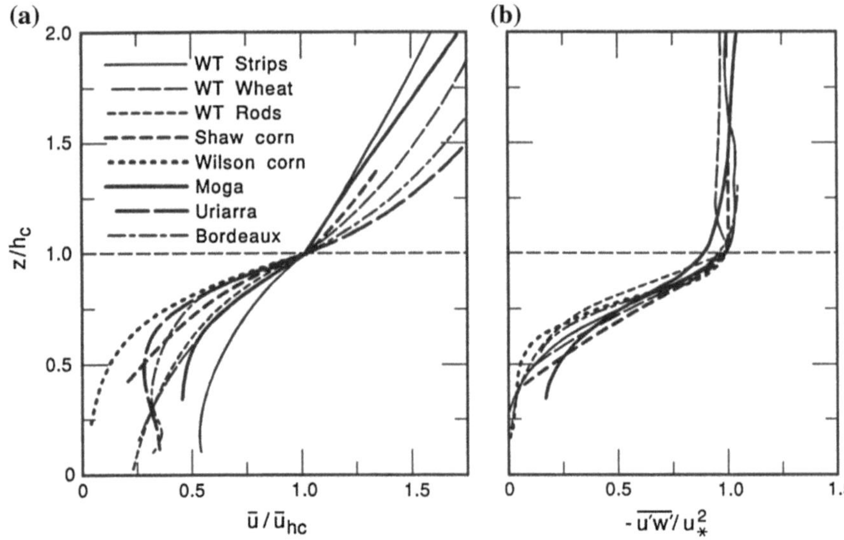

Fig. 2.8 Profile of the mean wind velocity and the momentum flux in the canopy normalized with the value at the top of the canopy from wind tunnel, corn and forest measurements by different authors (Kaimal and Finnigan 1994): Wind tunnel (WT) shrips, Raupach et al. (1986); wind tunnel wheat, Brunet et al. (1994); wind tunnel rods, Seginer et al. (1976); Shaw corn (Shaw et al. 1974); Wilson corn (Wilson et al. 1982); moga, Raupach et al. (1996); uriarra, Denmead and Bradley (1987); Bordeaux forest, Brunet personal communication (Published with kind permission of © Oxford University Press, 1994. All Rights Reserved)

and are partly ventilated. Some examples are given by Geiger et al. (2009). For tall vegetation and for forests, much more data is available. By normalizing the profiles with their respective values at the top of the canopy, the profiles are similar to those inside a wind tunnel, in low vegetation and even in forest canopies (Fig. 2.8). Cionco (1978) proposed a profile function which depends on canopy height h_c

$$\overline{u(z)} = \overline{u(h_c)} \cdot e^{\alpha(z/h_c - 1)} \tag{2.44}$$

and a coefficient α given for different plants in Table 2.7.

A direct calculation of the coefficient α that is vegetation type dependent is provided to us by Goudriaan (1977). The formulation of the coefficient is a function of both the mean distance of the leaves (l_m), and the leaf area index (LAI):

$$\alpha \cong \left(\frac{0.2 \cdot LAI \cdot h_c}{l_m} \right). \tag{2.45}$$

Table 2.7 Values of the profile parameter α in a plant canopy according to Eq. (2.45)

Plant canopy	Profile parameter α	References
Wheat	2.45	Cionco (1978)
	1.6	Brunet et al. (1994)
Corn	1.97	Cionco (1978)
	2.4	Shaw et al. (1974)
	4.1	Wilson et al. (1982)
Rice	1.62	Cionco (1978)
Sunflower	1.32	Cionco (1978)
Larch plantation	1.00	Cionco (1978)
Forest, 20 m	1.7	Denmead and Bradley (1987)

2.2.4 Roughness Sublayer

Above the canopy, the profiles of the state parameters are strongly influenced by the roughness of the surface and the ideal profile (Eqs. (2.24)–(2.26)) must be modified. This range is called the roughness sublayer and includes the canopy height. It is approximately three times the canopy height. The roughness sublayer was firstly found in laboratory experiments (Raupach et al. 1980) and later in the natural environment e.g. by Shuttleworth (1989). This variable is of considerable significance to flux measurements, especially when using profile functions. While over low vegetation, typical relations of the measuring height to the roughness length z/z_0 are 100–1,000. Above a forest canopy with a generally significant roughness sub-layer, that dimensionless value hovers around 5–10 (Garratt 1980).

Therefore, the Monin-Obukhov similarity assumption cannot be applied in the roughness sublayer of thickness z_*(Garratt 1978, 1980; Raupach et al. 1980; Raupach and Legg 1984), which according to Verhoef et al. (1997) is

$$z_* = h_c + cL_s \tag{2.46}$$

where h_c is the canopy height, c is 2 for momentum and 3 for heat exchange (Mölder et al. 1999) and L_s is the characteristic length scale (shear scale) of the mixing layer (Raupach et al. 1996; Finnigan 2000):

$$L_s = \frac{u(h_c)}{\left(\partial u / \partial z\right)_{z=h_c}}. \tag{2.47}$$

In their model, Rannik et al. (2003) assumed the roughness sublayer high as $h_c + d$, where d is the zero plane displacement of $d = 2/3\ h_c$. In this layer, weaker gradients are found but the turbulent transport occurs largely through the action of coherent structures in the mixing layer (Raupach et al. 1996; Finnigan 2000). Therefore, an additional function $\varphi_*(z/z_*)$ must be added to the profile equations

(Eqs. (2.24)–(2.26) to represent the effect of the roughness sublayer, since the latter increases the diffusion coefficient (Garratt 1992):

$$u_* = \sqrt{-\overline{u'w'}} = \frac{\kappa}{\varphi_{*u}(z/z_*) \cdot \varphi_m(\varsigma)} \cdot \frac{\partial u}{\partial \ln z} \tag{2.48}$$

$$\overline{w'T'} = \frac{\mathrm{Pr}_t^{-1} \cdot \kappa \cdot u_*}{\varphi_{*T}(z/z_*) \cdot \varphi_H(\varsigma)} \cdot \frac{\partial T}{\partial \ln z} \tag{2.49}$$

$$\overline{w'q'} = \frac{Sc_t^{-1} \cdot \kappa \cdot u_*}{\varphi_{*q}(z/z_*) \cdot \varphi_E(\varsigma)} \cdot \frac{\partial q}{\partial \ln z} \tag{2.50}$$

where $1/\varphi_*(z/z_*)$ is called the enhancement factor (Raupach and Legg 1984; Simpson et al. 1998). The universal function for the roughness sublayer for the wind variables is given by

$$\varphi_{*_u}(z/z_*) = \exp[-0.7\,(1 - z/z_*)] \tag{2.51}$$

(Garratt 1992; Graefe 2004). Another definition was given by Cellier and Brunet (1992)

$$\varphi_{*_u} = \left(\frac{z}{z_*}\right)^\eta \tag{2.52}$$

where $\eta = 0.6$, which was also found by Mölder et al. (1999). The functions for scalars are not well defined. Mölder et al. (1999) found a linear relation with height for humidity and temperature i.e.

$$\varphi_{*_{T,q}} = z/z_*. \tag{2.53}$$

Another more sophisticated method to describe this phenomenon which takes into account the coherent structures is the mixing-layer theory (Raupach et al. 1996; Finnigan 2000). This theory suggests that the reduced gradients above the top of the canopy can be attributed to the presence of Kelvin-Helmholtz instability present in strong shear flows and by the generation of disturbances and coherent structures. This approach has not yet been used in footprint analysis. A combination of both was given by Harman and Finnigan (2007, 2008), who defined the roughness sublayer for momentum and scalar fluxes dependent on the mixing layer length scale L_s, Eq. (2.47). According to this theory, both the displacement height and roughness length vary with stability.

2.2.5 Power Laws

For many applied purposes, power laws are used to determine the wind profile in the surface layer and the lower boundary layer (Doran and Verholek 1978; Sedefian 1980; Joffre 1984; Wieringa 1989; Hsu et al. 1994):

$$\frac{u_1}{u_2} = \left(\frac{z_1}{z_2}\right)^p. \tag{2.54}$$

In wind power applications, an exponent $p = {}^1/_7$ is often used (Peterson and Hennessey Jr 1978).

Differentiating Eq. (2.54), we obtain the expression by Huang (1979):

$$p = \frac{z}{u} \cdot \frac{\partial u}{\partial z}. \tag{2.55}$$

This method offers a more complicated approach including also a dependency on the roughness of the surface and the stability using universal functions of the Monin-Obukhov similarity theory (see Sect. 2.2.1). Irvin (1978) proposed the following simple equation:

$$p = \frac{u_*}{u \cdot \kappa} \cdot \varphi_m(\varsigma). \tag{2.56}$$

The factor $u \cdot \kappa \cdot u_*^{-1}$ can be expressed by the integrated form of the universal function given in Sedefian (1980):

$$p = \frac{\varphi_m\left(\frac{z}{L}\right)}{\left[\ln\left(\frac{z}{z_0}\right) - \psi_m\left(\frac{z}{L}\right)\right]}. \tag{2.57}$$

Huang (1979) used also this form but used the concrete universal functions by Webb (1970) and Dyer (1974) allowing for large roughness elements in contrast with the earlier integration provided by Paulson (1970). For the unstable case, it follows

$$\begin{aligned} p &= \frac{\left(1 - 16\frac{z}{L}\right)^{-1/4}}{\ln\frac{(\eta-1)(\eta_0+1)}{(\eta+1)(\eta_0-1)} + 2\tan^{-1}\eta - 2\tan^{-1}\eta_0} \\ \eta &= \left(1 - 16\frac{z}{L}\right)^{1/4} \quad \eta_0 = \left(1 - 16\frac{z_0}{L}\right)^{1/4} \end{aligned} \tag{2.58}$$

and for the stable case:

$$p = \frac{1 + 5\frac{z}{L}}{\ln\frac{z}{z_0} + 5\frac{z}{L}}. \tag{2.59}$$

This approach is used in the footprint model by Kormann and Meixner (2001). According to Högström (1988), the coefficients 16 and 5 should be replaced by 19.3 and 6.0.

The use of this method is fraught with difficulties. In the hours before noon, when in the presence of a developing convective boundary layer, the method works well. Later in the afternoon, due to the cooling by longwave upwelling radiation, the layer close to the surface becomes stable while the layers above are still unstable (see Sect. 2.3.2.). Therefore, the stability measured near the surface cannot be applied using the power law. This is because atmospheric stratification sets in first near the surface due to radiative cooling early in the late afternoon and evening while aloft, the upper layers of the atmosphere are still unstable (Foken 2008).

2.2.6 Dispersion Profiles

Standard deviations of the three wind components are necessary input parameters of analytical footprint models which are based on a Gaussian dispersion approach. Lagrangian footprint models also need parameterizations of the profiles of standard deviations in the canopy (in the case of tall vegetation) and above.

The similarity between fluxes and variances based on the equation of the turbulent kinetic energy or on analogue equations for sensible heat and other scalars (flux-variance similarity, see Foken 2008). In these equations, the standard deviations of the vertical wind component and the temperature or another scalar are included (Wyngaard and Coté 1971; Foken et al. 1991):

$$\sigma_w = \sqrt{\overline{w'^2}} \quad \textit{and} \quad \sigma_T = \sqrt{\overline{T'^2}}. \tag{2.60}$$

The normalized standard deviations are also called integral turbulence characteristics (Tillman 1972), because they characterize the atmospheric turbulence over the entire range of turbulence spectra. In the surface layer and in steady-state conditions, these characteristics of the three wind components in the neutral case (Lumley and Panofsky 1964; Panofsky 1984) can be expressed as:

$$\begin{aligned} \sigma_w/u_* &\cong 1.25 \\ \sigma_u/u_* &\cong 2.45 \\ \sigma_v/u_* &\cong 1.9. \end{aligned} \tag{2.61}$$

In the atmospheric surface layer, the turbulence is anisotropic. Therefore the standard deviations of the wind components are different, $\sigma_w < \sigma_v < \sigma_u$. For

Table 2.8 Integral turbulence characteristics for stable and unstable conditions (Foken 2008)

Parameter	z/L	c_1	c_2
σ_w/u_*	0 > z/L > −0.032	1.3	0
	−0.032 > z/L	2.0	1/8
σ_u/u_*	0 > z/L > −0.032	2.7	0
	−0.032 > z/L	4.15	1/8
σ_T/T_*	0.02 < z/L < 1	1.4	−1/4
	0.02 > z/L > −0.062	0.5	−1/2
	−0.062 > z/L > −1	1.0	−1/4
	−1 > z/L	1.0	−1/3

non-neutral conditions, a large number of parameterizations is given in the literature (Foken 2008). For the wind components, these follow the form

$$\sigma_{u,v,w}/u_* = c_1 \cdot \left(z/L\right)^{c_2} \tag{2.62}$$

and for the temperature or other scalars (with a different normalization instead of T_*)

$$\sigma_T/T_* = c_1 \cdot \left(z/L\right)^{c_2}. \tag{2.63}$$

An example of the integral characteristics for unstable (not free convection) and neutral conditions is given in Table 2.8. In the stable case, there are only a few parameterizations available. One can use the above given parameterizations only for the wind components of the unstable case also for the stable case as a first approach.

For the vertical wind component, most studies agree with one another with the parameterization given by Panofsky et al. (1977) is mainly used for a wide range of stratification, $-1 < z/L < 0$:

$$\sigma_w/u_* = 1.3 \cdot \left(1 - 2 \cdot \frac{z}{L}\right)^{1/3}. \tag{2.64}$$

The integral turbulence characteristics for temperature and other scalars are in the neutral case due to $T_* \rightarrow 0$ not well defined. In the unstable range, these turbulence properties are closely coupled to atmospheric stability.

Several authors also found a dependency on the mixed layer height (Panofsky et al. 1977; Peltier et al. 1996; Johansson et al. 2001; Thomas and Foken 2002). This dependency arises mostly in very unstable conditions. Other authors (Yaglom 1979; Tennekes 1982; Högström 1990; Smedman 1991) assumed a dependency on the Coriolis parameter, probably only for neutral conditions to be statistically significant. This was first found by Högström et al. (2002).

For free convective conditions ($z/L < -1$), the scaling parameter is the convective velocity (*Deardorff-velocity*)

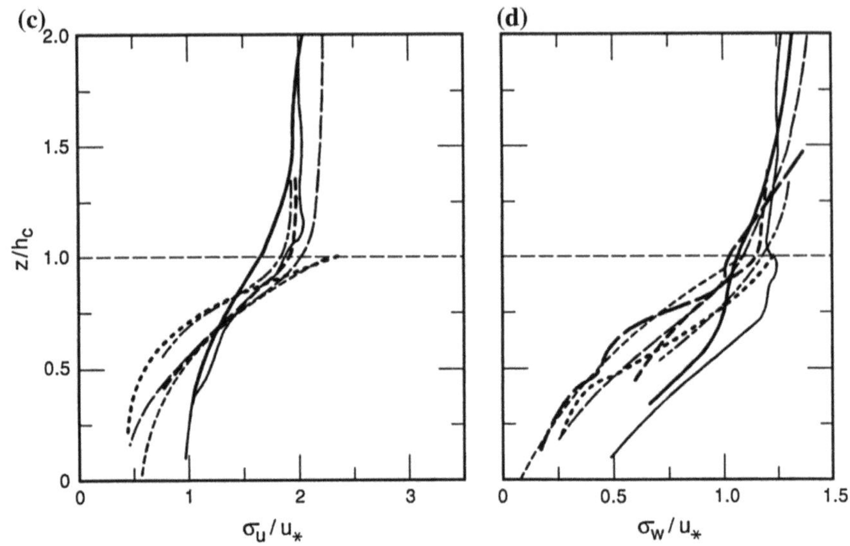

Fig. 2.9 Profiles of the standard deviation of the horizontal and vertical wind component for wind tunnel, corn and forest measurements by different authors (Kaimal and Finnigan 1994), for legend see Fig. 2.8. (Published with kind permission of © Oxford University Press, 1994. All Rights Reserved)

$$w_* = \left(\frac{g \cdot z_i}{\theta_v} \cdot \overline{\theta_v' w'} \right)^{1/3} \tag{2.65}$$

and partly the mixed layer height z_i (Garratt 1992). Such parameterizations must take into account the decrease of the characteristics with increasing height and an increase in the entrainment layer. One possible parameterization is given by Sorbjan (1989):

$$\sigma_w/_{w_*} = 1.08 \left(z/_{z_i} \right)^{1/3} \cdot \left(1 - z/_{z_i} \right)^{1/3} \tag{2.66}$$

$$\sigma_T/_{T_*} = 2 \left(z/_{z_i} \right)^{-2/3} \cdot \left(1 - z/_{z_i} \right)^{4/3} + 0.94 \left(z/_{z_i} \right)^{4/3} \cdot \left(1 - z/_{z_i} \right)^{-2/3}. \tag{2.67}$$

The profiles of the integral turbulence characteristics within the canopy are also of special interest in footprint modeling. These profiles are very similar for different types of canopies when normalized with their value in the height of the top of the canopy. This is illustrated in Fig. 2.9.

Inside the canopy, the profiles are strongly dependent on the leaf area index (Shaw et al. 1988). Furthermore, the profiles are also stability dependent (Shaw et al. 1988; Leclerc et al. 1990, 1991) and change with the coupling stage between the atmosphere and the canopy (Göckede et al. 2007). Also, the application of an

Table 2.9 Coefficients in Eqs. (2.68) and (2.69) for two forested sites

References	i	a_i	α_i	β_i	γ_i
Rannik et al. (2003),	u	2.30	1.0	1.0	−0.3
neutral, for Hyytiäla site	v	1.75	1.0	0.85	−0.2
(FI-Hyy)	w	1.25	0.9	1.2	−0.63
Foken et al. (2012),	u	2.01	8.97	1.37	0.29
for Waldstein-Weidenbrunnen	v	1.60	5.18	1.11	0.34
site (DE-Bay)	w	1.13	0.9	1.2	−0.63

analytical second-order closure model used a plant area profile (Massman and Weil 1999) to represent profiles of integral turbulence characteristics comparable with measured data (Göckede et al. 2007).

At a single site, the functional form of these relationships above the canopy is similar to that of measurements above low vegetation. Due to the lack of coupling between the canopy layer flow and the flow above forest canopies, a universal formulation of turbulence profiles is still conspicuously absent. Nevertheless, a site-specific parameterization is required to accurately model the footprint.

For measurements inside the canopy ($z < h_c$), a parameterization was proposed by Rannik et al. (2003)

$$\frac{\sigma_i}{u_*} = a_i \left\{ \exp\left[-\alpha_i \left(1 - \frac{z}{h_c} \right)^{\beta_i} \right] (1 - \gamma_i) + \gamma_i \right\} \qquad i = u, v, w; \quad z < h_c \tag{2.68}$$

and above the canopy constant values were assumed

$$\frac{\sigma_i}{u_*} = a_i \qquad i = u, v, w; \quad z > h_c. \tag{2.69}$$

The values are given in Table 2.9.

2.2.7 Relevance of Profile Parameterizations in Footprint Models

Many footprint models use profile functions for the parameterization of surface layer properties and, if the models are not limited to low vegetation, both the roughness sublayer and profile within the canopy must be parameterized. The discussion below deals with the specific parameterizations and differences between models based on their relevance (Chap. 3). Most footprint models use Monin-Obukhov similarity theory only with minor differences (Table 2.10)

Table 2.10 The use of surface layer parameterization for the stability influence in the widely distributed footprint models, "*italic*" not according the recently accepted modifications of the universal functions according to Högström (1988)

Footprint model	Use of	Remarks
Gash (1986)		Neutral
Schuepp et al. (1990), Horst and Weil (1992, 1994), Hsieh et al. (2000), Hsieh and Katul (2009)	Monin-Obukhov similarity theory	*Universal function by* Dyer (1974), but $\kappa = 0.4$
Leclerc and Thurtell (1990)	Monin-Obukhov similarity theory	*Universal function by Dyer (1974) for unstable and by* Businger et al. (1971)*for stable stratification,* but $\kappa = 0.4$
Schmid (1994, 1997)	Probably like Horst and Weil (1992; 1994)	
Leclerc et al. (1997)	Monin-Obukhov similarity theory	*Universal function by* Businger et al. (1971), but $\kappa = 0.4$
Kaharabata et al. (1997)	Monin-Obukhov similarity theory	*Universal function by* Businger et al. (1971), but $\kappa = 0.4$
Haenel and Grünhage (1999)	Monin-Obukhov similarity theory	*Universal function by* Dyer (1974)
Rannik et al. (2000, 2003)	Monin-Obukhov similarity theory, roughness sublayer	Universal function by Businger et al. (1971) in the re-evaluated form by Högström (1988)
Kormann and Meixner (2001)	Combination of power law and Monin-Obukhov similarity theory according to Huang (1979)	*Universal function by* Webb (1970) *and* Dyer (1974)
Kljun et al. (2002)	Monin-Obukhov similarity theory and convective boundary layer	See Rotach et al. (1996): universal function by Businger et al. (1971) in the re-evaluated form by Högström (1988)

between each other. Furthermore, most footprint models require a parameterization of the standard deviations of the wind components. In Lagrangian models, this parameterization is mainly those of vertical wind components, while, in the two-dimensional analytical case, the parameterization must include those of the lateral wind component (Table 2.11).

2.3 Internal Boundary Layers

The above given parameterizations of the atmospheric turbulence are based on the assumption of horizontal homogeneity. Landscapes are composed of a mosaic of typically heterogeneous surfaces with a change in surface characteristics within the

Table 2.11 The use of surface layer parameterization of the standard deviation of the wind components in the widely distributed footprint models

Footprint model	Use of	Remarks
Leclerc and Thurtell (1990)	Similar to Lumley and Panofsky (1964) in the neutral and stable case and Panofsky et al. (1977) and Hicks (1981) in the unstable case	Vertical wind component
Horst and Weil (1992, 1994)	Panofsky et al. (1977)	Vertical wind component
Hsieh et al. (2000) and Hsieh and Katul (2009)	See Leclerc and Thurtell (1990)	Vertical wind component
Rannik et al. (2000, 2003)	Similar to Panofsky et al. (1977)	

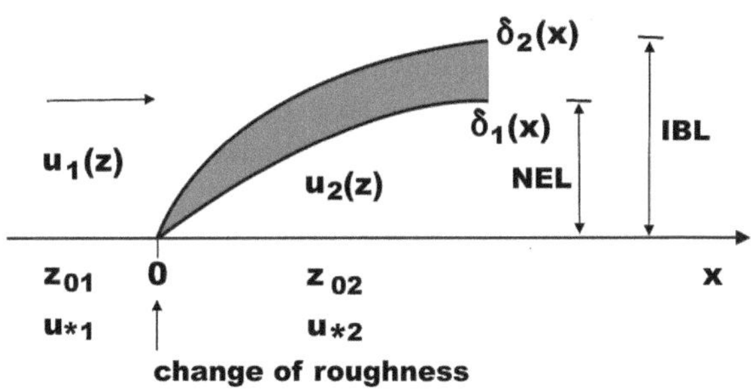

Fig. 2.10 Schematic structure of the internal boundary layer at a sudden change of the surface roughness according with the new equilibrium layer (NEL), the internal boundary layer (IBL), the fetch x, and the discontinuity layer between both according to the findings by Rao et al. (1974) from Foken (2008)

first 100 m. The wind profile develops depending on surface roughness, temperature profile-dependent on the surface temperature, etc. on the downwind site of such changes in the surface characteristics. Due to the horizontal wind field, the different profiles are shifted downwind. Therefore, internal boundary layers are significantly developed close to the surface. These arise in the presence of horizontal advection over discontinuities of surface properties (roughness, thermal properties, etc.). Overviews are given by Stull (1988), Garratt (1990, 1992) and Savelyev and Taylor (2001, 2005).

The internal boundary layer is a disturbed layer, which can be divided into different layers (Fig. 2.10). The layer below the discontinuity layer is called new equilibrium layer (NEL). Their properties come from the new surface. Above that layer (internal boundary layer, IBL), the layer is influenced by the surface on the upwind site. Above the new equilibrium layer, the discontinuity layer is not a sharp line but rather covers a range. For large fetches, the differences between both sides of an internal boundary layer decrease.

The concept of the internal boundary layer was used by Schmid and Oke (1990) to define the outer dimensions of the source area within the new equilibrium layer.

Typical fetch requirements for measurement levels located within the presence of internal boundary layers scale with the scale of footprint areas (Horst 2000). Nevertheless, the footprint concept provides an essential contribution to measurement sites, since most footprint models are not able to determine the effect of roughness changes. Some progress was made by Luhar and Rao (1994) and further on by Klaassen and Sogatchev (2006) for footprints at a forest edge and by Markkanen et al. (2010) for thermal heterogeneous surfaces. The practical application of the internal boundary-layer concept is discussed in Sect. 8.1.

2.3.1 Mechanical Internal Boundary Layer

The development of a mechanical internal boundary layer is caused by mechanical inhomogeneities both upwind and downwind (roughness length). In the simplest case, the height of an internal boundary layer can be determined by extrapolating the wind profiles above and within the internal boundary layer (Elliott 1958; Raabe 1983):

$$u_1(\delta) = u_2(\delta). \tag{2.70}$$

This method has the disadvantage that the point of intersection may be above the internal boundary layer or within the new equilibrium layer. More successful is the assumption that the undisturbed wind profile can be well fixed below and above the internal boundary layer and the height is between the upper and lower point of disturbance:

$$\delta = \frac{\delta_1 + \delta_2}{2}. \tag{2.71}$$

For practical reasons, it can be useful to use the lower level of the layer of disturbances as the height of the internal boundary layer, $\delta = \delta_l$, because the new equilibrium layer can assumed to be as undisturbed above the new surface (Rao et al. 1974) enabling the experimentalist to make measurements within that layer that reflect the properties of the surface beneath.

The mechanical internal boundary layer occurs for the flow from rough to smooth as well as from smooth to rough. Since different wind gradients differ between above smooth and rough surfaces, there is a characteristic development of internal boundary layers (Fig. 2.11).

The dependency of the height of an internal boundary layer on the fetch x was found in hydrodynamical investigations and is given by a 4/5-exponential law (Shir 1972; Garratt 1990; Savelyev and Taylor 2001):

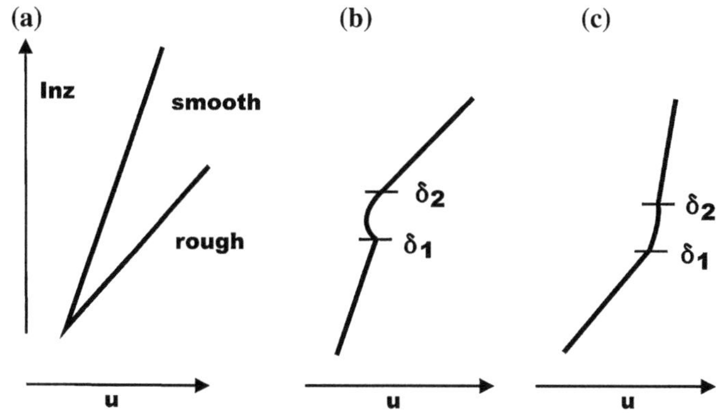

Fig. 2.11 The schematic wind profile at an internal boundary layer for neutral stratification: **a** typical profile for rough and smooth surfaces, **b** Change of the surface roughness from rough to smooth, **c** Change of the surface roughness from smooth to rough (Foken 2008)

Table 2.12 Experimental results for the coefficients in Eq. (2.73) to depend the height of the internal boundary layer (new equilibrium layer according to Rao et al. 1974), for more data see Savelyev and Taylor (2005)

Author	a	b	Conditions
Bradley (1968), Shir (1972)	0.11	0.8	z_{01}/z_{02} = 125 and 0.08 artificial roughness
Antonia and Luxton (1971, 1972)	0.28 0.04	0.79 0.43	$x \leq 10$ m, rough–smooth $x \leq 10$ m, smooth–rough wind tunnel
Raabe (1983)	0.30 ± 0.05	0.50 ± 0.05	Beach, on- and off-shore winds, 5 m < x < 1,000 m

$$\delta = f_1\left({}^{z_{01}}/_{z_{02}}\right) \cdot x^{4/5 + f_2\left(z_{01}/z_{02}\right)}. \tag{2.72}$$

A lot of experiments were done to verify this equation. Because of the large scatter in experimental results, most of the authors assume a simplified dependency

$$\delta = a \cdot x^b \tag{2.73}$$

for which some data are given in Table 2.12.

The height of the internal boundary layer normalized by the upwind roughness length was found to be higher for smooth to rough transition than for rough to smooth transition according to model calculations (Garratt 1990; Savelyev and Taylor 2001) and also the internal boundary layer is higher in the unstable case than in the stable case (Savelyev and Taylor 2005). But in the case of experimental data, no significant differences in the height of the internal boundary layer could be found due to the large scatter in the experimental data (Jegede and Foken 1999). Therefore, simple parameterizations of the internal boundary layer or equilibrium layer height according to Eq. (2.73) with the coefficients by Raabe (1983) are a good approach.

Large fetch requirements of internal boundary layers have constituted the basis of micrometeorological measurements for decades. Micrometeorological measurements were typically done at height/fetch ratios of 1/100 of undisturbed fetch. This requirement for the measuring height $z_m = 100x$ gives similar results as Eq. (2.73). Leclerc and Thurtell (1990) found that the 1:100 ratio used by micrometeorologists agrees with footprint calculations for short crop canopies in unstable conditions.

2.3.2 Thermal Internal Boundary Layer

In analogy to the mechanical internal boundary layer formed by a sudden change in surface roughness, a thermal internal boundary layer develops as a result of a change in surface temperature due to different land-use characteristics. Furthermore, other surface characteristics like different surface moisture or gas exchange conditions can lead to a scalar internal boundary layer. Few if any experimental results are available on the subject since this layer is typically combined with the mechanical internal boundary layer.

The height of the thermal internal boundary layer is given by Raynor et al. (1975):

$$\delta_T = c\left(\frac{u_*}{u}\right)\left[\frac{x\,(\theta_1-\theta_2)}{\left|{}^{\partial T}/_{\partial z}\right|}\right]^{1/2}. \tag{2.74}$$

The temperature gradient is measured on the upwind side or above the internal boundary layer, all other parameters in the reference level. The coefficient c depends on the reference level and is in the order of 1 (Arya 2001). Obviously, such parameterizations are similarly robust as in Eq. (2.73) for the mechanical internal boundary layer.

A special case is the thermal internal boundary layer during the afternoon mainly due to the "oasis effect" (Stull 1988). Shortly after noon above an evaporating surface, the temperature near the surface decreases and the stratification becomes stable. The height of the deflection point between stable stratification near the surface and the unstable conditions in the higher layers grows over time. This inversion layer close to the surface is also called a thermal internal boundary layer. Below the inversion, the sensible heat flux is downward and above upward. The height increases up to 50–100 m after sunset and is then identical with the stable boundary layer (see Sect. 3.1). According to this finding, the thermal internal boundary layer exists over a period of several hours at typical micrometeorological levels. Also in the early morning, this effect can be found, but with a much shorter duration, i.e. of the order of minutes. The problem is of particular relevance when the stability near the surface is used e.g. to determine the power law in the lower atmospheric boundary layer. The consequence on footprint models in the afternoon has not yet been investigated.

2.3.3 Blending Height Concept

According to the structure of internal boundary layers, it can be assumed that the internal boundary layers can only develop up to a certain level. The layers merge with one another far away from the change in surface roughness. Above this height, an area-averaged flux can be assumed. That means the properties near the surface fade (Taylor 1987). This idea is the basis of the so-called blending height concept according to Mason (1988) and its updated formulations by Claussen (1991) and Claussen and Walmsley (1994). The blending height is assumed to be at heights ranging between approximately 30–100 m with a close dependence on the magnitude of the underlying surface roughness and atmospheric conditions. The concept considers especially larger scale changes in surface roughness with characteristically horizontal distances of $L_x > 1$ km. The blending height l_b can be estimated as (Mahrt 1996)

$$l_b = 2\left(\frac{u_*}{u}\right)^2 L_x \approx 2\left(\frac{\sigma_w}{u}\right)^2 L_x \tag{2.75}$$

or as a simple approximation $l_b = L_x/200$.

The blending height concept has a large practical evidence for area averaging in numerical models (see Sect. 2.4), because it can be assumed for the model level in the height of approximately the blending height the fluxes above an heterogeneous surface are area averaged (Claussen 1995).

From the experimental standpoint, this concept is controversial. In an atmospheric boundary layer, conditions of free convection exist for $z/L < -1$, and for example for $z/L = -0.1$ at 2 m height, free convection starts already above 20 m (Eigenmann et al. 2009). According to Andreas and Cash (1999), the conditions for free convection are given for $\delta/L < -1$ in a growing internal boundary layer. This is a level where internal boundaries can be easily detected. The conditions of single surfaces can be also detected by aircraft measurements in the whole boundary layer if the single areas are large enough that convection can be developed. This is the case for horizontal extensions larger 200 m (Shen and Leclerc 1994). The convection areas can be typically found several hundreds of meters on the downwind side of the roughness change due to the development of internal boundary layers.

2.4 Modeling Concepts

In the following chapter, modeling concepts used to describe the family of footprint models will be presented. The specific application of these model types is presented in Chap. 3.

2.4.1 Diffusion Model

The diffusion model based on Pasquill (1972) was the most widely used type of models used for footprint modelling as most analytical models applied this method. The simplest models often used inair pollution applications are Gaussian plume models (Pasquill 1972; Pasquill and Smith 1983; Blackadar 1997; Arya 1999). The dispersion of air pollution in a three-dimensional volume can be described with probability density functions for the distribution of pollutants or particles in the three directions $F(x)$, $G(y)$, and $H(z)$. The three-dimensional distribution becomes, according to the continuity principle

$$\int_{-\infty}^{\infty}\int_{-\infty}^{\infty}\int_{-\infty}^{\infty} F(x)G(y)H(z)\, dx\, dy\, dz = 1. \tag{2.76}$$

For a point source with constant emission rate $Q\mathrm{d}t$ and constant horizontal wind velocity, the distribution density function is:

$$F(x) = \frac{1}{\overline{u}\, dt}. \tag{2.77}$$

For the transverse horizontal and vertical distributions, the Gaussian distribution functions are used:

$$G(y) = \frac{1}{\sqrt{2\pi}\,\sigma_v} \exp\left(-\frac{y^2}{2\,\sigma_v}\right) \tag{2.78}$$

$$H(z) = \frac{1}{\sqrt{2\pi}\,\sigma_w} \exp\left(-\frac{z^2}{2\,\sigma_w}\right) \tag{2.79}$$

where σ_v and σ_w are the standard deviations of the lateral and vertical wind component.

The concentration distribution can be also calculated using Fick's diffusion law:

$$\frac{\partial \chi}{\partial t} + \overline{u}\,\frac{\partial \chi}{\partial x} = \frac{\partial}{\partial x}\left(K_x\,\frac{\partial \chi}{\partial x}\right) + \frac{\partial}{\partial y}\left(K_y\,\frac{\partial \chi}{\partial y}\right) + \frac{\partial}{\partial z}\left(K_z\,\frac{\partial \chi}{\partial z}\right). \tag{2.80}$$

The parameterization of the diffusion coefficients is made with error functions:

$$\sigma_u^2 = 2\,K_x\,t \qquad \sigma_v^2 = 2\,K_y\,t \qquad \sigma_w^2 = 2\,K_z\,t. \tag{2.81}$$

The concentration distribution for a constant source strength Q and the mean horizontal wind speed $\overline{u}$ in the x-direction is given by

$$\chi(x, y, z) = \frac{Q}{2\,\pi\,\overline{u}\,\sigma_v\,\sigma_w}\exp\left(-\frac{y^2}{2\,\sigma_v} - \frac{z^2}{2\,\sigma_w}\right). \tag{2.82}$$

In the absence of meteorological data, the standard deviations of the wind components can be parameterized using the micrometeorological approach described in Sect. 2.2.6. Gryning et al. (1987) use a plume crosswind dispersion based on Draxler (1976), which includes the Lagrangian time scale (see below) for crosswind dispersion.

The most widely applied footprint models are based on the analytical solution of the vertical diffusion by van Ulden (1978) and Gryning et al. (1983, 1987) determined solutions for Eq. (2.82) for individual atmospheric scenarios. For the vertical diffusion, it follows

$$\frac{\chi_z(x, z)}{Q} = \frac{A}{\overline{z}(x)}exp\left[-\left(\frac{B \cdot z}{\overline{z}}\right)^r\right] \tag{2.83}$$

and for crosswind diffusion

$$\frac{\chi_y(x, y)}{Q} = \frac{A}{\overline{z}(x)\overline{u}}exp\left[-\left(\frac{B \cdot z}{\overline{z}}\right)^r\right] \tag{2.84}$$

where $A = r \cdot \Gamma(2/r)/\Gamma^2(1/r)$ and $\mathrm{B} = \Gamma(2/r)/\Gamma(1/r)$ are functions of the shape parameter r and Γ is the gamma function and $\overline{z}$ is the mean height of the plume. In some cases, a Gaussian distribution is commonly used to include the diffusion in the lateral direction:

$$\frac{\chi_y(x, y)}{Q} = \frac{1}{\sqrt{2\pi} \cdot \sigma_v}exp\left[-\frac{1}{2}\left(\frac{y}{\sigma_v}\right)^2\right] \tag{2.85}$$

where A and B are functions of the exponent (shape parameter) r. Gryning et al. (1983) provided an approximate formula for r in terms of the mass-weighted mean plume height, $\overline{z}(x)$ and stability. $\overline{u}$ is the mass-weighted mean plume velocity. Although van Ulden's solution is analytical, it is implicitly in x by

Table 2.13 The use of diffusion model parameterizations in the widely distributed footprint models

Footprint model	Use of
Gash (1986)	Pasquill (1961)
Schuepp et al. (1990),	Gash (1986)
Horst and Weil (1992, 1994)	van Ulden (1978), Horst (1979)
Schmid (1994, 1997)	van Ulden (1978), Gryning et al. (1987)
Haenel and Grünhage (1999)	van Ulden (1978), Horst (1999)
Kormann and Meixner (2001)	van Ulden (1978), Horst and Weil (1992)

$$\bar{z}(x) = \frac{\int_0^\infty z\chi_y(x,y)dz}{\int_0^\infty \chi_y(x,y)dz}. \tag{2.86}$$

Using K-theory, van Ulden (1978) expressed the evolution of the centroid of the plume

$$\frac{d\bar{z}}{dx} = \frac{K(p\bar{z})}{\bar{u}(p\bar{z})p\bar{z}} \tag{2.87}$$

where K is the eddy diffusivity, and p is a weak function of r. The wind profile above the canopy is given by the logarithmic wind profile in Eq. (2.39), while the mean wind profile inside a canopy is given in Eq. (2.44).

An overview of the different uses of diffusion model parameterizations in analytical footprint models is given in Table 2.13.

2.4.2 Lagrangian Model

The spectrum of atmospheric turbulence scales (Frisch 1995) for state parameters and fluxes in the range of micrometeorological processes (periods lower than approx. 30 minutes depending on the site properties, altitude, etc.) is divided into three regions. The range of energy transfer from the mean motion into turbulent flow is characterized by the integral turbulent length scale Λ, which is approx. 10^1–$5\cdot10^2$ m (Kaimal and Finnigan 1994). The typical range of frequencies is f ~ 10^{-4} Hz. High frequencies follow the inertial sub range with isotropic turbulence. This range follows Kolmogrorov's law (Kolmogorov 1941a, b) with a defined decrease in energy density with increasing frequency in a manner proportional to $f^{-5/3}$. At higher frequencies (f ~ 10–30 Hz), eddies disappear through viscous dissipation ε. The scale is the Kolmogorov's micro-scale of about 10^{-3} m:

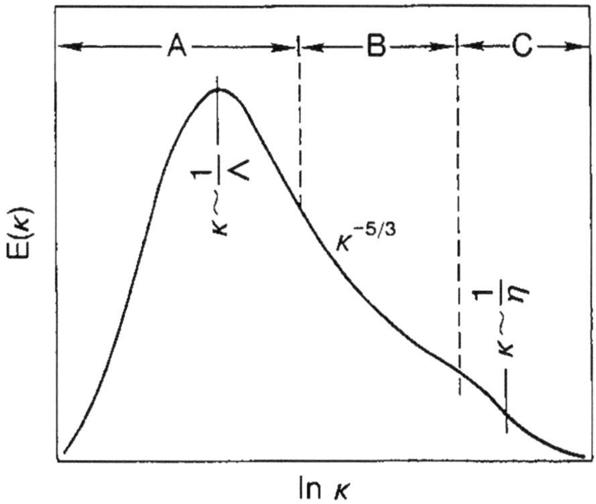

Fig. 2.12 Schematic illustration of the turbulence spectra with the range of energy production (**a**), Energy dissipation (**c**) and the inertial subrange (**b**) depending on the wave number κ (Kaimal and Finnigan 1994, Published with kind permission of © Oxford University Press, 1994. All Rights Reserved)

$$\eta = \left(\frac{\nu^3}{\varepsilon}\right)^{1/4} \tag{2.88}$$

The three ranges in the turbulence spectra in micrometeorology are illustrated in Fig. 2.12 as a function of wave length. The spectral peak corresponds to an integral turbulent length scale (exact: $\kappa = \pi/\Lambda$, Λ: Eulerian length scale). This length scale can be determined for wind components and scalars. According to Taylor's hypothesis on frozen turbulence (Taylor 1923, 1938), for which the relation

$$\kappa = 2\pi \cdot f / \overline{u} \tag{2.89}$$

is valid, this length scale can be combined with the integral turbulent time scale using the mean wind velocity.

The Lagrangian integral time scale τ can be determined from the autocorrelation function ρ (Monin and Yaglom 1973, 1975; Schlichting and Gersten 2003; Wyngaard 2010). Because the autocorrelation function is usually an exponential function, the integral time scale of ξ is $\rho(\xi) = 1/e \sim 0.37$. This is illustrated in Fig. 2.13. For the horizontal length scale follows with the horizontal wind velocity:

$$\Lambda_u = \overline{u} \cdot \tau_u = \overline{u} \int_0^\infty \rho_u(\xi)\, d\xi = \overline{u} \int_0^\infty \frac{\overline{u'(t)\, u'(t+\xi)}}{\sigma_u^2}\, d\xi. \tag{2.90}$$

The transport of a conserved passive scalar be it of carbon dioxide, water vapor or the likes is predicated on the state of the atmosphere. The atmosphere near the

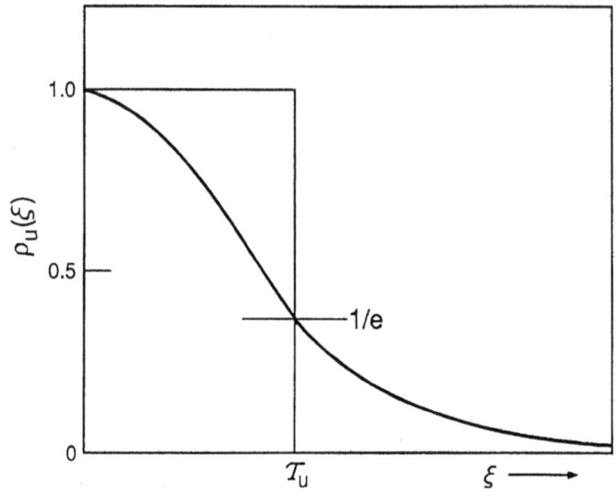

Fig. 2.13 Autocorrelation function and its dependency on the integral time scale. The value 1/e is a good approximation for which the square of the rectangle is identical with the square below the exponential graph (Kaimal and Finnigan 1994, Published with kind permission of © Oxford University Press, 1994. All Rights Reserved)

ground is characterized by vertical inhomogeneity in the flow, i.e. $\tau = \tau(z)$, the Lagrangian time scale of the turbulence, is thought to be a function of height. Earlier, seminal studies of Wilson et al. (1981) have found this value to $\tau \sim 0.5z$ in the neutral atmospheric surface layer above a smooth surface. The Lagrangian timescale plays an important role in Lagrangian modeling (Koeltzsch 1999).

The atmospheric surface layer is also generally the layer of air where anisotropy $(\sigma_u > \sigma_v \ > \sigma_w)$is significant, $\sigma_{u_i} = \sigma_{u_i}(z)$, where $\sigma_{u_i}$is the turbulence velocity scale in dimensions i.e. the streamwise, crosswind, and vertical directions increases with distance from the surface in the surface layer (see Sect. 2.2.6). It is customarily assumed that the Eulerian and the Lagrangian turbulent velocity scales of the turbulence are equivalent, thus greatly simplifying our prescription of input variables. In the neutral case, Wilson et al. (1982) found for the vertical and horizontal wind velocity, the time scales can be defined as

$$\tau_w = \frac{0.1\ z}{\sigma_w(z)}, \quad \tau_u \equiv \tau_L = \frac{0.5\ z}{\sigma_w(z)}, \tag{2.91}$$

which were used by Leclerc and Thurtell (1990) in the original Lagrangian footprint model. In the neutral atmospheric surface layer, $\sigma_w(z)$ in the Lagrangian timescale can also be replaced by $1.25u_*$ according to Eq. (2.61).

The integral time scale used in Lagrangian footprint models is given in Table 2.14.

The trajectory of the fluid element is given as

$$du_i = u_i\, dt \tag{2.92}$$

where u_i is the instantaneous Lagrangian velocity in the x_i direction and where $d\underline{t}$ is the instantaneous time increment, typically taken to be generally 0.1 τ:

Table 2.14 The use of surface layer parameterization for the Lagrangian integral time scale in the widely distributed footprint models, "*italic*" not according the present state

Footprint model	Use of	Remarks
Leclerc and Thurtell (1990)	$\tau(z) = \frac{0.5\,z}{\sigma_w(z)}$	Wilson et al. (1982)
Baldocchi (1997)	$\tau = \frac{0.3h_c}{u_*}$	*No height dependence*
Rannik et al. (2000, 2003)	$\tau = \frac{2\sigma_w^2}{C_0\varepsilon}$	$C_0 = 4$ (Kolmogorov constant)

$$u_i = a_i(x_i, t)dt + b_i(x_i, t)r. \tag{2.93}$$

With the first term representing the 'memory' term and the second term the 'random' term. Both a_i and b_i are a function of position, time and velocity while r is a random process with Gaussian statistics, exhibiting a mean of zero and a variance of 1. It is the respective magnitude of each of the coefficients that dictate the relative weight of the 'memory' term and that of the 'random' term. In inhomogeneous turbulence as are the cases treated here, an additional term must be added. The characteristics of the latter but a drift term is generally sufficient to treat the diffusion near a simple surface in the atmospheric surface layer. That original method has often yielded to the generalized method used by Thomson (1987) to deal with inhomogeneous turbulence.

An asset of Lagrangian stochastic models over analytical solutions lies in their applicability to model the dispersion from a very close range (near field), a subject of particular importance inside vegetation i.e. the region of sources and sinks. Lagrangian simulations intrinsically account for the characteristics of diffusion both in the near-field and in the far field as particles travel away from their source; this key feature allows for a proper description of the physics within vegetation (Denmead and Bradley 1985; Thurtell 1988).

2.4.3 Higher-Order Closure Model

Over the last decade, also classical numerical models based on the Navier-Stokes equations have also been used in modeling atmospheric footprints. The transformation of streamwise flow components into the equations of motion to the equation for turbulent flow is necessary. This leads to a system of differential equations with more unknown parameters than equations. To solve the system of equations, assumptions have to be made to calculate the unknown parameters. This is often referred to as closure techniques.

The order of the closure refers to the highest order of the parameters that must be calculated with the prognostic equations. Therefore, the moments of the next higher order must be determined (Stull 1988). A first-order closure generally use either the K-closure approach or stability functions linked to Monin-Obukhov similarity theory. To calculate state variables like the wind velocity or temperature, the *K*-approach must be applied to determine fluxes according to Eqs. (2.24)–(2.26). With a second-order closure, turbulent fluxes can be determined using the prognostic equation, but a parameterization for the triple correlation is necessary. A 1.5 order closure uses the equation of the turbulent kinetic energy (Stull 1988) to determine variance terms, see Sect. 2.2.6.

The benefit of closure techniques larger than the first-order closure allows the modeling of counter gradients, a frequent occurrence inside canopies (Denmead and Bradley 1985). In the case of counter gradient diffusion, the direction of the flux does not follow the direction of the gradient due to possible exchange by coherent structures, while the proportionality between flux and gradient is given for the *K*-approach (1st order closure). The models by Sogachev et al. (2002, 2008), Sogachev and Panferov (2006), and Sogachev and Leclerc (2011) are of 1.5 order closure and by Hsieh and Katul (2009), Hsieh et al. (2000), and Luhar and Rao (1994) of 2nd order closure.

2.4.4 Large-Eddy Simulation Model

This Large-Eddy Simulation (LES) method, applied for the first time to the atmosphere by Deardorff (1972) and Moeng and Wyngaard (1988), is considered the *ne plus ultra* approach for complex flows not otherwise within the realm of most models; LES can also incorporate pressure gradients and other challenging flow/surface scenarios conferring it a definite advantage.

The LES approach is based on the fact that most of the flux is contained in the large eddies, which are directly resolved. Therefore, a parameterization is necessary to account for the contribution of smaller eddies to fluxes. This method provides a high level of realism of the flow despite complex boundary conditions. This powerful type of simulations has been used extensively in atmospheric flow modeling and in particular in convective boundary layers (Mason 1989).

The LES computes the three-dimensional, time-dependent turbulence motions, and only parameterizes the subgrid-scale motions (SGS). Using the Navier-Stokes equations, LES resolves the large eddies with scales equal to or greater than twice the grid size, while parameterizing SGS processes.

The LES approach is free of the drawback of prescribing a turbulence field, hence the importance of initial and boundary conditions. Typically, LES determines the three-dimensional velocity field, pressure, and turbulent kinetic energy. The LES can also contain a set of cloud microphysical and thermodynamic equations and can predict the temperature and mixing ratios. It can also simulate the turbulent transport of moisture, carbon dioxide, and pollutants.

The first seminal study using the LES to model the turbulence inside a forest canopy was performed by Shaw and Schumann (1992). That study revealed the feasibility of using the LES to model correctly the three-dimensional structure of the turbulence in the canopy layer. In the absence of experimental data, LES is often a substitute, providing a realistic turbulence structure. Canopy LES simulations were made for homogeneous canopy flows (Patton et al. 2001; Shaw and Patton 2003; Yue et al. 2007; Mao et al. 2008; Shen and Leclerc 1997; Su et al. 1998) and recently to the canopy layer near the edge (Dupont and Brunet 2009).

There are several parameterizations available in treating the sub-grid scales. One of the most widely used simulations is that originally developed by Moeng (1984) and Moeng and Wyngaard (1988) and later adapted for flux footprints applications by Leclerc et al. (1997) and Mao et al. (2008). Often, the in canopy SGS are parameterized using the 1.5 order of closure scheme. Sullivan et al. (2003) have discussed and proposed realistic closure schemes.

Some LES also include a terrain-following coordinate system. A spatial cross-average and temporal average is most often applied to the simulated 'data' once the simulation has reached quasi steady-state equilibrium. Typical boundary conditions are periodic with a rigid lid applied to the top of the domain so that waves are absorbed and reflection from the upper portion of the domain is decreased. The LES is computationally very expensive and limited by the number of grid points in flow simulations. As computers' performance and speed keep going up, this becomes less and less of an issue though still of significance when footprint modeling in moderately stable conditions is required.

Despite the many advantages of the present method, moderately stable boundary layers remain the Achilles' heel of the LES, with errors due to an imperfect SGS becoming more pronounced in these stable conditions since the characteristic eddy size is notably smaller.

The LES technique is generally prized amongst other numerical modeling approaches in part because, as is the case for a canopy layer, they have reproduced key features of the canopy turbulence structure such as vertically distributed shear levels, high turbulence intensities changing rapidly with depth inside the canopy layer, scalar microfronts and inflection point near treetop in the velocity profile (Su and Leclerc 1998; Su et al. 1998; Fitzmaurice et al. 2004; Watanabe 2004; Yue et al. 2007; Mao et al. 2008).

The LES technique is applied for footprint modeling either in the way that the LES model produces the necessary input parameters e.g. for a Lagrangian footprint model (Leclerc et al. 1997) or by embedding an Lagrangian footprint model directly into the LES model (Steinfeld et al. 2008). The footprint application of LES models is still an ongoing issue.

2.5 Averaging Surface Characteristics

Most footprint models are based on atmospheric transport patterns over regions with spatially uniform flux sources and surface characteristics (e.g. roughness length, leaf area index, or surface moisture). Lagrangian backward models (Kljun et al. 2002), Large-Eddy Simulation (e.g. Steinfeld et al. 2008) and higher-order closure models (e.g. Sogachev and Lloyd 2004) provide complex footprint descriptions over heterogeneous surfaces. Simpler footprint models needs the averaging of the surface characteristics of a heterogeneous surface to determine averaged input parameters.

The net impact of spatially heterogeneous surface characteristics on atmospheric transport, and therefore on footprint computations is governed by highly non-linear processes. Despite the above, many approaches that sacrifice complex physics are available for the sake of mathematical simplicity. The simplest approach available, called the parameter aggregation, linearly averages parameters such as the aerodynamic roughness length over an area characterized by surface parcels of distinct properties

$$\overline{z_0} = \frac{1}{N}\sum_{i=1}^{N} z_{0i}. \tag{2.94}$$

Parameter aggregation is easy to apply and commonly used in footprint studies. This is however achieved at the expense of a significant over simplification of the underlying physics. The error of such averaging can be considerable (Stull and Santoso 2000). For example, in the case of a heterogeneous landscape composed of equal parts of water ($z_0 = 0.001$ m) and of forest ($z_0 = 1.0$ m), the linearly averaged roughness length would suggest shrub land ($z_0 = 0.5$ m); however, the flow characteristics over shrubs differ significantly from those over a forested area with lakes.

The reason for the failure of the parameter aggregation is the fact that the interaction of the atmosphere with the underlying surface takes parties intertwined with fluxes such as the momentum flux

$$\tau = \rho \cdot u_*^2 = \frac{\rho}{N} \cdot \sum_{i=1}^{N} \left(\frac{\kappa \cdot u_i(z)}{\ln z - \ln z_{0i}} \right)^2. \tag{2.95}$$

The relationship between surface characteristics (e.g. roughness length) and resulting fluxes (e.g. momentum flux) is non-linear. Therefore, averaged surface parameters lead to a construction of a regional flux that differs from that created using a superposition of fluxes from the individual patches. Accordingly, these fluxes must be averaged (flux aggregation) to yield more realistic results. Besides the momentum flux, also the stability (momentum and sensible heat flux) and the evaporation (latent heat flux) must be averaged for special applications (Chap. 6). Due to the involvement of additional flux equations and their interdependencies, it is evident that the flux aggregation is complex than parameter aggregation.

2.5.1 Averaging Using Effective Parameters

A common approach to arrive at spatially averaged parameter sets that produce representative area-averaged fluxes without the use of complex algorithms consists in selecting effective parameters. A well-known example is that of the 'effective roughness lengths' presented by Fiedler and Panofsky (1972). These effective roughness lengths are not describing the roughness of a particular underlying surface per se, but instead focus on the effect on momentum fluxes on the landscape scale (flat terrain: 0.42 m; low hills: 0.99 m; high mountains: 1.42 m). Generally speaking, the determination of the effective roughness length needs the normalization with the friction velocity inherent to a particular area (Taylor 1987; Schmid and Bünzli 1995a, b; Mahrt 1996; Hasager and Jensen 1999) in the following form:

$$z_{0eff} = \frac{\overline{u_* \cdot \ln z_0}}{\overline{u_*}} \tag{2.96}$$

Such a simple averaging method is helpful for countless practical applications, such as the micrometeorological characterization of the boundary layer above settlement areas (Grimmond et al. 1998).

A more empirical form for the determination of an effective roughness length was presented in the European Wind Atlas (Troen and Peterson 1989). The authors classified only four types (0–3) of the surface roughness (water, flat meadows, landscape with bushes, forest), each of which are assigned a basic roughness length. To include subgrid-scale heterogeneities, the model area is divided up into quarters, each of which are assigned individualist own land-cover class. In Table 2.15 the different portions of the roughness classes (water, flat meadows, landscape with bushes, forest) to the entire area are given. Unfortunately no algorithm is available or published. The effective roughness length is an "empirical" function of the contribution of each surface area to the whole area. Göckede et al. (2004) used this concept to determine an effective roughness length for each grid cell in a heterogeneous landscape. The authors assumed that a simple parameter averaging of these effective values was sufficient to subsequently average the roughness length of all grid cells within the footprint area.

2.5.2 Flux-Averaging Models in Inhomogeneous Terrain

The simplest form of a flux aggregation approach is the tile approach, where land cover characteristics within a grid cell accumulate (Fig. 2.14a) according to the proportional use of each land-use type. To arrive at representative averaged parameters, fluxes must be determined for each land use type, and averaged afterwards (Beyrich et al. 2006). However, this method neglects the interaction between the different grid cells, e.g. flow transitions from smooth to rough surfaces or vice versa, local circulation or large water bodies, which can influence the flux systematically. Both experimental (Panin et al. 1996; Klaassen et al. 2002; Klaassen and Sogatchev 2006) and numerical studies (Schmid and Bünzli 1995a;

Table 2.15 Averaging schema of the roughness length in the European Wind Atlas (Troen and Peterson 1989) used to generate a "flux-averaged" effective roughness length

Type	Water (%)	Flat meadow (%)	Landscape with bushes (%)	Forest (%)	Effective roughness length (m)
z_0 (m)	0.0002	0.03	0.1	0.4	
	75	25			0.001
	75		25		0.002
	75			25	0.003
	50	50			0.004
	50	25	25		0.006
	50	25		25	0.010
	50		50		0.009
	50		25	25	0.015
	50			50	0.025
	25	75			0.011
	25	50	25		0.017
	25	50		25	0.027
	25	25	50		0.024
	25	25	25	25	0.038
	25	25		50	0.059
	25		75		0.033
	25		50	25	0.052
	25		25	50	0.079
	25			75	0.117
		75	25		0.042
		75		25	0.064
		50	50		0.056
		50	25	25	0.086
		50		50	0.127
		25	75		0.077
		25	50	25	0.113
		25	25	50	0.163
		25		75	0.232
			75	25	0.146
			50	50	0.209
			25	75	0.292

Friedrich et al. 2000; Sogachev et al. 2008) found that in a heterogeneous landscape, significantly higher fluxes are found close to such roughness changes.

This problem can be circumvented using the sub-grid scale approach (Mölders et al. 1996; Wang et al. 2006), should be used: For each surface element, a separate multilayer model (often of the Surface-Vegetation-Atmosphere-Transfer type: SVAT) interacting with the neighbourhood grid cell by horizontal fluxes and advection can then be calculated (Fig. 2.14b). The structure of the surface will be

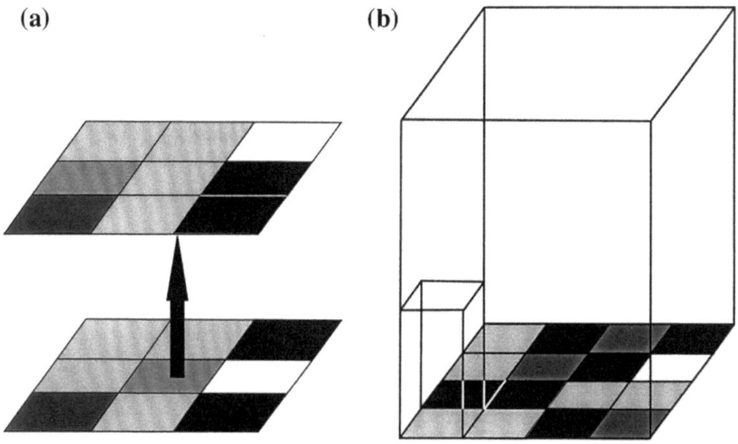

Fig. 2.14 **a** Schematic figure of the tile approach. The initial distribution of the surface structures will be combined according their contributions for further calculation. **b** Schematic view of the sub-grid scale (Foken 2008)

not changed and for each surface type a multilayer model will be used which interacts each other by horizontal flow.

A more sophisticated approach to aggregate roughness lengths under consideration of local advection effects was developed by Hasager and Jensen (1999). This microscale aggregation model accounts for the adjustment of the flow to roughness change in arbitrary surface conditions. The physics consists of a linearized version of the atmospheric momentum equation in which only the advective term and the vertical flux divergence are assumed to be of importance, while all other terms such as the Coriolis term are neglected (Hasager et al. 2003). The algorithms are solved by Fast Fourier Transform to allow the time-efficient computation of the effective roughness parameter in accordance with average stress for a given background flow. Terrain information is provided using high-resolution two-dimensional land-use maps, with a fixed roughness length assigned to each land-use class. The application of this method for footprint modeling was demonstrated by Göckede et al. (2006).

Since area averaging of representative surface parameters is usually a pre-processing step to the actual footprint computation, the problem of finding a suitable averaging scheme is more a responsibility of the model user than of the model developer. Accordingly, the use of different aggregation approaches has been published mostly in application papers (Chap. 8). The experimentalist using footprint models must keep in mind that most footprint models are only valid for homogeneous surfaces, and any application in heterogeneous terrain may compromise the model output. The use of averaging techniques to produce aggregated, homogeneous surface fluxes opens footprint models to larger areas of application, but the neglect of high variability of turbulent fluxes at subgrid-scale in the

Table 2.16 Flux-averaging approaches used in footprint models

Averaging technique	Used in
Homogeneous surface	Schuepp et al. (1990), Leclerc and Thurtell (1990), Horst and Weil (1992, 1994), Schmid (1994, 1997), Flesch (1996), Leclerc et al. (1997), Haenel and Grünhage (1999), Rannik et al. (2000, 2003), Kormann and Meixner (2001) and others
Effective roughness length	Applicable to most models
Effective roughness length according to a momentum flux averaging (Hasager and Jensen 1999)	Göckede et al. (2006) using the model by Rannik et al. (2000, 2003)
Heterogeneous surface	Kljun et al. (2002), Steinfeld et al. (2008), Sogachev and Lloyd (2004)

heterogeneous terrain may significantly bias the computations. Table 2.16 separates the footprint model regarding use for homogeneous or heterogeneous surfaces and summarizes flux aggregation approaches used for footprint models.

References

Andreas EL (2002) Parametrizing scalar transfer over snow and ice: a review. J Hydrometeorol 3:417–432

Andreas EL and Cash BA (1999) Convective heat transfer over wintertime leads and polynyas. J Geophys Res. 104:15.721–725.734

Antonia RA, Luxton RE (1971) The response of turbulent boundary layer to a step change in surface roughness, part 1. J Fluid Mech 48:721–761

Antonia RA, Luxton RE (1972) The response of turbulent boundary layer to a step change in surface roughness, part 2. J Fluid Mech 53:737–757

Arya SP (1999) Air pollution meteorology and dispersion. Oxford University Press, New York, Oxford, 310 pp

Arya SP (2001) Introduction to micrometeorology. Academic Press, San Diego, 415 pp

Aubinet M, Vesala T and Papale D (2012) Eddy covariance: a practical guide to measurement and data analysis. Springer, Dordrecht, 438 pp

Baldocchi D (1997) Flux footprints within and over forest canopies. Bound-Layer Meteorol 85:273–292

Behrendt A, Wulfmeyer V, Riede A, Wagner G, Pal S, Bauer H, Radlach M, Späth F (2009) 3-Dimensional observations of atmospheric humidity with a scanning differential absorption lidar. In: Picard RH et al (eds) Remote sensing of clouds and the atmosphere XIV, SPIE conference proceeding, vol 7475, Art No 74750L. doi:74710.71117/74712.835143

Beyrich F et al (2006) Area-averaged surface fluxes over the LITFASS region on eddy-covariance measurements. Bound-Layer Meteorol 121:33–65

Blackadar AK (1997) Turbulence and Diffusion in the Atmosphere. Springer, Berlin, 185 pp

Blöschl G, Sivapalan M (1995) Scale issues in hydrological modelling: a review. Hydrol Processes 9:251–290

Bowen IS (1926) The ratio of heat losses by conduction and by evaporation from any water surface. Phys Rev 27:779–787

Bradley EF (1968) A micrometeorological study of velocity profiles and surface drag in the region modified by change in surface roughness. Quart J Roy Meteorol Soc 94:361–379
Brunet Y, Finnigan JJ, Raupach MR (1994) A wind tunnel study of air flow in waving wheat: single-point velocity statistics. Bound-Layer Meteorol 70:95–132
Businger JA, Yaglom AM (1971) Introduction to Obukhov's paper "Turbulence in an atmosphere with a non-uniform temperature". Bound-Layer Meteorol 2:3–6
Businger JA, Wyngaard JC, Izumi Y, Bradley EF (1971) Flux-profile relationships in the atmospheric surface layer. J Atmos Sci 28:181–189
Cellier P, Brunet Y (1992) Flux-gradient relationships above tall plant canopies. Agric Forest Meteorol 58:93–117
Cionco RM (1978) Analysis of canopy index values for various canopy densities. Bound-Layer Meteorol 15:81–93
Claussen M (1991) Estimation of areally-averaged surface fluxes. Bound-Layer Meteorol 54:387–410
Claussen M (1995) Flux aggregation at large scales: on the limits of validity of the concept of blending height. J Hydrol 166:371–382
Claussen M, Walmsley JL (1994) Modification of blending procedure in a proposed new PBL resistance law. Bound-Layer Meteorol 68:201–205
Davenport AG, Grimmond CSB, Oke TR and Wieringa J (2000) Estimating the roughness of cities and shelterred country. In: 12th conference on applied climatology, Ashville, NC2000. American Meteorological Society, pp 96–99
Deardorff JW (1972) Numerical investigation of neutral und unstable planetary boundary layer. J Atmos Sci 29:91–115
Denmead DT, Bradley EF (1985) Flux-gradient relationships in a forest canopy. In: Hutchison BA, Hicks BB (eds) The forest-atmosphere interaction. D. Reidel Publishing Company, Dordrecht, pp 421–442
Denmead DT, Bradley EF (1987) On scalar transport in plant canopies. Irrig Sci 8:131–149
Doran JC, Verholek MG (1978) A note on vertical extrapolation formulas for Weibull velocity distribution parameters. J Climate Appl Meteorol 17:410–412
Draxler RR (1976) Determination of atmospheric diffusion parameters. Atmos Environ 10:99–105
Dupont S, Brunet Y (2009) Coherent structures in canopy edge flow: a large-eddy simulation study. J Fluid Mech 630:93–128
Durden DJ, Nappo CJ, Leclerc MY, Duarte HF, Zhang G, Parker MJ, Kurzeja RJ (2013) On the impact of wave-like disturbances on turbulent fluxes and turbulence statistics in night-time conditions: a case study. Biogeosci 10:8433–8443
Dyer AJ (1974) A review of flux-profile-relationships. Bound-Layer Meteorol 7:363–372
Eigenmann R, Metzger S, Foken T (2009) Generation of free convection due to changes of the local circulation system. Atmos Chem Phys 9:8587–8600
Elliott WP (1958) The growth of the atmospheric internal boundary layer. Trans Am Geophys Union 39:1048–1054
ESDU (1972) Characteristics of wind speed in the lowest layers of the atmosphere near the ground: strong winds. Engineering Science Data Unit Limited. Regent street, London
Fiedler F, Panofsky HA (1972) The geostrophic drag coefficient and the 'effective' roughness length. Quart J Roy Meteorol Soc 98:213–220
Finnigan J (2000) Turbulence in plant canopies. Ann Rev Fluid Mech 32:519–571
Finnigan JJ, Clement R, Malhi Y, Leuning R, Cleugh HA (2003) A re-evaluation of long-term flux measurement techniques, part I: Averaging and coordinate rotation. Bound-Layer Meteorol 107:1–48
Fitzmaurice L, Shaw RH, Paw UKT, Patton EG (2004) Three-dimensional scalar microfront systems in a large-eddy simulation of vegetation canopy flow. Bound-Layer Meteorol 112:107–127
Flesch TK (1996) The footprint for flux measurements, from backward Lagrangian stochastic models. Bound-Layer Meteorol 78:399–404

Foken T (1990) Turbulenter Energieaustausch zwischen Atmosphäre und Unterlage: Methoden, meßtechnische Realisierung sowie ihre Grenzen und Anwendungsmöglichkeiten. Ber Dt Wetterdienstes 180:287 pp
Foken T (2006) 50 years of the Monin-Obukhov similarity theory. Bound-Layer Meteorol 119:431–447
Foken T (2008) Micrometeorology. Springer, Berlin, 308 pp
Foken T, Wichura B (1996) Tools for quality assessment of surface-based flux measurements. Agric Forest Meteorol 78:83–105
Foken T, Skeib G, Richter SH (1991) Dependence of the integral turbulence characteristics on the stability of stratification and their use for Doppler-Sodar measurements. Z Meteorol 41:311–315
Foken T et al (2012) Coupling processes and exchange of energy and reactive and non-reactive trace gases at a forest site: results of the EGER experiment. Atmos Chem Phys 12:1923–1950
Friedrich K, Mölders N, Tetzlaff G (2000) On the influence of surface heterogeneity on the Bowen-ratio: a theoretical case study. Theor Appl Clim 65:181–196
Frisch U (1995) Turbulence. Cambridge University Press, Cambridge, 296 pp
Garratt JR (1978) Flux profile relations above tall vegetation. Quart J Roy Meteorol Soc 104:199–211
Garratt JR (1980) Surface influence upon vertical profiles in the atmospheric near surface layer. Quart J Roy Meteorol Soc 106:803–819
Garratt JR (1990) The internal boundary layer: a review. Bound-Layer Meteorol 50:171–203
Garratt JR (1992) The atmospheric boundary layer. Cambridge University Press, Cambridge, 316 pp
Gash JHC (1986) A note on estimating the effect of a limited fetch on micrometeorological evaporation measurements. Bound-Layer Meteorol 35:409–414
Geiger R, Aron RH and Todhunter P (2009) The climate near the ground. Rowman & Littlefield Publishers, Lanham, 623 pp
Göckede M, Rebmann C, Foken T (2004) A combination of quality assessment tools for eddy covariance measurements with footprint modelling for the characterisation of complex sites. Agric Forest Meteorol 127:175–188
Göckede M, Markkanen T, Hasager CB, Foken T (2006) Update of a footprint-based approach for the characterisation of complex measuring sites. Bound-Layer Meteorol 118:635–655
Göckede M, Thomas C, Markkanen T, Mauder M, Ruppert J, Foken T (2007) Sensitivity of Lagrangian stochastic footprints to turbulence statistics. Tellus 59B:577–586
Goudriaan J (1977) Crop micrometeorology: A simulation study. Center for Agricultural Publishing and Documentation, Wageningen, 249 pp
Graefe J (2004) Roughness layer corrections with emphasis on SVAT model applications. Agric Forest Meteorol 124:237–251
Grimmond CSB, Oke TR (1999) Aerodynamic properties of urban areas derived from analysis of surface form. J Appl Meteorol 38:1262–1292
Grimmond CSB, King TS, Roth M, Oke TR (1998) Aerodynamic roughness of urban areas derived from wind observations. Bound-Layer Meteorol 89:1–24
Gryning S-E, van Ulden AP, Larsen S (1983) Dispersions from a ground level source investigated by a K model. Quart J Roy Meteorol Soc 109:355–364
Gryning SE, Holtslag AAM, Irvin JS, Sivertsen B (1987) Applied dispersion modelling based on meteorological scaling parameters. Atmos Environ 21:79–89
Haenel H-D, Grünhage L (1999) Footprint analysis: a closed analytical solution based on height-dependent profiles of wind speed and eddy viscosity. Bound-Layer Meteorol 93:395–409
Handorf D, Foken T, Kottmeier C (1999) The stable atmospheric boundary layer over an Antarctic ice sheet. Bound-Layer Meteorol 91:165–186
Harman IN, Finnigan JJ (2007) A simple unified theory for flow in the canopy and roughness sublayer. Bound-Layer Meteorol 123:339–363
Harman IN, Finnigan JJ (2008) Scalar concentration profiles in the canopy and roughness sublayer. Bound-Layer Meteorol 129:323–351

Hasager CB, Jensen NO (1999) Surface-flux aggregation in heterogeneous terrain. Quart J Roy Meteorol Soc 125:2075–2102

Hasager CB, Nielsen NW, Jensen NO, Boegh E, Christensen JH, Dellwik E, Soegaard H (2003) Effective roughness calculated from satellite-derived land cover maps and hedge-information used in a weather forecasting model. Bound-Layer Meteorol 109:227–254

Hatfield JL and Baker JM (eds) (2005) Micrometeorology in agricultural systems. American Society of Agronomy, Madison, 584 pp

Hicks BB (1981) An examination of the turbulence statistics in the surface boundary layer. Bound-Layer Meteorol 21:389–402

Högström U (1988) Non-dimensional wind and temperature profiles in the atmospheric surface layer: a re-evaluation. Bound-Layer Meteorol 42:55–78

Högström U (1990) Analysis of turbulence structure in the surface layer with a modified similarity formulation for near neutral conditions. J Atmos Sci 47:1949–1972

Högström U (1996) Review of some basic characteristics of the atmospheric surface layer. Bound-Layer Meteorol 78:215–246

Högström U, Hunt JCR, Smedman A-S (2002) Theory and measurements for turbulence spectra and variances in the atmospheric neutral surface layer. Bound-Layer Meteorol 103:101–124

Horst TW (1979) Lagrangian similarity modeling of vertical diffusion from a ground level source. J Appl Meteorol 18:733–740

Horst TW (1999) The footprint for estimation of atmosphere-surface exchange fluxes by profile techniques. Bound-Layer Meteorol 90:171–188

Horst TW (2000) An intercomparison of measures of special inhomogeneity for surface fluxes of passive scalars. In: 14th symposium on boundary layer and turbulence, Aspen CO, 7–11 August 2000. American Meteorological Society, pp 11–14

Horst TW, Weil JC (1992) Footprint estimation for scalar flux measurements in the atmospheric surface layer. Bound-Layer Meteorol 59:279–296

Horst TW, Weil JC (1994) How far is far enough?: the fetch requirements for micrometeorological measurement of surface fluxes. J Atm Oceanic Tech 11:1018–1025

Hsieh C-I, Katul G (2009) The Lagrangian stochastic model for estimating footprint and water vapor fluxes over inhomogeneous surfaces. Int J Biometeorol 53:87–100

Hsieh C-I, Katul G, Chi T-W (2000) An approximate analytical model for footprint estimation of scalar fluxes in thermally stratified atmospheric flows. Adv Water Res 23:765–772

Hsu SA, Meindl EA, Gilhousen DB (1994) Determination of power-law wind-profile exponent under near: neutral stability conditions at sea. J Appl Meteorol 33:757–765

Huang CH (1979) A theory of dispersion in turbulent shear flow. Atmos Environ 13:453–463

Irvin JS (1978) A theoretical variation of the wind profile power-law exponent as a function of surface roughness and stability. Atmos Environ 13:191–194

Jegede OO, Foken T (1999) A study of the internal boundary layer due to a roughness change in neutral conditions observed during the LINEX field campaigns. Theor Appl Clim 62:31–41

Joffre SM (1984) Power laws and the empirical representation of velocity and directional sheer. J Clim Appl Meteorol 23:1196–1203

Johansson C, Smedman A, Högström U, Brasseur JG, Khanna S (2001) Critical test of Monin-Obukhov similarity during convective conditions. J Atmos Sci 58:1549–1566

Kader BA, Yaglom AM (1972) Heat and mass transfer laws for fully turbulent wall flows. Int J Heat Mass Transf 15:2329–2350

Kaharabata SK, Schuepp PH, Ogunjemiyo S, Shen S, Leclerc MY, Desjardins RL, MacPherson JI (1997) Footprint considerations in BOREAS. J Geophys Res 102(D24):29113–29124

Kaimal JC and Finnigan JJ (1994) Atmospheric boundary layer flows: their structure and measurement. Oxford University Press, New York, 289 pp

Kaimal JC, Gaynor JE (1991) Another look to sonic thermometry. Bound-Layer Meteorol 56:401–410

Kanda M, Moriwaki R, Roth M, Oke T (2002) Area-averaged sensible heat flux and a new method to determine zero-plane displacement length over an urban surface using scintillometry. Bound-Layer Meteorol 105:177–193

Klaassen W, Sogatchev A (2006) Flux footprint simulation downwind of a forest edge. Bound-Layer Meteorol 121:459–473
Klaassen W, van Breugel PB, Moors EJ, Nieveen JP (2002) Increased heat fluxes near a forest edge. Theor Appl Clim 72:231–243
Kljun N, Rotach MW, Schmid HP (2002) A three-dimensional backward Lagrangian footprint model for a wide range of boundary layer stratification. Bound-Layer Meteorol 103:205–226
Koeltzsch K (1999) On the relationship between the Lagrangian and Eulerian time scale. Atmos Environ 33:117–128
Kolmogorov AN (1941a) Lokalnaja struktura turbulentnosti v neschtschimaemoi schidkosti pri otschen bolschich tschislach Reynoldsa (The local structure of turbulence in incompressible viscous fluid for very large Reynolds numbers). Dokl AN SSSR 30:299–303
Kolmogorov AN (1941b) Rassejanie energii pri lokolno-isotropoi turbulentnosti (Dissipation of energy in locally isotropic turbulence). Dokl AN SSSR 32:22–24
Kormann R, Meixner FX (2001) An analytical footprint model for non-neutral stratification. Bound-Layer Meteorol 99:207–224
Leclerc MY, Thurtell GW (1990) Footprint prediction of scalar fluxes using a Markovian analysis. Bound-Layer Meteorol 52:247–258
Leclerc MY, Beissner KC, Shaw RH, den Hartog G, Neumann HH (1990) The influence of atmospheric stability on the budgets of the Reynolds stress and turbulent kinetic energy within and above a deciduous forest. J Appl Meteorol 29:916–933
Leclerc MY, Beissner KC, Shaw RH, den Hartog G, Neumann HH (1991) The influence of bouyancy on third-order turbulent velocity statistics within d deciduous forest. Bound-Layer Meteorol 55:109–123
Leclerc MY, Shen S, Lamb B (1997) Observations and large-eddy simulation modeling of footprints in the lower convective boundary layer. J Geophys Res 102(D8):9323–9334
Luhar AK, Rao KS (1994) Source footprint analysis for scalar fluxes measured in flows over an inhomogeneous surface. In: Gryning SE, Millan MM (eds) Air pollution modeling and its applications. Plenum Press, New York, pp 315–323
Lumley JL and Panofsky HA (1964) The structure of atmospheric turbulence. Interscience Publishers, New York, 239 pp
MacDonald RW, Griffiths RF, Hall DJ (1998) An improved method for the estimation of surface roughness of obstacle arrays. Atmos Environ 32:1857–1864
Mahrt L (1996) The bulk aerodynamic formulation over heterogeneous surfaces. Bound-Layer Meteorol 78:87–119
Mao S, Leclerc MY, Michaelides EE (2008) Passive scalar flux footprint analysis over horizontally inhomogeneous plant canopy using large-eddy simulation. J Atmos Sci 42:5446–5458
Markkanen T, Steinfeld G, Kljun N, Raasch S, Foken T (2010) A numerical case study on footprint model performance under inhomogeneous flow conditions. Meteorol Z 19:539–547
Mason PJ (1988) The formation of areally-averaged roughness length. Quart J Roy Meteorol Soc 114:399–420
Mason PJ (1989) Large-eddy simulation of convective atmospheric boundary layer. J Atmos Sci 46:1492–1516
Massman WJ, Weil JC (1999) An analytical one-dimensional second-order closure model of turbulence statistics and the Lagrangian time scale within and above plant canopies of arbitrary structure. Bound-Layer Meteorol 91:81–107
Moene AF, van Dam JC (2014) Transport in the atmosphere-vegetation-soil continuum. Cambridge University Press, Cambridge, 436 pp
Moeng C (1984) A large-eddy simulation model for the study of planetary boundary-layer turbulence. J Atmos Sci 41:2052–2062
Moeng C, Wyngaard JC (1988) Spectral analysis of large-eddy simulations of the convective boundary layer. J Atmos Sci 45:3573–3587
Mölder M, Grelle A, Lindroth A, Halldin S (1999) Flux-profile relationship over a boreal forest: roughness sublayer correction. Agric Forest Meteorol 98–99:645–648

Mölders N, Raabe A, Tetzlaff G (1996) A comparison of two strategies on land surface heterogeneity used in a mesoscale ß meteorological model. Tellus 48A:733–749
Monin AS, Obukhov AM (1954) Osnovnye zakonomernosti turbulentnogo peremesivanija v prizemnom sloe atmosfery (Basic laws of turbulent mixing in the atmosphere near the ground). Trudy geofiz inst AN SSSR 24(151):163–187
Monin AS and Yaglom AM (1973) Statistical fluid mechanics: mechanics of turbulence, Volume 1. MIT Press, Cambridge, 769 pp
Monin AS and Yaglom AM (1975) Statistical fluid mechanics: mechanics of turbulence, Volume 2. MIT Press, Cambridge, 874 pp
Monteith JL, Unsworth MH (2008) Principles of environmental physics, 3rd edn. Elsevier, Academic Press, Amsterdam, Boston, 418 pp
Obukhov AM (1946) Turbulentnost' v temperaturnoj: neodnorodnoj atmosfere (Turbulence in an atmosphere with a non-uniform temperature). Trudy Inst Theor Geofiz AN SSSR 1:95–115
Obukhov AM (1960) O strukture temperaturnogo polja i polja skorostej v uslovijach konvekcii (Structure of the temperature and velocity fields under conditions of free convection). Izv AN SSSR, ser Geofiz, 1392–1396
Obukhov AM (1971) Turbulence in an atmosphere with a non-uniform temperature. Bound-Layer Meteorol 2:7–29
Orlanski I (1975) A rational subdivision of scales for atmospheric processes. Bull Am Meteorol Soc 56:527–530
Panin GN, Tetzlaff G, Raabe A, Schönfeld H-J, Nasonov AE (1996) Inhomogeneity of the land surface and the parametrization of surface fluxes: a discussion. Wiss Mitt Inst Meteorol Univ Leipzig und Inst Troposphärenforschung Leipzig 4:204–215
Panofsky HA (1984) Vertical variation of roughness length at the boulder atmospheric observatory. Bound-Layer Meteorol 28:305–308
Panofsky HA, Tennekes H, Lenschow DH, Wyngaard JC (1977) The characteristics of turbulent velocity components in the surface layer under convective conditions. Bound-Layer Meteorol 11:355–361
Pasquill F (1961) Estimation of the dispersion of windborne material. Meteorol Mag 90:33–49
Pasquill F (1972) Some aspects of boundary layer description. Quart J Roy Meteorol Soc 98:469–494
Pasquill F, Smith FB (1983) Atmospheric diffusion, 3rd edn. Wiley, Chichester
Patton EG, Davis KJ, Barth MC, Sullivan PP (2001) Decaying scalars emitted by a forest canopy: a numerical study. Bound-Layer Meteorol 100:91–192
Paulson CA (1970) The mathematical representation of wind speed and temperature profiles in the unstable atmospheric surface layer. J Clim Appl Meteorol 9:857–861
Peltier LJ, Wyngaard JC, Khanna S, Brasseur JG (1996) Spectra in the unstable surface layer. J Atmos Sci 53:49–61
Peterson EW, Hennessey Jr JP (1978) On the use of power laws for estimates of wind power potential. J Clim Appl Meteorol 17:390–394
Prandtl L (1925) Bericht über Untersuchungen zur ausgebildeten Turbulenz. Z angew Math Mech 5:136–139
Raabe A (1983) On the relation between the drag coefficient and fetch above the sea in the case of off-shore wind in the near shore zone. Z Meteorol 33:363–367
Rannik Ü, Aubinet M, Kurbanmuradov O, Sabelfeld KK, Markkanen T, Vesala T (2000) Footprint analysis for measurements over heterogeneous forest. Bound-Layer Meteorol 97:137–166
Rannik Ü, Markkanen T, Raittila T, Hari P, Vesala T (2003) Turbulence statistics inside and above forest: influence on footprint prediction. Bound-Layer Meteorol 109:163–189
Rao KS, Wyngaard JC, Coté OR (1974) The structure of the two-dimensional internal boundary layer over a sudden change of surface roughness. J Atmos Sci 31:738–746
Raupach MR (1994) Simplified expressions for vegetation roughness lenght and zero-plane displacement as functions of canopy height and area index. Bound-Layer Meteorol 71:211–216

Raupach MR, Legg BJ (1984) The uses and limitations of flux-gradient relationships in micrometeorology. Agric Water Manag 8:119–131
Raupach MR, Thom AS, Edwards I (1980) A wind-tunnel study of turbulent flow close to regularly arrayed rough surface. Bound-Layer Meteorol 18:373–379
Raupach MR, Coppin PA, Legg BJ (1986) Experiments on scalar dispersion within a model plant canopy. Part I: The turbulence structure. Bound-Layer Meteorol 35:21–52
Raupach MR, Finnigan JJ, Brunet Y (1996) Coherent eddies and turbulence in vegetation canopies: the mixing-layer analogy. Bound-Layer Meteorol 78:351–382
Raynor GS, Michael P, Brown RM, SethuRaman S (1975) Studies of atmospheric diffusion from a nearshore oceanic site. J Clim Appl Meteorol 14:1080–1094
Reithmaier LM, Göckede M, Markkanen T, Knohl A, Churkina G, Rebmann C, Buchmann N, Foken T (2006) Use of remotely sensed land use classification for a better evaluation of micrometeorological flux measurement sites. Theor Appl Clim 84:219–233
Richardson LF (1920) The supply of energy from and to atmospheric eddies. Proc Roy Soc A 97:354–373
Rotach MW, Gryning S-E, Tassone C (1996) A two-dimensional Lagrangian stochastic dispersion model for daytime conditions. Quart J Roy Meteorol Soc 122:367–389
Roth M, Salmond J, Satyanarayana A (2006) Methodological considerations regarding the measurement of turbulent fluxes in the urban roughness sublayer: the role of scintillometry. Bound-Layer Meteorol 121:351–375
Savelyev SA, Taylor PA (2001) Notes on an internal boundary-layer height formula. Bound-Layer Meteorol 101:293–301
Savelyev SA, Taylor PA (2005) Internal boundary layers: I. Height formulae for neutral and diabatic flow. Bound-Layer Meteorol 115:1–25
Schlichting H and Gersten K (2003) Boundary-layer theory. McGraw Hill, New York, XXIII, 799 pp
Schmid HP (1994) Source areas for scalars and scalar fluxes. Bound-Layer Meteorol 67:293–318
Schmid HP (1997) Experimental design for flux measurements: matching scales of observations and fluxes. Agric Forest Meteorol 87:179–200
Schmid HP, Bünzli D (1995a) The influence of the surface texture on the effective roughness length. Quart J Roy Meteorol Soc 121:1–21
Schmid HP, Bünzli D (1995b) Reply to comments by E. M. Blyth on 'The influence of surface texture on the effective roughness length'. Quart J Roy Meteorol Soc 121:1173–1176
Schmid HP, Oke TR (1990) A model to estimate the source area contributing to turbulent exchange in the surface layer over patchy terrain. Quart J Roy Meteorol Soc 116:965–988
Schoonmaker PK (1998) Paleoecological perspectives on ecological scales. In: Peterson DL, Parker VT (eds) Ecological scale. Columbia University Press, New York, pp 79–103
Schuepp PH, Leclerc MY, MacPherson JI, Desjardins RL (1990) Footprint prediction of scalar fluxes from analytical solutions of the diffusion equation. Bound-Layer Meteorol 50:355–373
Sedefian L (1980) On the vertical extrapolation of mean wind power density. J Clim Appl Meteorol 19:488–493
Seginer I, Mulhearn PJ, Bradley EF, Finnigan JJ (1976) Turbulent flow in a model plant canopy. Bound-Layer Meteorol 10:423–453
Seibert P, Beyrich F, Gryning S-E, Joffre S, Rasmussen A, Tercier P (2000) Review and intercomparison of operational methods for the determination of the mixing height. Atmos Environ 34:1001–1027
Shaw RH, Patton EG (2003) Canopy element influences on resolved- and subgrid-scale energy within a large-eddy simulation. Agric Forest Meteorol 115:5–17
Shaw RH, Schumann U (1992) Large-eddy simulation of turbulent flow above and within a forest. Bound-Layer Meteorol 61:47–64
Shaw RH, Silversides RH, Thurtell GW (1974) Some observations of turbulence and turbulent transport within and above plant canopies. Bound-Layer Meteorol 5:429–449

Shaw RH, den Hartog G, Neumann HH (1988) Influence of foliar density and thermal stability on profiles of Reynolds stress and turbulence intensity in a deciduous forest. Bound-Layer Meteorol 45:391–409

Shen S, Leclerc MY (1994) Large-eddy simulation of small scale surface effects on the convective boundary layer structure. Atmos Ocean 32:717–731

Shen S, Leclerc MY (1995) How large must surface inhomogeneous be before they influence the convective boundary layer structure? a case study. Quart J Roy Meteorol Soc 121:1209–1228

Shen S, Leclerc MY (1997) Modelling the turbulence structure in the canopy layer. Agric Forest Meteorol 87:3–25

Shir CC (1972) A numerical computation of the air flow over a sudden change of surface roughness. J Atmos Sci 29:304–310

Shuttleworth WJ (1989) Micrometeorology of temperate and tropical forest. Phil Trans R Soc London B 324:299–334

Simpson IJ, Thurtell GW, Neumann HH, Hartog GD, Edwards GC (1998) The validity of similarity theory in the roughness sublayer above forests. Bound-Layer Meteorol 87:69–99

Smedman A-S (1991) Some turbulence characteristics in stable atmospheric boundary layer flow. J Atmos Sci 48:856–868

Sogachev A, Leclerc MY (2011) On concentration footprints for a tall tower in the presence of a nocturnal low-level jet. Agric Forest Meteorol 151:755–764

Sogachev A, Lloyd J (2004) Using a one-and-a-half order closure model of atmospheric boundary layer for surface flux footprint estimation. Bound-Layer Meteorol 112:467–502

Sogachev A, Panferov O (2006) Modification of two-equation models to account for plant drag. Bound-Layer Meteorol 121:229–266

Sogachev A, Menzhulin G, Heimann M, Lloyd J (2002) A simple three dimensional canopy-planetray boundary layer simulation model for scalar concentrations and fluxes. Tellus 54B:784–819

Sogachev A, Leclerc MY, Zhang G, Rannik U, Vesala T (2008) CO_2 fluxes near a forest edge: a numerical study. Ecol Appl 18:1454–1469

Sorbjan Z (1989) Structure of the atmospheric boundary layer. Prentice Hall, New York, 317 pp

Steinfeld G, Raasch S, Markkanen T (2008) Footprints in homogeneously and heterogeneously driven boundary layers derived from a Langrangian stochastic particle model embedded into large-eddy simulation. Bound-Layer Meteorol 129:225–248

Stull RB (1988) An introduction to boundary layer meteorology. Kluwer Acadamic Publisher, Dordrecht, 666 pp

Stull RB (2000) Meteorology for scientists and engineers. Brooks/Cole, Pacific Grove, 502 pp

Stull R, Santoso E (2000) Convective transport theory and counter-difference fluxes. In: 14th symposium on boundary layer and turbulence, Aspen, CO., 7–11 August 2000. American Meteorological Society, Boston, pp 112–113

Su H-B, Leclerc MY (1998) Large-eddy simulation of trace gas footprints from infinite crosswind line sources inside a forest canopy. In: 23th symposium on agricultural and forest meteorology1998. American Meteorological Society, Boston, pp 388–391

Su H-B, Shaw RH, Paw UKT, Moeng C-H, Sullivan PP (1998) Turbulent statistics of neutrally stratified flow within and above sparse forest from large-eddy simulation and field observations. Bound-Layer Meteorol 88:363–397

Sullivan PP, Horst TW, Lenschow DH, Moeng C-H, Weil JC (2003) Structure of subfilter-scale fluxes in the atmospheric surface layer with application to large-eddy simulation modelling. J Fluid Mech 482:101–139

Taylor GI (1923) Stability of a viscous liquid contained between two rotating cylinders. Phil Trans R Soc London 223:289–343

Taylor GI (1938) The spectrum of turbulence. Proc Roy Soc London A 164:476–490

Taylor PA (1987) Comments and further analysis on the effective roughness length for use in numerical three-dimensional models: a research note. Bound-Layer Meteorol 39:403–418

Tennekes H (1982) Similarity relations, scaling laws and spectral dynamics. In: Nieuwstadt FTM, Van Dop H (eds) Atmospheric turbulence and air pollution modelling. D. Reidel Publishing Company, Dordrecht, pp 37–68

Thomas C, Foken T (2002) Re-evaluation of integral turbulence characteristics and their parameterisations. In: 15th conference on turbulence and boundary layers, Wageningen, 15–19 July 2002. American Meteorological Society, pp 129–132

Thomson DJ (1987) Criteria for the selection of stochastic models of particle trajectories in turbulent flows. J Fluid Mech 189:529–556

Thurtell GW (1988) Canopy transport processes: commentary. In: Steffen WL, Denmead OT (eds) Flow and transport in the natural environment: advances and applications. Springer, Berlin, pp 128–132

Tillman JE (1972) The indirect determination of stability, heat and momentum fluxes in the atmospheric boundary layer from simple scalar variables during dry unstable conditions. J Clim Appl Meteorol 11:783–792

Troen I, Peterson EW (1989) European wind Atlas. Risø National Laboratory, Roskilde, 656 pp

van Ulden AP (1978) Simple estimates for vertical diffusion from sources near the ground. Atmos Environ 12:2125–2129

Verhoef A, McNaughton KG, Jacobs AFG (1997) A parameterization of momentum roughness lenght and displacement height for a wide range of canopy densities. Hydrol Earth Syst Sci 1:81–91

Vogel H-J, Roth K (2003) Moving through scales of flow and transport in soil. J Hydrol 272:95–106

Wang WG, Davis KJ, Cook BD, Butler MP, Ricciuto DM (2006) Decomposing CO2 fluxes measured over a mixed ecosystem at a tall tower and extending to a region: a case study. J Geophys Res 111:G02005.02001–G02005.02014

Watanabe T (2004) Large-eddy simulation of coherent turbulence structures associated with scalar ramps over plant canopies. Bound-Layer Meteorol 112:307–341

Webb EK (1970) Profile relationships: the log-linear range, and extension to strong stability. Quart J Roy Meteorol Soc 96:67–90

Wieringa J (1980) A revaluation of the Kansas mast influence on measurements of stress and cup anemometer overspeeding. Bound-Layer Meteorol 18:411–430

Wieringa J (1989) Shapes of annual frequency distribution of wind speed observed on high meteorological masts. Bound-Layer Meteorol 47:85–110

Wieringa J (1992) Updating the Davenport roughness classification. J Wind Eng Ind Aerodyn 41:357–368

Wilczak JM, Oncley SP, Stage SA (2001) Sonic anemometer tilt correction algorithms. Bound-Layer Meteorol 99:127–150

Wilson JD, Thurtell GW, Kidd GE (1981) Numerical simulation of particle trajectories in inhomogeneous turbulence. II. Systems with variable turbulent velocity scale. Bound-Layer Meteorol 21:423–441

Wilson JD, Ward DP, Thurtell GW, Kidd GE (1982) Statistics of atmospheric turbulence within and above a corn canopy. Bound-Layer Meteorol 24:495–519

Wyngaard JC (2010) Turbulence in the atmosphere. Cambridge University Press, Cambridge, 393 pp

Wyngaard JC, Coté OR (1971) The budgets of turbulent kinetic energy and temperature variance in the atmospheric surface layer. J Atmos Sci 28:190–201

Yaglom AM (1979) Similarity laws for constant-pressure and pressure-gradient turbulent wall flow. Ann Rev Fluid Mech 11:505–540

Yue W, Parlange MB, Meneveau C, Zhu W, van Hout R, Katz J (2007) Large-eddy simulation of canopy flow using plant-scale representation. Bound-Layer Meteorol 124:183–203

Chapter 3
Classification of Footprint Models

Footprint models constitute quality-assurance tools that can be used to support the interpretation of scalar, flux, and concentration measurements. The purpose of the present chapter is to provide the reader with an overview of key developments in footprint modeling. This chapter also highlights the key features and limitations inherent to each model. Here, the authors seek to provide sufficient information to empower the reader to select the most appropriate model for the purpose at hand. The reader is thus strongly encouraged to read the original papers for a more exhaustive and thorough description of these models.

In a nutshell, five approaches are presently available to calculate the flux footprint to the experimentalists and modelers alike: (1) Numerous forms of analytical solutions with varying degrees of complexity, whether 1D, 2D or even with the full 3D formulation, (2) Lagrangian simulations based on stochastic modeling of inhomogeneous turbulence, (3) Higher-order closure models, (4) Finally, in the hierarchy of footprint models, Large-Eddy Simulation, and (5) A combination of the above methods. Each category will be reviewed briefly. Table 1.3 presents some of the most important footprint models.

In the selection of a footprint model, the experimentalist is advised to consider several important criteria: (1) Complexity of the site as defined by roughness characteristics, vegetation types and biological, chemical and physical characteristics encompassed within the footprint area, topography, (2) Extent and depth of the experimental database for the site in question, (3) Range of atmospheric stabilities covered by the flux measurement campaign, (4) Purpose of application of the footprint model to determine the degree of accuracy needed in deriving credible footprint estimates, and (5) Level of familiarity and ease of use of each

M. Y. Leclerc and T. Foken, *Footprints in Micrometeorology and Ecology*,
DOI: 10.1007/978-3-642-54545-0_3, © Springer-Verlag Berlin Heidelberg 2014

footprint model, along with a working knowledge and understanding of the underlying hypotheses and assumptions.

Stationarity assumption aside ($d/dt = 0$ over the period of time), most models here represent adequately the state of the atmosphere. The turbulent transport of a scalar during the day over flat terrain is also generally well represented while rapid progress is being made to incorporate forcings such as that induced by the presence of hydrostatic pressure gradient arising from sloping terrain and discontinuities in roughness characteristics of the terrain over the area encompassed by an atmospheric exchange measurement. Moreover, recent progress is also being made quantifying and interpreting, not over a 2D "surface", but over a 3D surface complete with a vertical distribution of sources and sinks in addition to that of the horizontal distribution of the same sources. This is a great step forward in the analysis of surface-atmosphere exchange over tall forest canopies. The following overview ranges from the simplest, most field-ready footprint models such as the analytical solutions to the sophisticated CPU-intensive Large-Eddy Simulation.

3.1 Analytical Footprint Models

Analytical flux footprint models based either on exact or approximate solutions to the advection-diffusion equation aim at providing the first description for the vertical diffusion of material released at the surface. The majority of the models described here assumes an infinite crosswind line source of passive scalars and often resort to the use of power laws of height-dependent wind speed and eddy diffusivity. Some of the historical developments, concepts, and equations were described earlier in Chap. 2.

The analytical approach to footprint modeling is attractive in that the calculations are relatively few, easy, and quick to perform. However, analytical solutions to the advection-diffusion equation, despite their level of refinement, still have a limited ability to reproduce the diffusion process correctly is limited in many cases. Present analytical solutions for ground-level releases generally require a smooth surface, precluding their use immediately above orchards or forest canopies. The vast majority of these solutions ignore both the effect of atmospheric stability on the flow field and the height-dependence of eddy diffusivity. A notable exception is the work of Horst and Slinn (1984) whose predictions (precursors to its reformulation in terms of flux footprint functions by Horst and Weil 1992) compare well with experimental results in near-neutral conditions. However, the discrepancy between their solution and experiments increases as stability departs from neutral conditions. While powerful, its use is also not straight forward and requires the inclusion of ill-defined constants. These analytical solutions assume that the streamwise diffusion is negligible when compared with the advective component of the flow. This fact should be kept in mind for calm conditions. None of these footprint models work very well in the roughness sub-layer, inside canopies, for conditions of free convection and moderately stable conditions, and outside the atmospheric surface layer. They are thus expected to provide the highest degree of realism for cases where the mean flow is much larger than the turbulence intensity, i.e. $\overline{u(z)} \gg u'$ with $\overline{u(z)}$ for mean horizontal wind velocity in the height z und u' for its turbulent fluctuations.

3.1.1 The Schuepp et al. (1990) Approach

With the intent of incorporating a footprint analysis as part of a flux analysis package, Schuepp et al. (1990) formulated a simple approximate one-dimensional analytical solution to the diffusion equation proposed by Gash (1986) in terms of ***footprint***. That approximate solution is based on Calder's (1952) early work. At that time, Schuepp et al. (1990) ignored the van Ulden (1978) approach. For comparison, Schuepp et al. (1990) proposed an approach with a shape parameter $r = 1$ (and $A = B = 1$, see Sect. 2.4.1). A brief review is presented here.

If one considers an infinite crosswind area source of uniform flux density which satisfies the flux boundary condition of 0 for $x \leq 0$ (outside the source area) and Q_0 for $x > 0$, the concentration $\chi(x, z)$ at horizontal distance x and height z distribution is given by

$$\chi(x,z) = -\frac{Q_0}{\kappa u_* x} exp^{-\frac{uz}{\kappa u_* x}}, \qquad (3.1)$$

compare with the more general Eq. (2.82). The flux distribution can be calculated with Eq. (2.3) using the turbulent diffusion coefficient Eq. (2.7) and the concentration gradient by differentiating Eq. (3.1) with respect to z

$$\frac{\partial \chi(x,z)}{\partial z} = -\frac{Q_0}{\kappa u_* x} exp^{-\frac{uz}{\kappa u_* x}}\Big|_0^x. \qquad (3.2)$$

The total sum of the contributions from all elements of the upwind surface flux with the footprint is the spatial weighted elemental emission

$$Q_\chi(z) = \sum_{i=1}^{n} Q_i exp^{-\frac{uz}{\kappa u_* x}}\Big|_{x_{i-1}}^{x_i}. \qquad (3.3)$$

Equation (3.3) thus defines the one-dimensional "footprint". The relative contribution to the vertical flux ($\frac{1}{Q_0}\frac{dQ}{dx}$) from $x = 0$ to infinity can be used to determine the position of the peak of the footprint ($x_{\max}$), i.e. the area to which the observation at is most sensitive is,

$$x_{\max} = \frac{u}{u_*}\frac{(z-d)}{2\kappa}. \tag{3.4}$$

The above solution, however compact and simplified, does not include the influence of atmospheric stability on the behavior of the spatial extension of the footprint; in addition, this solution is limited to applications over smooth terrain.

3.1.2 The Schmid and Oke (1990) approach

Building upon the work of Pasquill (1972), Schmid and Oke (1990) developed a ***reverse plume source-area model*** (SAM) in a study focusing mostly on the identification of the maximum effect source location to a reference location on the basis of Fick's diffusion law for scalars (Eq. 2.76). Using the probability density function (pdf) plume of Gryning et al. (1987), Schmid and Oke (1990) derived an equation to establish an a priori selected most important source areas ω to a concentration measurement at a point, and to describe the influence of that particular source area to a reference (e.g. sensor) location. The identification of the sensitive regions to a concentration measurement is given with the function

$$P = \iint_{\omega=\omega_P} Q(x,y)dxdy \Big/ \int_{-\infty}^{\infty}\int_{0}^{\infty} Q(x,y)\,dx\,dy \tag{3.5}$$

with P being the portion of the total integrated effect controlled by a P-criterion source area bounded by the weight distribution function isopleths $\omega = \omega_P$, Schmid and Oke (1990) determined the concentration distribution from a continuous point source of a passive scalar. The schema of this reverse plume source-area model is illustrated in Fig. 3.1.

The Schmid and Oke (1990) study incorporates the influence of non-neutral atmospheric stability on resulting ***concentration footprints***. The approach of that study is based on the use of self-similar profiles of wind velocity and eddy diffusivity expressed by power laws matched to Monin-Obukhov similarity surface-layer profiles (Gryning et al. 1987).

3.1.3 The Family of Horst and Weil's (1992) Analytical Solution

This 'family' includes analytical solutions to the advection-diffusion equation originating from Horst and Weil (1992) with subsequent improvements (Horst and Weil 1994, 1995; Schmid 1994; Finn et al. 1996; Schmid 1997; Haenel and Grünhage 1999; Horst 1999). A leapfrog step was made subsequently by Kormann

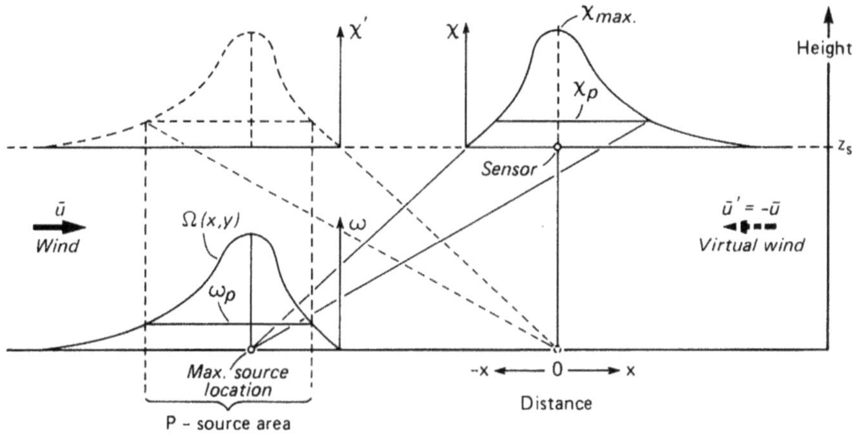

Fig. 3.1 Schematic cross-section of a P-criterion source area according to Schmid and Oke (1990): the source weight distribution ω is equivalent to the plane projection of the effect-level projection, at a height z_m of the virtual source beneath the sensor (and with a virtual wind in the reverse direction), Published with kind permission of © Royal Meteorological Society, 1990. All Rights Reserved

and Meixner (2001) which also used the same Horst and Weil model as the starting point of their derivation. The description of these approaches follow below.

As alluded to in Chap. 1, two important limitations of earlier solutions such as Calder's (1952), used by Gash's (1986) general solution to the diffusion equation and later applied by Schuepp et al. (1990) to quantify the flux footprint function include the constant u/u_* and the limitation that these solutions are mostly applicable in neutral stability. The studies in this section represent the many efforts by the footprint community to include a height-dependent and stability-dependent wind and diffusivity profiles.

3.1.3.1 The Horst and Weil (1992, 1994) Approach

Most earlier exact or approximate solutions were hampered by either or both the limitation of a constant height-independent wind speed profile and restricted to neutral stability. Following van Ulden (1978) and Horst (1979), the arrival of the Horst and Weil (1992, 1994) analytical solutions describing footprint functions provided a welcome alternative. Oversimplifications that had previously crippled solutions to the diffusion equations, i.e. that of a constant u/u_* and the limitation to neutral stability, were circumvented later by Horst and Weil (1992) using a numerical solution. Horst and Weil (1994) extends their model's use to special cases of non-passive scalar transfer and became an approximate analytical solution to the diffusion equation.

Horst and Weil (1992) used the approximate vertical concentration profile equation proposed by van Ulden (1978) and Horst (1979) to formulate the crosswind-integrated flux footprint $\overline{f}^y(x, z_m)$ as briefly described below.

Imagining a fluid element released at a point (x_r, y_r, z_r) detected at a downwind position (x, y, z) with $Q_0(x_r, y_r, 0)$ source strength per unit area at ground-level at position $(x_r, y_r, 0)$, the flux density measured at measurement height z_m at point (x, y, z_m), $F(x, y, z_m)$ is given by:

$$F(x,y,z_m) = \int_{-\infty}^{\infty}\int_{0}^{\infty} Q_0(x_r,y_r,0)f(x',y',z_m)dx_r dy_r \tag{3.6}$$

where $f(x' = x - x_r, y' = y - y_r, z_m)$ is the ***source probability density*** (***footprint***) ***function*** and where it is assumed that the flow field of turbulence is horizontally homogeneous. Assuming that the wind is in the x-direction (from higher to lower values), the flux footprint f therefore depends only on the streamwise direction separation distance $x - x_r$ and the crosswind separation distance $y - y_r$. The sum (discrete case) or the integral (continuous) of the flux footprint values f is unity.

Horst and Weil (1994, 1995) gave the special case of a surface point of emission rate Q where $F_0(x_r, y_r) = Q\delta(x_r)\delta(y_r)$ with the result that the flux footprint equals the vertical flux downwind of a unit surface point source:

$$f(x,y,z_m) = \frac{F_m(x,y,z_m)}{Q} \tag{3.7}$$

Defining $\overline{f}^y(x, z_m)$ as the crosswind-integrated footprint, follows

$$\overline{f}^y(x,y,z_m) = \int_{-\infty}^{\infty} f(x,y,z_m)dy \tag{3.8}$$

and integrating the two-dimensional advection-diffusion equation from the surface to the flux measurement height z_m:

$$\overline{\eta_y}(x,z_m) = -\frac{\partial}{\partial x}\int_0^{z_m} \overline{u(z)\chi_y(x,z)}dz \tag{3.9}$$

where $\overline{u(z)}$ is the mean wind speed profile and $\overline{\chi_y(x,z)}$ is the crosswind-integrated concentration distribution downwind of a unit surface point source. Thus, the relationship between upwind surface source or sink distributions and measurements of concentration or flux at some point z_m above the surface is defined by the footprint function.

Horst and Weil (1992) presented a normalized integrated crosswind integrated footprint, η, which strongly depends on $\overline{z}/z_m$ and exhibits only very weak remaining dependence on stability and surface roughness. For averaging the mean height, please see van Ulden (1978) or Eq. (2.82).

Horst and Weil (1992) demonstrated this universality by a comparison with a Lagrangian simulation footprint model. To circumvent the challenge that their model can only be evaluated numerically, Horst and Weil (1994) provided an approximate analytical expression for the normalized crosswind-integrated footprint Φ (for A and $b = B^{-1}$ see Sect. 2.3.1), which is the exact solution of Eq. (3.8) for power law wind profiles (Horst 1999),

$$\Phi = \frac{z_m \bar{f}^y(x, z_m)}{d\bar{z}/dx} \approx A \left[\frac{z_m}{\bar{z}}\right]^2 \frac{\overline{u(z_m)}}{u(\bar{z})} exp^{-(z_m/b\bar{z})^r}. \tag{3.10}$$

In a tracer flux experiment aimed at validating the various footprint models available at that time, Finn et al. (1996) evaluated the parameter r experimentally for a range of atmospheric stabilities at four diffusion distances over a total of 136 cases spanning a total of 485 hypothetical towers, free of edge effects and with quality data; they did so by running their analytical model for each tracer flux period at each of the four tower distances and allowed shape parameter r to vary over the range of 0.3–5.0, encompassing the theoretical limit of the Gryning et al. (1987) model. The empirical value of r was then chosen as that value for which the predicted flux F was equal to the measured flux.

The Horst and Weil (1992) solution has been widely used worldwide and provided invaluable insight in the analysis of field campaigns with fluxes measured over a wide variety of environmental conditions.

3.1.3.2 The Schmid (1994, 1997) Approaches

A notable step forward was provided by the use of the flux footprint analytical solution originally attributed Horst and Weil (1992, 1994), see Eq. (3.10), with the arrival of the extension of the Horst and Weil formulation by Schmid (1994) to a two-dimensional flux footprint analysis, thus generating much additional insight into the interpretation of experimental data collected over patchy surfaces. Furthermore, the expanded model extended the solutions to a range of atmospheric stabilities. A useful two-dimensional source area of level P defined as the integral of *hitherto* the source weight function of the smallest possible domain comprising the fraction P of the total surface influence in the measured signal and based on Horst and Weil (1992) was described in Schmid (1994). The latter presented simple expressions to provide parameterization formulae for the principal source areas as functions of the measurement level, atmospheric stability, and crosswind turbulence. These formulae have ease of use but are limited in their range of applications. That model provides P information.

On this basis, the original model of Schmid and Oke (1990), SAM, was extended to the two-dimensional case and not only used for the concentration footprint but also as a flux footprint model FSAM by Schmid (1994).

A later paper by Schmid (1997) discussed criteria and guidelines to be used in field campaigns over a variety of surfaces (short crops, agricultural areas, and urban areas) with particular emphasis given to the need of matching the choice of the measurement system to the spatial scale of measured surface. Schmid (1997) cautions the limits of the model regarding the range of atmospheric stability, vertical range of the applicability of the model being limited to the atmospheric surface layer, and states that flux footprints obtained from concentration profile or from Bowen ratio measurements should refer to the model of Horst (1999). For more details see Sect. 7.1.2.

The Schmid (1994) model has been one of the most useful footprint models to date owing to its inherent relevance to deal with interpretation of real-terrain sources and sinks. As is the case for all other analytical solutions, this flux footprint calculation algorithm can be tagged to signal processing packages for online analysis of flux outputs.

Unfortunately, the model is numerical unstable for $z_m/z_0 < 12$ and in highly unstable and in stable conditions. Therefore, for field applications where a wide range of atmospheric conditions is sought, the Kormann and Meixner (2001) model should be applied.

3.1.3.3 The Kaharabata et al. (1997) Approach

The model by Kaharabata et al. (1997), based on the previous works by Horst and Weil (1992, 1994) and Schmid (1994), is like the later model (Kaharabata et al. 1999) a 2D approach. According to van Ulden (1978), they us the logarithmical wind profile Eq. (2.8), but multiplied the aerodynamical height in the logarithm and in the universal function with the Euler-Mascheroni constant of 0.5772. The model was mainly applied for aircraft measurements during the BOREAS experiment by Chen et al. (1999) and Ogunjemiyo et al. (2003) and further on for the emission of VOCs from forest canopies (Kaharabata et al. 1999).

3.1.3.4 The Haenel and Grünhage (1999) approach

Haenel and Grünhage (1999) presented a slightly different analytical solution, which, unlike existing analytical solutions for crosswind-integrated flux footprint, normalizes the footprint using a closed analytical formula based on height-dependent profiles of wind speed and eddy diffusivity. They pointed out that the implementation of the expression above causes Φ, the normalized crosswind integrated footprint, to overshoot its theoretical constraint of unity at large diffusion distances $\overline{z}(x)$.

The major difference between the Haenel and Grünhage's (1999) description and the earlier analytical flux footprint models is that the parameter r shaping the vertical plume dispersion is set constant as opposed to being a function of upwind distance. As a result of this change, their model satisfies the condition that the cumulative normalized footprint approaches unity for an infinite upwind distance.

The authors proposed to keep the use of power laws expressions longer in the use of their derivation of the crosswind integration of the flux footprint expression and prescribed r as independent of $\bar{z}$ and thus of x. Since r is a constant, the crosswind integrated flux footprint expression can be integrated numerically and becomes, in normalized form:

$$\Phi = AB'\left[\left(\frac{z_m}{b\bar{z}}\right)^{(3+r)/2}\right]\left(\frac{z_m}{b\bar{z}}\right)^{r} \tag{3.11}$$

where B' is an analytical function of r in Haenel and Grünhage (1999):

$$B' = b^{(1-r)/2}\Gamma(1/r)\{\Gamma[(1+r)/2r]\}^{-1} \tag{3.12}$$

Applying Gram-Schmidt's conjugate powers for the power laws of wind speed and diffusivity, they finally reintroduced Monin-Obukhov similarity theory at this stage by expressing r as a function of stability z_m/L and measurement height and roughness z_m/z_0.

Thus, the Horst and Weil (1992, 1994) models and the Haenel and Grünhage (1999) model follow the van Ulden (1978) use of the Monin-Obukhov similarity theory for profiles of K and $\bar{u}$, and ignore the weak dependence of p on r . The result is a similarity relation for $d\bar{z}/dx$ that can only be solved for $\bar{z}$ numerically.

The cumulative normalized footprint function approach unit asymptotically if and when the constant m is defined, normally using the Gram-Schmidt's conjugate power law. This value can be defined by applying Gram-Schmidt's conjugate power law.

Haenel and Grünhage (1999) state that their model is both less complex and computationally more effective than the Horst and Weil (1994) approximate and Horst (1999) profile models.

3.1.3.5 The Kormann and Meixner (2001) Approach

Both the approach of Kormann and Meixner (2001) and that of Haenel and Grünhage aim at avoiding the apparent inconsistent behavior of the Horst and Weil (1992) model and decreasing the computational time to evaluate footprint functions. Unlike Haenel and Grünhage (1999), Kormann and Meixner (2001) used the power law profiles for both K and $\bar{u}$ in the solution for $d\bar{z}/dx$. It therefore allows an analytical integration.

Kormann and Meixner (2001) used two different approaches to circumvent this difficulty in solving the power-law profile, the first one resorting to the purely analytical description of Huang (1979) and a simple numerical one, which minimizes the deviations between the two different profiles. The reader is referred to Sect. 2.2.5, Eqs. (2.54)–(2.58) for details regarding the Huang (1979) method. That solution matches the power law for $u(z)$ and $K(z)$ and the Monin-Obukhov profiles for the stability dependence of the exponents in the power laws at a certain height. The numerical approach is simple when compared to that of Schmid (1994) and requires essentially a one-dimensional numerical root finding. Finally, in order to relate the two different types of profiles, we have to specify the Businger–Dyer relationships. Kormann and Meixner (2001) note further that the reference height for the exponents p and p' and the proportionality constants u and K need not be the same and that this approach generally overestimates the velocity near the ground, especially for unstable conditions and large roughness length values. In the same way, this solution tends to overestimate the eddy diffusivity in stable conditions.

In the Kormann and Meixner (2001), the original version by Huang (1979) was modified in the following form (identical with Eqs. 2.55 and 2.56)

$$p = \frac{z}{u}\frac{\partial u}{\partial z} = \frac{u_*}{\kappa u}\varphi_m(\zeta) \tag{3.13}$$

and

$$p' = \frac{z}{K}\frac{\partial K}{\partial z} = \begin{cases} \frac{1}{1+5z/L}, & L > 0 \\ \frac{1-24z/L}{1-16z/L}, & L<0 \end{cases}. \tag{3.14}$$

In its simplest form where $p = 1$ and p is the power exponent in the diffusivity expression, the Kormann and Meixner (2001) method is equivalent to that of Schuepp et al. (1990) but when Gram-Schmidt's conjugate laws must be applied ($p + p' = 1$), it is equivalent to the analytical solution of Haenel and Grünhage (1999).

To date, the Kormann and Meixner (2001) analytical solution is an algebraic expression in x and z and thus constitutes the only truly analytical flux footprint model based on realistic profiles of K and u. It is one of the most desired solutions due to a combination of attributes including its ease of use, its wide range of stability and its numerical stability. An applicable version of the model was published by Neftel et al. (2008)—see also Supplement 3.1.

Supplement 3.1: The Kormann and Meixner (2001) in the Version by Neftel et al. (2008)

The tool by Neftel et al. (2008) is available on the WEB-page:
http://www.agroscope.admin.ch/art-footprint-tool/
and includes an instruction and an EXCEL sheet. The user has to copy his input data into the EXCEL Sheet and receives as the output data for the

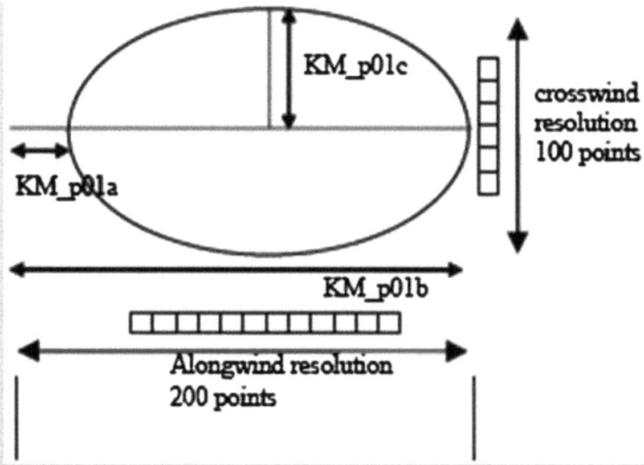

Fig. 3.S1 Output parameter KM_p01 a, b. c of model by Neftel et al. (2008) for the 1 % effect level

footprint ellipse as given in Fig. 3.S1. The input data are given in Table 3.S1. No input of the roughness length is necessary. This is done with an internal calculation and the roughness length as an output parameter. Further output parameters are the model functions. Details are given in an instruction file.

Table 3.S1 Input parameter of model by Neftel et al. (2008)

	x, y Sensor coordinates	u_*	L	σ_v	Wind direction	z_m	u
Dimension	m	m s^{-1}	m	m s^{-1}	°	m	m s^{-1}
Restriction	1 m resolution		>1 m	0.01–5.0 m s^{-1}		Above zero plane displ.	

3.1.4 Analytical Solutions Based on Lagrangian Models

Analytical solutions have also resorted to the results and insights provided by Lagrangian simulations. Nonetheless, given that Lagrangian simulations are used to construct analytical solutions. Their linkage to analytical solutions is briefly described. Due to the release of a large number of trajectories required to obtain stable solutions, the long computing time needed to produce statistically reliable results is an unavoidable weakness of Lagrangian stochastic footprint models. This can be partly overcome using the method proposed by Hsieh et al. (2000) which sought to bypass this difficulty by using an analytical model derived from Lagrangian model results.

Hsieh et al. (2000) developed a Lagrangian stochastic dispersion model based on a Markov process and the application of the well-mixed criterion by Thomson (1987). They used the scaling and fetch analysis of previous models (Schuepp et al. 1990; Horst and Weil 1994; Luhar and Rao 1994; Hsieh et al. 1997). Finally they found that the result was very close to Gash's (1986) analytical value but describes also the flux change due to a change of the Obukhov length.

In addition, given the need for real-time footprint information during the data collection period, Kljun et al. (2004) proposed a simple parameterization based on a Lagrangian footprint model. Their parameterization is highly valuable to an experimentalist as it allows the determination of the footprint from atmospheric variables obtained during flux measurements. Kljun et al. (2004) investigated the Lagrangian backward model by Kljun et al. (2002) with an analysis of dimensionless parameters according to Buckingham's Π-Theorem (Kantha and Clayson 2000). By ensemble averaging of model runs the parameterizatious where found to be a function of following dimensionless parameters $\Pi_1 = z_m \bar{f}_y$, $\Pi_2 = x/z_m$, $\Pi_3 = z_i/z_m$, and $\Pi_4 = \sigma_w/u_*$, which could be combined finally to two parameters

$$X_* = \Pi_4^{\alpha_1} \Pi_2 = \left(\frac{\sigma_w}{u_*}\right)^{\alpha_1} \frac{x}{z_m}, \tag{3.15}$$

$$F_* = \Pi_4^{\alpha_2} \Pi_3' \Pi_1 = \left(\frac{\sigma_w}{u_*}\right)^{\alpha_2} \left(1 - \frac{z_m}{z_i}\right)^{-1} z_m \overline{f_y} \tag{3.16}$$

By ensemble averaging model runs, the parameterizations were found as functions of both dimensionless parameters (Fig. 3.2)

$$\widehat{F}_* = a\left(\frac{\widehat{X}_* + d}{c}\right)^b exp\left\{b\left(1 - \frac{\widehat{X}_* + d}{c}\right)\right\}, \tag{3.17}$$

where $\widehat{F}_*$ and $\widehat{X_*}$ are the ensemble averaged functions, $\alpha_{1,2}$ are free parameters, and a, b, c, d are coefficients. These parameterizations are similar to those of Hsieh et al. (2000) and are well comparable with the Kormann and Meixner (2001) analytical approach.

In comparison to the Lagrangian model the approximation is only applicable for homogeneous terrain but in comparison to many other models it can be used also outside the surface layer and was tested for a wide range of meteorological conditions: $-200 \le z_m/L \le 1$, $u_* \ge 0.2$ m s^{-1}, $z_m > 1$ m. The model is available online (Supplement 3.2).

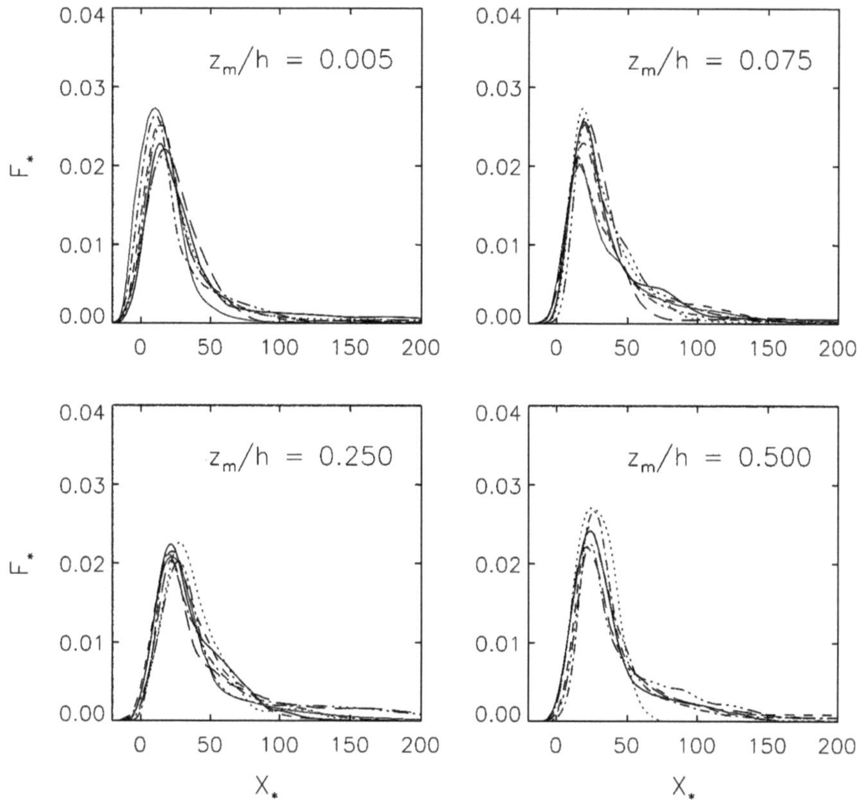

Fig. 3.2 Parameterisation according to Kljun et al. (2004) of the ensemble of scaled flux footprints (Eq. 3.16 *solid line*). The *dashed lines* indicate the scaled footprint estimates with the backward Lagrangian model (Kljun et al. 2002). These footprint estimates range from strongly convective to strongly stable with receptor heights of $z_m/z_i = \{0.005, 0.075, 0.25, 0.50\}$. Roughness length as indicated in each panel

Supplement 3.2: Online Version of the Model by Kljun et al. (2002) in the Analytical Version by Kljun et al. (2004)

The program by Kljun et al. (2004) is online available on the WEB-page as executable online version:
http://footprint.kljun.net/index.php
The input parameters are given in Table 3.S2. Furthermore, the effect level (up to 90 %) should be given for the calculation. The output is a visual presentation (Fig. 3.S2) and a data set of the master footprint according to in the dimensionless function

Table 3.S2 Input parameter of model by Kljun et al. (2004)

	σ_v	u_*	z_m	z_i	L	z_0
Dimension	m s^{-1}	m s^{-1}	m	m	m	m
Restriction		>0.2 m s^{-1}	>1 m above zero plane displ.	$z_i > z_m$	$-200 \le z_m/L \le 1$	

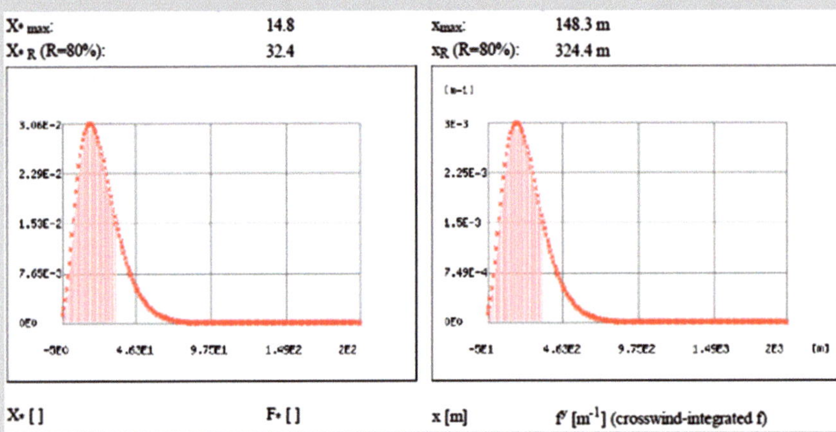

Fig. 3.S2 Output graph of the Kljun et al. (2004) model

$$X_* = \left(\frac{\sigma_w}{u_*}\right)^{\alpha_1} \frac{x}{z_m} \tag{3.S2a}$$

and

$$F_* = \left(\frac{\sigma_w}{u_*}\right)^{\alpha_2} \left(1 - \frac{z_m}{h}\right)^{-1} z_m \overline{f^y}, \tag{3.S2b}$$

with $\alpha_1 = -\alpha_2 = 0.8$, as well as the usual crosswind integrated footprint function dependent on the distance from the measuring point. Furthermore the location of the maximum of the footprint and the extension of the footprint for the given effect level are calculated.

3.2 Lagrangian Simulations

The stochastic Lagrangian approach represents one of the most natural methods for simulating the motions of molecules advected in a turbulent flow to a point measurement; its approach is simple and lends itself particularly well in numerous

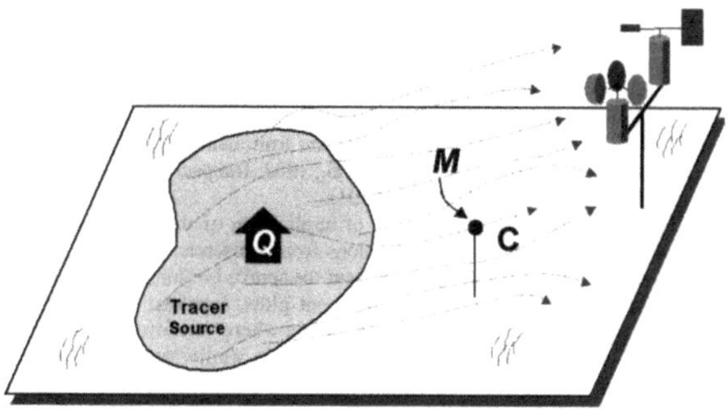

Fig. 3.3 The inverse-dispersion method for estimating tracer emission rates (Q). Average tracer concentration C measured at pint M. A dispersion model predicts the ratio of concentration at M to the emission rate (C/Q) (Flesch and Wilson 2005, Published with kind permission of © American Society of Agronomy, 2005. All Rights Reserved)

footprint applications in flows ranging from homogeneous turbulence to sheared, anisotropic inhomogeneous turbulent flows. Once the form of the parameterization is chosen, the stochastic Langevin type equation is solved (e.g. Sawford 1985; Thomson 1987; Sabelfeld and Kurbanmuradov 1990). The Lagrangian approach needs only the one-point probability density function (pdf) of the Eulerian velocity field. The Lagrangian stochastic trajectory simulation, together with appropriate simulation methods and corresponding estimators for concentration or flux footprints, are then merged into a Lagrangian footprint model. For a detailed overview of the estimation of concentration and flux footprints in particular, the reader is referred to Kurbanmuradov et al. (2001).

The Lagrangian simulation (LS) method used for footprint applications involving a myriad of other atmospheric turbulent diffusion problems is based on a stochastic differential equation. That equation, the Langevin equation, determines the evolution of fluid particles in space and time. With the LS, the approach typically consists of releasing millions of fluid particles of infinitesimal mass at the surface point source and tracking their trajectories in a fluid to which a turbulent flow field is assigned and downwind of this source towards the measurement location forward in time (Fig. 3.3, Leclerc and Thurtell 1990; Rannik et al. 2000, 2003). An ensemble of particle trajectories then reproduces the dispersion process. This has the advantage that the small time behavior, i.e. the diffusion of particles for short travel times following their release, can be accounted for, something not otherwise possible in an Eulerian frame of reference (Sawford 1985; Nguyen et al. 1997). Such footprint models require a prescribed turbulence field, often obtained using scaling laws such as Monin-Obukhov similarity theory or atmospheric boundary layer scaling laws. The approach is stochastic in nature and is often treated as a Gaussian process. The idea has its origin in the "drunkard's walk", first coined by Einstein (1905) to describe the behavior of molecular diffusion. This reflects well the behavior of an infinitesimally small fluid particle embedded

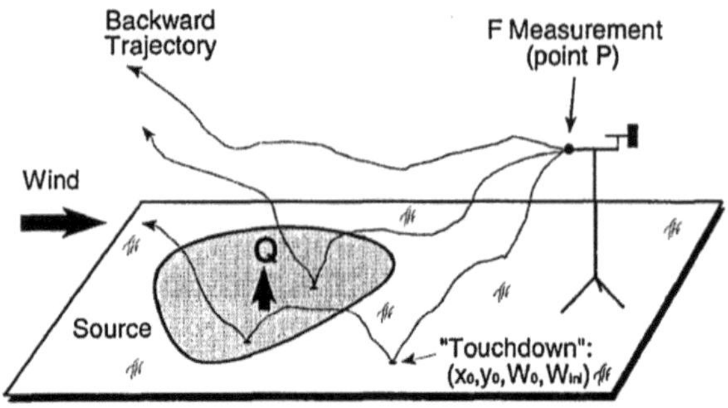

Fig. 3.4 Idealization of the backward Lagrangian simulation methodology: Particles are released from flux measurement location (Point P) and followed upstream. A touchdown catalogue stores touchdown locations (x_0,y_0), vertical touchdown velocities (W_0) and vertical velocities at release (W_{ini}) for all particles (Flesch 1996)

in a fluid in motion. In this frame of reference and in contrast with its Eulerian counterpart, the fluid particle moves with the flow.

Lagrangian footprint models have also been run in the backward mode (Fig. 3.4), i.e. tracking the trajectories from their point of measurements back, using a negative time step, to their point of origin on the surface; this has been done for both flux and concentration footprints (Flesch and Wilson 1992; Flesch et al. 1995, 2004; Flesch 1996; Kljun et al. 2002; Cai and Leclerc 2007).

The treatment of upper and lower boundaries must be treated carefully and reflection scheme near the lower boundary developed. The literature is replete with different formulations of Lagrangian simulations for inhomogeneous turbulence, a thorny topic with theoreticians. Reviews of the myriad of formulations have been given to us by Rodean (1996), Wilson and Sawford (1996) and Kurbanmuradov and Sabelfeld (2000). Given the formulation of necessary conditions to obtain a correct simulation of diffusion in inhomogeneous turbulence, the main criterion for robust simulations is the well-mixed condition and so is correct within the most rigorous Lagrangian formulation *ab extensio*, of Lagrangian footprint models Thomson (1987). It also should be pointed out that this well-mixed criterion condition can be fulfilled. Yet, the stochastic model is not necessarily uniquely defined for atmospheric flow conditions (called the uniqueness problem). Rannik et al. (2012) point out that, even in the case of homogeneous but anisotropic turbulence, there are several stochastic models which satisfy the well-mixed condition (Thomson 1987; Sabelfeld and Kurbanmuradov 1998). In addition to the well-mixed condition by Thomson (1987), the trajectory curvature has also been proposed as the additional criterion to select the most appropriate Lagrangian stochastic model (Wilson and Flesch 1997), but this additional criterion does not define the unique model (Sawford 1999).

While many if not most Lagrangian footprint models run in the forward mode, the backward Lagrangian method has been gaining in popularity. In this scheme, particle trajectories are tracked from their point of measurements backward, using a negative timestep, to their point of origin on the surface. This has been done for both flux and concentration footprints (Flesch et al. 1995; Flesch 1996; Kljun et al. 2002; Flesch and Wilson 2005; Cai and Leclerc 2007; Hsieh and Katul 2009; Sogachev and Leclerc 2011). The basic equations for concentration and flux footprints are (Flesch 1996)

$$\chi(x,y,z) = \frac{2}{N}\sum_{i=1}^{N}\sum_{j=1}^{n_i}\frac{1}{W_{ij}}Q(x_{ij},y_{ij},z_0), \tag{3.18}$$

$$F(x,y,z) = \frac{2}{N}\sum_{i=1}^{N}\sum_{j=1}^{n_i}\frac{W_{i0}}{W_{ij}}Q(x_{ij},y_{ij},z_0), \tag{3.19}$$

with the initial velocity W_{i0} and the touchdown velocity W_{ij} for all particles.

The forward and backward methods used to derive footprints are theoretically equivalent. In practice, the forward LS models are generally applicable in horizontally homogeneous conditions. The attribute intrinsic in the (Flesch et al. 1995) backward trajectory approach that neither horizontal homogeneity nor stationarity of the turbulence field is required makes it in principle a powerful method to construct footprint estimates over non-homogeneous terrain. The reader is referred to Sogachev and Leclerc (2011) as an illustration.

The Lagrangian stochastic approach can be applied to any turbulence regime, thus allowing footprint calculations for various atmospheric boundary layer flow regimes. For example, in the convective boundary layer, turbulence statistics are typically non-Gaussian and for realistic dispersion simulations, a non-Gaussian trajectory model has to be applied. An indication of the departure from Gaussianity is often obtained using the turbulence velocity skewness; for instance, in convective boundary-layers, the vertical velocity skewness is typically 0.3 while a neutral canopy layer can exhibit negative vertical velocity skewness as large as −2.0 (Leclerc and Thurtell 1990; Finnigan 2000). However, most Lagrangian trajectory models fulfill the main criterion for construction of Lagrangian stochastic models, the well-mixed condition (Thomson 1987) for only one given turbulence regime.

In the case of tall forest canopies, similarity laws that work well within the atmospheric surface layer break down within the canopy, i.e. in a region characterized by fluxes rapidly changing within the canopy layer even over short distances and so the assumption of a 'constant' flux layer cannot be assumed. A theoretical framework, such as the Monin-Obukhov similarity theory, which describes time-averaged wind relations as a function of measurement height z_m above the surface and atmospheric stability z_m/L at any given level as long as it is contained within the atmospheric surface layer, adds a layer of complexity in the assessment of the contribution of individual source signatures to a point flux measurement, whether located within or above a canopy layer. With both thermal and mechanical turbulence most often co-arising inside a canopy layer, a sound framework to predict turbulence statistics used to identify the local footprint in

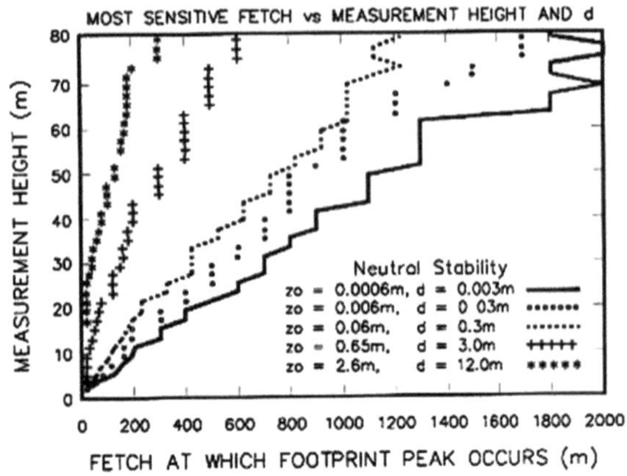

Fig. 3.5 Dependence of the footprint peak dependent on the fetch and the measuring height (Leclerc and Thurtell 1990)

non-isothermal conditions is still needed, though recent work to palliate this deficiency is gradually emerging (Zhang et al. 2010). Also of paramount importance, knowledge of Lagrangian timescales inside a canopy is also required and, despite their capital importance in Lagrangian simulations, is still needed. Some of the Lagrangian simulations (Rannik et al. 2000; Mölder et al. 2004; Poggi and Katul 2008) also use a Kolmogorov constant C_0 whose model results are sensitive to the absolute value of the constant (Rannik et al. 2003; Mölder et al. 2004). Poggi and Katul (2008) revealed that C_0 may vary nonlinearly inside the canopy while the LS model predictions were not sensitive to gradients of C_0 inside canopy.

Göckede et al. (2004) solved this problem of long computing time needed to produce statistically reliable results by pre-calculation of look-up tables in a wide range of atmospheric and surface characteristics. These simplifications allow the determination of the footprint from atmospheric variables usually measured during flux observation programs.

3.2.1 The Leclerc and Thurtell (1990) Approach

As one of the first two-paper series describing the behavior of flux footprint above natural surfaces, Leclerc and Thurtell (1990) investigated the signatures of individual sources contributing to a point flux measurement using a 2-D Lagrangian stochastic dispersion model parameterized by effective roughness length and displacement height (Fig. 3.5). Leclerc and Thurtell (1990) describe a two-dimensional Markovian (random walk) simulation of the respective contribution of upwind sources to a point flux measurement at height z_m.

This early study was the first to highlight the prominent influence of atmospheric stability on the upwind footprint; it also demonstrated the role of measurement level and surface roughness with particular attention to the location of the peak source contribution. While such results are now well accepted, their early simulations demonstrated how measurements obtained during unstable daytime conditions represent fluxes from upwind sources closer to the observation point than those measurements made during stable nighttime conditions. They also demonstrated the sensitivity of the footprint peak to measurement level and surface properties. Despite the fact that this is the first Lagrangian footprint model, its robustness was tested by several extensive turbulence tracer flux experiments and proved to be describing accurately the flux footprint over surfaces ranging from smooth surfaces to above forest canopies (Finn et al. 1996; Leclerc et al. 2003a, b)

3.2.2 *The Sabelfeld-Rannik Approach*

The formalism of the Lagrangian simulation used by Rannik et al. (2000) to model footprints in the canopy layer is based on the work of Kurbanmuradov et al. (1999). The Kurbanmuradov-based simulation satisfies the well-mixed condition (Thomson 1987; Sabelfeld and Kurbanmuradov 1998). The Sabelfeld and Kurbanmuradov (1998) approach was compared with that of Thomson (1987) and will be shown below.

Rannik et al. (2000) evaluated both approaches, the one given by Thomson (1987) and that of Kurbanmuradov et al. (1999). A comparison between the stochastic footprint flux model using the formalism of Thomson (1987) and that of Kurbanmuradov et al. (1999) is shown in Fig. 5.1 with flux footprints found to be virtually identical to one another. Finally, the basis of this model stems from Sabelfeld and Kurbanmuradov (1990), which includes the well-mixed conditions by Thomson (1987). For the wind profile they applied the Monin-Obukhov similarity theory with a modification for the roughness sublayer according to Cellier and Brunet (1992) and an in-canopy profile according to Kaimal and Finnigan (1994). The model was tested at the FLUXNET site Vielsam (BE-Vie).

Rannik et al. (2003) improved their model by applying it to FLUXNET site Hyytiälä (FI-Hyy) data, for which they made a site specific parameterization for the in-canopy profile (see Eq. 2.68). Göckede et al. (2007) used this model for the Waldstein-Weidenbrunnen site (DE-Bay) data and found in comparison with the approach by Massman and Weil (1999) that the footprint can be significantly improved by using site specific in-canopy profile parameterizations. A further improvement is possible according to Göckede et al. (2007), if the in-canopy parameterization can be used in a specific form dependent on the coupling between the forest and the atmosphere (Thomas and Foken 2007).

3.2.3 The Kljun et al. (2002) 3D Backward Lagrangian Footprint Model

The study of Kjun et al. (2002) is credited for the first use a 3D backward Lagrangian footprint model (LPDM-B) to determine flux footprints using the three-dimensional model of de Haan and Rotach (1998). The latter, based on Rotach et al. (1996) uses the approach of backward Lagrangian dispersion method first attributed to Flesch et al. (1995). The Kljun et al. (2002) model accommodates a wide spectrum of atmospheric stabilities and satisfies the well-mixed condition throughout a wide range of stabilities. It also can be used in three-dimensional footprint calculations above the surface layer, something particularly useful in the interpretation of observations from airborne flux platforms (Leclerc et al. 1997). Following Rotach et al. (1996), Kljun et al. (2002) approximated a skewed probability density function of the vertical velocity to model the footprint in the convective boundary layer using a scheme proposed by Baerentsen and Berkowicz (1984) in their LPDM-B. This approximation was done by adding the sum of two Gaussian distributions one for the updrafts and one for the downdrafts respectively. Gibson and Sailor (2012) found some mathematical inconsistencies in the Rotach et al. (1996) and therefore also in the Kljun et al. (2002) approach. The correction would make the model more stable.

Based on Flesch's method (personal comm., 2001), Kljun et al. (2002) also introduced a spin-up procedure in the model. The simulated flux footprint depends strongly on the particle's initial velocities since these are explicitly included in the footprint calculation. According to the authors, Lagrangian particle models need to incorporate the correlation of the streamwise and vertical velocity components resulting in unrealistic individual particle velocities produced. Assuming this to be the case, when calculated over hundreds of thousands of particles, this effect is non-negligible and biases the trajectories, and thus the resulting concentrations. While the distributions without spin-up were almost symmetric (neglected correlation of u and w) in the quadrant analysis (not shown here), the spin-up procedure leads to a more realistic distribution between the flow quadrants. The figures below depict the footprint flux and concentration results of the LPDM-B for contrasting stability conditions (Figs. 3.6 and 3.7).

Also based on the approach by Flesch et al. (1995) Wang and Rotach (2010) developed a backward model for undulating surfaces in a non-flat topography. They found that the topographic influence on the footprint depends on the stratification, the wind speed and the wind direction in relative to the orientation of the topography.

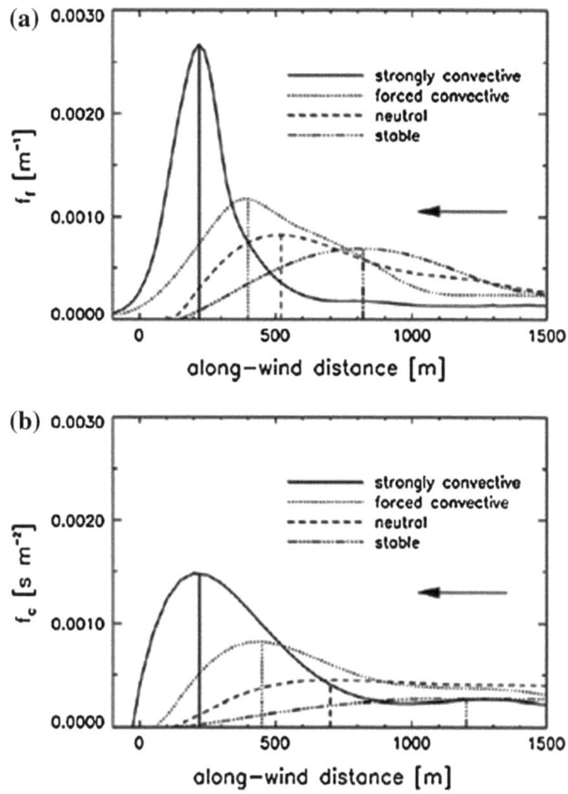

Fig. 3.6 **a** Crosswind integrated footprint for flux and **b** concentration measurements for four different cases of stability. The location of respective peaks are indicated by *vertical lines* (Kljun et al. 2002)

3.3 Higher-Order Closure Footprint Models

An alternative to analytical solutions or to Lagrangian formalism arises in the form of higher-order closure models. These can be used to describe a step change in contrasting scalar flux in two adjoining fields with dissimilar scalar and aerodynamic properties provided the change from an upwind mixing length to a downwind equilibrium value is gradual. Amongst these, the second-order closure model is the closure order most often sought.

One recent member of this family of higher-order closure models, SCADIS, uses one and half order closure scheme: this two-equation model bypasses a predefined mixing length and includes a new parameterization for the drag term (Sogachev and Lloyd 2004). The numerical atmospheric boundary-layer (ABL) SCADIS model based on E-ω scheme (where E is turbulent kinetic energy and ω is specific dissipation of E) is a model that has been rapidly gaining ground because of its versatility as described in Sogachev et al. (2002, 2005a), Sogachev and Lloyd (2004) and in Sogachev and Leclerc (2011).

Model equations and details for SCADIS numerical schemes and boundary conditions and further improvements to the parameterization can be found in

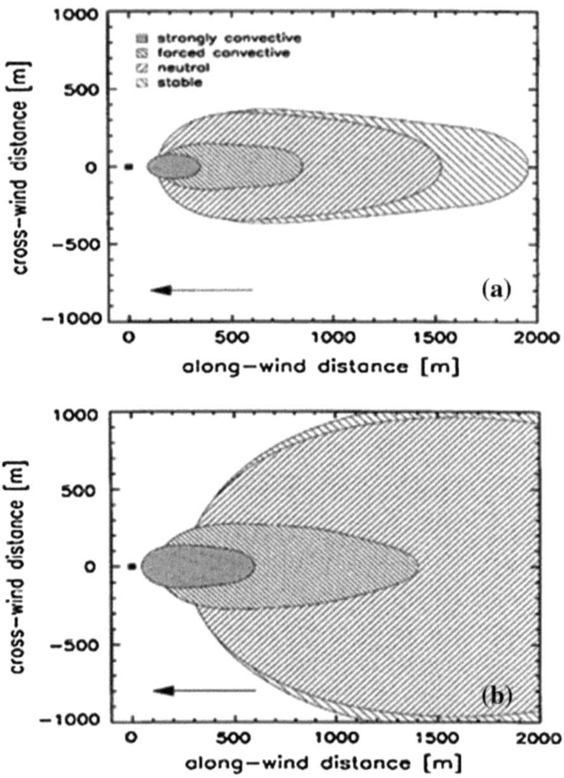

Fig. 3.7 **a** 50 % source area flux and **b** concentration measurements for four different cases of stability. The *square* indicates the receptor location (Kljun et al. 2002)

Sogachev and Lloyd (2004) and Sogachev et al. (2002, 2005a, b, 2008). In SCADIS, the two-dimensional governing equations solved are those for mass and momentum conservation (Navier-Stokes). To date, SCADIS has been used in footprint quantification over short crops, tall forest canopies, urban canopies, in downwind of clearcuts in a forest canopy. It has also recently been coupled to a Lagrangian Particle Dispersion Model (LPDM) ran in an inverse mode to determine the concentration footprint from tall towers (Sogachev and Leclerc 2011).

3.4 Large-Eddy Simulation Models

The advantage of LES compared to conventional footprint models lies in its ability to determine turbulence statistics, scalar fluxes and concentrations and thus to evaluate the corresponding footprints without the use of externally-derived turbulence statistics. Numerous workers have used this method to further their insight into footprints and apply it to a range of surface and flow properties (Leclerc et al. 1997; Cai and Leclerc 2007; Prabha et al. 2008; Steinfeld et al. 2008). Figure 3.8 presents the comparison of Langrangian and LES simulation for the concentration

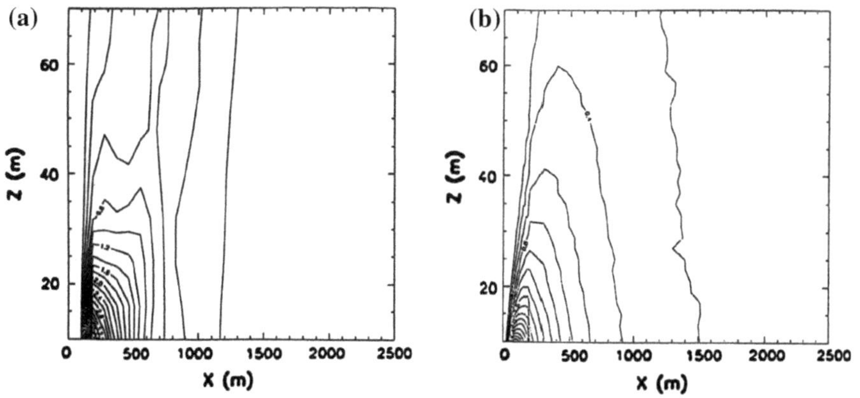

Fig. 3.8 Contours of the normalized crosswind integrated concentration in the surface layer as a function of the downwind distance x. **a** Large eddy simulation and **b** Lagrangian stochastic model (Leclerc and Thurtell 1990), L $= -32$ m in both models (Leclerc et al. 1997, Published with kind permission of © American Geophysical Union (Wiley), 2012. All Rights Reserved)

field according to Leclerc et al. (1997). Furthermore, the LES method has been applied to simulate footprints in the convective boundary layer (Leclerc et al. 1997; Guo and Cai 2005; Cai and Leclerc 2007; Peng et al. 2008; Prabha et al. 2008; Steinfeld et al. 2008).

In recent studies (Cai and Leclerc 2007; Steinfeld et al. 2008) the LES simulation was used in conjunction with the Lagrangian simulation at the sub-grid scale to model convective boundary layer turbulence and infer concentration footprints (Cai and Leclerc 2007; Cai et al. 2008, 2010); Steinfeld et al. (2008) used LES to describe the footprint in boundary layers of different complexities. The Steinfeld et al. (2008) study compared their results with the Finn et al. (1996) tracer flux footprint study and the LES flux footprint of Leclerc et al. (1997). More details can be found in Sect. 3.5.

3.5 Hybrid Footprint Models

A new breed of models which we will call 'hybrid' models is becoming increasingly popular owing to the increased computational speed. There is a rapidly growing proliferation of models seeking to harness the sophistication and advantages of different models, most often combining Eulerian and Lagrangian-generated statistics to solve practical problems in difficult atmospheric flows.

3.5.1 *LES-Driven Lagrangian Stochastic Models*

3.5.1.1 The Prabha et al. (2008) Approach

Examples include the work of Prabha et al. (2008) who first used the turbulence statistics obtained using the LES to drive a Lagrangian stochastic footprint model with a coupling in an offline mode. This strategy can be advantageous when LES is used as a standard technique for turbulence simulation allowing an evaluation of the performance of the Lagrangian simulation of footprints inside canopies. This point is interesting specially when considering that there is a paucity of tracer experiments in-canopy combining turbulence measurements with tracer information.

The Prabha et al. (2008) study solves the conservation equations for mass and momentum and TKE in a three-dimensional domain following Shaw and Patton (2003). The Lagrangian footprint model follows the Thomson criteria (1987) with algorithms for the coefficients of the Fokker-Planck equation based on Flesch and Wilson (1992) accounting for inhomogeneous anisotropic turbulence. Prabha et al. (2008) use on offline coupling of the two models whereby the LES data are saved at a certain timestep and from there, the data are used to run the LES.

They then used the LES-derived turbulent flow statistics as an input to a stochastic footprint model of Flesch and Wilson (1992). That study used several different Lagrangian timescale formulations in the Lagrangian simulations and compared their sensitivity to the resulting flux footprints and compared the results against those obtained with the LES simulations.

3.5.1.2 The Cai and Leclerc (2007) and Cai et al. (2008) Approach

Hybrid models have also been used successfully by Cai and Leclerc (2007) to drive a Lagrangian stochastic model with LES data. The authors used a turbulence field derived from the LES of a passive tracer to drive both forward and backward models in an effort to derive convective boundary layer concentration footprints. They derived concentration footprints at four levels in the convective boundary layer using both forward and backward models. They also used the two models in the reverse direction, i.e. using the stochastic simulation to parameterize sub-grid scale turbulence in the LES. Cai and Leclerc (2007) noted that there is equivalence between the results in horizontally homogeneous turbulence and that while the forward method agreed with laboratory experimental results (Willis and Deardorff 1976, 1978, 1981) for different release heights in the convective boundary layer results from backward dispersion are asymmetric in contrast with the forward method. The authors point out that the backward dispersion results show crosswind-integrated concentration footprints in a generalized sense, i.e. where concentration from all sources, ground and elevated, are included, not just those at the

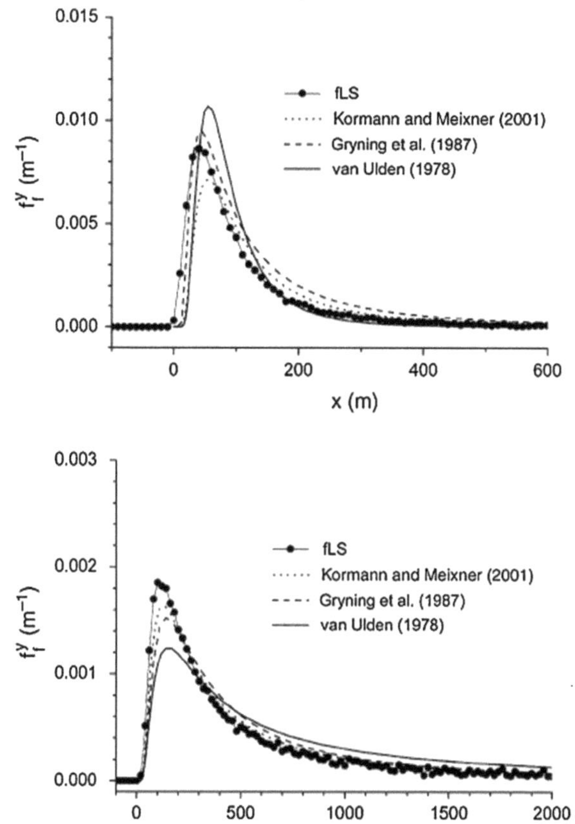

Fig. 3.9 Crosswind-integrated flux footprints of three analytical models and that of the forward LS model. Two stability cases with Obukhov length L equals (*above*) –16 m (*below*) 40 m. The *horizontal* coordinate denotes upwind distances (Cai et al. 2008)

surface. Furthermore Cai et al. (2008) show that the proposed model is in a good agreement with analytical models for concentration and flux footprints (Fig. 3.9).

3.5.2 LES-Embedded Lagrangian Stochastic Models: The Steinfeld et al. (2008) Approach

Steinfeld et al. (2008) also used a combination of LES coupled to a Lagrangian dispersion model to calculate footprints in both homogeneously- and heterogeneously-driven boundary layers. In their case, it is the Large-Eddy Simulation which is driven by the output of the Lagrangian model (Fig. 3.10). They documented positive and negative flux footprints in the convective boundary layer, as had been reported previously by Prabha et al. (2008) inside a forest canopy.

Results from these two studies are consistent with those of Finnigan's (2004) conclusion that the flux footprint function is a function of the concentration footprint function and in complex flows there is no guarantee that the flux footprint is positive, bounded by zero and one.

What really sets the Steinfeld et al. (2008) approach apart from the other LS-LES coupled models is (i) that the coupling of the two is done online and (ii) that the Lagrangian simulation of trajectories is embedded as a set of sub-routines in the LES code. The authors use that approach to evaluate the footprints obtain over a heterogeneously-heated convective boundary layer. The Steinfeld et al. (2008) LES code follows that of Raasch and Etling (1998) and Raasch and Schröter (2001) while the version of the Lagrangian model follows the method proposed by Thomson (1987).

3.5.3 Higher-Order Closure-Driven Lagrangian Simulation

The first such study was done by Luhar and Rao (1994), followed by Kurbanmuradov et al. (2003), and recently by Hsieh and Katul (2009) who applied a stochastic model to estimate footprint and water vapor fluxes over inhomogeneous surfaces. The latter derived the turbulence field of the two-dimensional flow over a change in surface roughness using a combination of both closure model and performed Lagrangian simulations to evaluate the footprint functions.

3.5.3.1 The Luhar and Rao (1994) Approach

The work presented by Luhar and Rao (1994) is also a coupled footprint model. They used a one dimensional second-order closure model of Rao et al. (1974) for the atmospheric boundary layer to account for the effects of vegetation on surface energy and water balance through a 'big-leaf' approach to determine the footprint for latent heat fluxes measured at various locations and heights near the surface. That model has a timescale determined by the model itself, not specified a priori, since the model includes a dynamical equation for the energy dissipation. These flow fields thus obtained were then used to drive the Lagrangian simulation to calculate footprints for cases where the flow is transitioning from an arid region to an irrigated crop field, changing the partitioning of net radiation into sensible and latent heat fluxes (Fig. 3.11).

3.5.3.2 The Hsieh and Katul (2009) Approach

While Hsieh and Katul (2009) resort to the use of stand-alone Lagrangian simulation to model the footprint over homogeneous surfaces. They used a Lagrangian simulation driven by a second-order closure scheme to determine the footprint flux

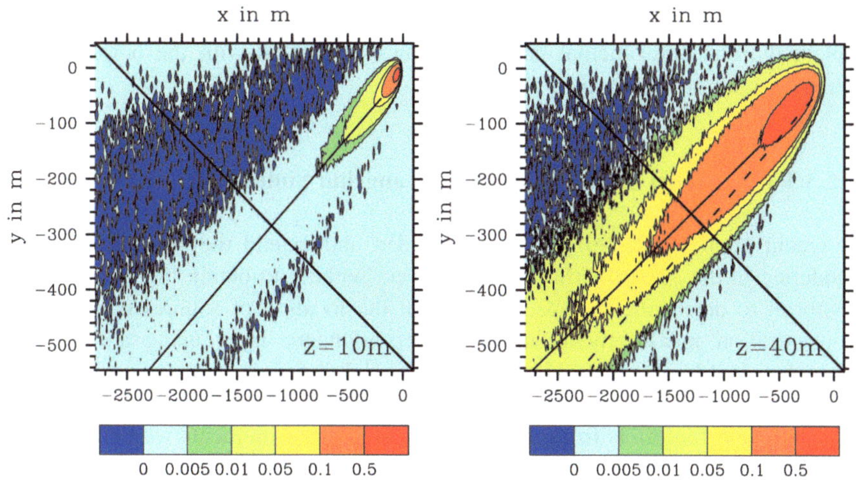

Fig. 3.10 Flux footprints normalized by the respective maximum value evaluated for the period between 7 and 9 a.m. local time after the start of the simulation for measurement heights of 10 and 40 m in the conventional neutral boundary layer (Steinfeld et al. 2008)

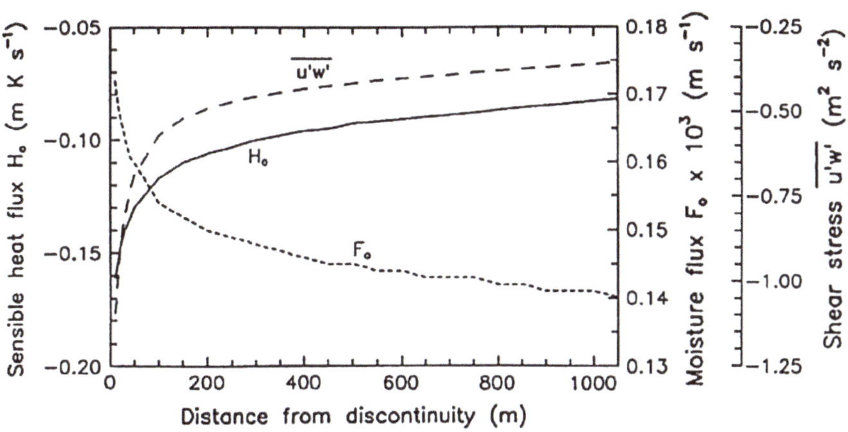

Fig. 3.11 Variation of surface fluxes of momentum ($u'w'$), sensible heat (H_0) at latent heat (E_0) flux over a wet grassy surface (Luhar and Rao 1994)

over inhomogeneous terrain involving a step change in surface source. Their Lagrangian model incorporates the fluctuating component of the streamwise velocity component and is two-dimensional, a feature that few Lagrangian stochastic footprint models incorporate. The Lagrangian simulation is particularly useful in planar inhomogeneous terrain. Given that the step change in temperature and moisture conditions over the domain precludes the use of Monin-Obukhov similarity theory to generate the fields of temperature and moisture, a second-order model was used. The closure formulations used are those of Wichmann and

Schaller (1986). Chap. 4 discusses results of comparisons of water vapor fluxes obtained from experimental data that shows agreement with the output of this Eulerian-Lagrangian coupled model.

3.5.3.3 E-ω Model Closure-Driven Lagrangrian Simulation

A recent approach is being used to broaden and extend the scope of footprint modeling in the atmospheric boundary layer. Seeking to obtain the concentration footprint to determine the spatial extent of the location of sources upwind from concentration measurements in the upper boundary layer using a tall-tower, Sogachev and Leclerc (2011) used a E-ω model (1.5 order closure model) called SCADIS, and embedded it into a Lagrangian simulation. They then examined concentration footprints from nocturnal tall tower measurements with and without the presence of a low-level jet (Sogachev and Leclerc 2011). The hybridization for these models is done in an offline mode. The model was run in the forward mode to examine the evolution of the spread of marked fluid particles and in the backward mode to determine the concentration footprint.

SCADIS incorporates meteorological variables which are dependent on net radiation and incoming solar radiation, surface roughness, sky conditions and initial air temperature profiles. The full description of SCADIS can be found in Sogachev et al. (2002, 2008). In their paper, Sogachev and Leclerc (2011) used SCADIS to drive a Lagrangian model in backward mode to determine the concentration footprints from a tall tower at a 500 m level.

The Lagrangian simulation used by the authors is based on the work of Legg and Raupach (1982) and in backward mode. The Lagrangian simulation used by Sogachev and Leclerc (2011) uses a spin up procedure as per Kljun et al. (2002). The coupled model is then used to create concentration footprints, over a series of terrain conditions, eddy diffusivity and atmospheric stability throughout a 12-h period.

References

Baerentsen JH, Berkowitz R (1984) Monte Carlo simulation of plume dispersion in the convective boundary layer. Atmos Environm 18:701–712

Cai X, Peng G, Guo X, Leclerc MJ (2008) Evaluation of backward and forward Lagrangian footprint models in the surface layer. Theor Appl Climat 93:207–233

Cai X, Leclerc MY (2007) Forward-in-time and backward-in-time dispersion in the convective boundary layer: the concentration footprint. Boundary-Layer Meteorol 123:201–218

Cai X, Chen J, Desjardins R (2010) Flux Footprints in the convective boundary layer: large-eddy simulation and lagrangian stochastic modelling. Boundary-Layer Meteorol 137:31–47

Calder KL (1952) Some recent British work on the problem of diffusion in the lower atmosphere. In: Proceedings of the US technology conference on air pollution. McGraw Hill, New York, pp 787–792

Cellier P, Brunet Y (1992) Flux-gradient relationships above tall plant canopies. Agric Forest Meteorol 58:93–117
Chen JM, Leblanc SG, Cihlar J, Desjardins RL, MacPherson IJ (1999) Extending aircraft- and tower-based CO_2 flux measurements to a boreal region using a Landsat thematic mapper land cover map. J Geophys Res 104(D14):16859–816877
de Haan P, Rotach MW (1998) A novel approach to atmospheric dispersion modelling: the Puff-Particle Model (PPM). Q J Roy Meteorol Soc 124:2771–2792
Einstein A (1905) Über die von der molekularkinetischen Theorie der Wärme geforderte Bewegung von in ruhenden Flüssigkeiten suspendierten Teilchen. Ann Phys 17:549–560
Finn D, Lamb B, Leclerc MY, Horst TW (1996) Experimental evaluation of analytical and Lagrangian surface-layer flux footprint models. Boundary-Layer Meteorol 80:283–308
Finnigan J (2000) Turbulence in plant canopies. Ann Rev Fluid Mech 32:519–571
Finnigan J (2004) The footprint concept in complex terrain. Agric Forest Meteorol 127:117–129
Flesch TK, Wilson JD (1992) A two-dimensional trajectory-simulation model for non-Gaussian inhomogeneous turbulence within plant canopies. Boundary-Layer Meteorol 61:349–374
Flesch TK, Wilson JD, Yee E (1995) Backward-time Lagrangian stochastic dispersion models and their application to estimate gaseous emissions. J Appl Meteorol 34:1320–1332
Flesch TK (1996) The footprint for flux measurements, from backward Lagrangian stochastic models. Boundary-Layer Meteorol 78:399–404
Flesch TK, Wilson JD, Harper LA, Crenna BP, Sharpe RR (2004) Deducing ground-air emissions from observed trace gas concentrations: a field trial. J Appl Meteorol 43:487–502
Flesch TK, Wilson JD (2005) Estimating tracer emissions with backward Lagrangian stochastic technique. In: Hatfield JL, Baker JM (eds) Micrometeorology in agricultural systems. American Society of Agronomy, Madison, pp 513–531
Gash JHC (1986) A note on estimating the effect of a limited fetch on micrometeorological evaporation measurements. Boundary-Layer Meteorol 35:409–414
Gibson M, Sailor D (2012) Corrections to the mathematical formulation of a backwards Lagrangian particle dispersion model. Bound-Layer Meteorol 145:399-406
Göckede M, Rebmann C, Foken T (2004) A combination of quality assessment tools for eddy covariance measurements with footprint modelling for the characterisation of complex sites. Agric Forest Meteorol 127:175–188
Göckede M, Thomas C, Markkanen T, Mauder M, Ruppert J, Foken T (2007) Sensitivity of Lagrangian stochastic footprints to turbulence statistics. Tellus 59B:577–586
Gryning SE, Holtslag AAM, Irvin JS, Sivertsen B (1987) Applied dispersion modelling based on meteorological scaling parameters. Atmos Environ 21:79–89
Guo XF, Cai XH (2005) Footprint characteristics of scalar concentration in the convective boundary layer. Adv Atmos Sci 22:821–830
Haenel H-D, Grünhage L (1999) Footprint analysis: a closed analytical solution based on height-dependent profiles of wind speed and eddy viscosity. Boundary-Layer Meteorol 93:395–409
Horst TW (1979) Lagrangian similarity modeling of vertical diffusion from a ground level source. J Appl Meteorol 18:733–740
Horst TW, Slinn WGN (1984) Estimates for pollution profiles above finite area-sources. Atmos Environm 18:1339–1346
Horst TW, Weil JC (1992) Footprint estimation for scalar flux measurements in the atmospheric surface layer. Boundary-Layer Meteorol 59:279–296
Horst TW, Weil JC (1994) How far is far enough?: the fetch requirements for micrometeorological measurement of surface fluxes. J Atmos Ocean Technol 11:1018–1025
Horst TW, Weil JC (1995) Corrigenda: how far is far enough?: the fetch requirements for micrometeorological measurement of surface fluxes. J Atmos Ocean Technol 12:447
Horst TW (1999) The footprint for estimation of atmosphere-surface exchange fluxes by profile techniques. Boundary-Layer Meteorol 90:171–188
Hsieh C-I, Katul GG, Schieldge J, Sigmon JT, Knoerr KK (1997) The Lagrangian stochastic model for fetch and latent heat flux estimation above uniform and non-uniform terrain. Water Resour Res 33:427–428

Hsieh C-I, Katul G, Chi T-W (2000) An approximate analytical model for footprint estimation of scalar fluxes in thermally stratified atmospheric flows. Adv Water Res 23:765–772

Hsieh C-I, Katul G (2009) The Lagrangian stochastic model for estimating footprint and water vapor fluxes over inhomogeneous surfaces. Int J Biometeorol 53:87–100

Huang CH (1979) A theory of dispersion in turbulent shear flow. Atmos Environ 13:453–463

Kaharabata SK, Schuepp PH, Ogunjemiyo S, Shen S, Leclerc MY, Desjardins RL, MacPherson JI (1997) Footprint considerations in BOREAS. J Geophys Res 102(D24):29113–29124

Kaharabata SK, Schuepp PH, Fuentes JD (1999) Source footprint considerations in the determination of volatile organic compound fluxes from forest canopies. J Appl Meteorol 38:878–884

Kaimal JC, Finnigan JJ (1994) Atmospheric boundary layer flows: their structure and measurement. Oxford University Press, New York 289 pp

Kantha LH, Clayson CA (2000) Small scale processes in geophysical fluid flows. Academic Press, San Diego 883 pp

Kljun N, Rotach MW, Schmid HP (2002) A three-dimensional backward Lagrangian footprint model for a wide range of boundary layer stratification. Boundary-Layer Meteorol 103:205–226

Kljun N, Calanca P, Rotach M, Schmid HP (2004) A simple parameterization for flux footprint predictions. Boundary-Layer Meteorol 112:503–523

Kormann R, Meixner FX (2001) An analytical footprint model for non-neutral stratification. Boundary-Layer Meteorol 99:207–224

Kurbanmuradov O, Rannik U, Sabelfeld KK, Vesala T (1999) Direct and adjoint Monte Carlo for the footprint problem. Monte-Carlo Meth Appl 5:85–111

Kurbanmuradov O, Sabelfeld KK (2000) Lagrangian stochastic models for turbulent dispersion in atmospheric boundary layers. Boundary-Layer Meteorol 97:191–218

Kurbanmuradov O, Rannik Ü, Sabelfeld KK, Vesala T (2001) Evaluation of mean concentration and fluxes in turbulent flows by Lagrangian stochastic models. Math Comput Simul 54:459–476

Kurbanmuradov O, Levykin A, Rannik Ü, Sabelfeld K, Vesala T (2003) Stochastic Langrarian footprint calculations over a surface with an abrupt change of roughness height. Monte-Carlo Methods Appl 9:167–188

Leclerc MY, Thurtell GW (1990) Footprint prediction of scalar fluxes using a Markovian analysis. Boundary-Layer Meteorol 52:247–258

Leclerc MY, Shen S, Lamb B (1997) Observations and large-eddy simulation modeling of footprints in the lower convective boundary layer. J Geophys Res 102(D8):9323–9334

Leclerc MY, Karipot A, Prabha T, Allwine G, Lamb B, Gholz HL (2003a) Impact of non-local advection on flux footprints over a tall forest canopy: a tracer flux experiment (Special issue: Advances in micrometeorology: Tribute to G. W. Thurtell). Agric Forest Meteorol 115:19–30

Leclerc MY, Meskhidze N, Finn D (2003b) Comparison between measured tracer fluxes and footprint modeling predictions over a homogeneous canopy of intermediate roughness. Agric Forest Meteorol 117:145–158

Legg BJ, Raupach MR (1982) Markov-chain simulation of particle dispersion in inhomogeneous flows: the mean drift velocity induced by a gradient in Eulerian velocity variance. Boundary-Layer Meteorol 24:3–13

Luhar AK, Rao KS (1994) Source footprint analysis for scalar fluxes measured in flows over an inhomogeneous surface. In: Gryning SE, Millan MM (eds) Air pollution modeling and its applications. Plenum Press, New York, pp 315–323

Massman WJ, Weil JC (1999) An analytical one-dimensional second-order closure model of turbulence statistics and the Lagrangian time scale within and above plant canopies of arbitrary structure. Boundary-Layer Meteorol 91:81–107

Mölder M, Klemedtsson L, Lindroth A (2004) Turbulence characteristics and dispersion in a forest—verification of Thomson's random-flight model. Agric Forest Meteorol 127:203–222

Neftel A, Spirig C, Ammann C (2008) Application and test of a simple tool for operational footprint evaluations. Environ Pollut 152:644–652

Nguyen AT, Budker D, DeMille D, Zolotorev M (1997) Search for parity nonconservation in atomic dysprosium. Phys Rev A56:3453–3463

Ogunjemiyo SO, Kaharabata SK, Schuepp PH, MacPherson IJ, Desjardins RL, Roberts DA (2003) Methods of estimating CO_2, latent heat and sensible heat fluxes from estimates of land cover fractions in the flux footprint. Agric Forest Meteorol 117:125–144

Pasquill F (1972) Some aspects of boundary layer description. Q J Roy Meteorol Soc 98:469–494

Peng G, Cai X, Zhang H, Li A, Hu F, Leclerc MY (2008) Heat flux apportionment to heterogeneous surfaces using flux footprint analysis. Adv Atm Sci 25:107–116

Poggi D, Katul GG (2008) Turbulent intensities and velocity spectra for bare and forested gentle hills: flume experiments. Boundary-Layer Meteorol 129:25–46

Prabha T, Leclerc MY, Baldocchi D (2008) Comparison of in-canopy flux footprints between Large-Eddy Simulation and the Langrangian simulation. J Appl Meteorol Climatol 47:2115–2128

Raasch S, Etling D (1998) Modeling deep ocean convection: large eddy simulation in comparison with laboratory experiments. J Phys Ocean 28:1796–1802

Raasch S, Schröter M (2001) PALM—a large-eddy simulation model performing on massively parallel computers. Meteorol Z 10:363–372

Rannik Ü, Aubinet M, Kurbanmuradov O, Sabelfeld KK, Markkanen T, Vesala T (2000) Footprint analysis for measurements over heterogeneous forest. Boundary-Layer Meteorol 97:137–166

Rannik Ü, Markkanen T, Raittila T, Hari P, Vesala T (2003) Turbulence statistics inside and above forest: influence on footprint prediction. Boundary-Layer Meteorol 109:163–189

Rannik Ü, Sogachev A, Foken T, Göckede M, Kljun N, Leclerc MY, Vesala T (2012) Footprint analysis. In: Aubinet M et al (eds) Eddy covariance: a practical guide to measurement and data analysis. Springer, Berlin, pp 211–261

Rao KS, Wyngaard JC, Coté OR (1974) The structure of the two-dimensional internal boundary layer over a sudden change of surface roughness. J Atmos Sci 31:738–746

Rodean HC (1996) Stochastic Lagrangian models of turbulent diffusion. Am Meteorol Soc, Boston 84 pp

Rotach MW, Gryning S-E, Tassone C (1996) A two-dimensional Lagrangian stochastic dispersion model for daytime conditions. Q J Roy Meteorol Soc 122:367–389

Sabelfeld KK, Kurbanmuradov OA (1990) Numerical statistical model of classical incompressible isotropic turbulence. Sov J Numer Anal Math Model 5:251–263

Sabelfeld KK, Kurbanmuradov OA (1998) One-particle stochastic Lagrangian model for turbulent for turbulent dispersion in horizontally homogeneous turbulence. Monte-Carlo Meth Appl 4:127–140

Sawford BL (1985) Lagrangian statistical simulation of concentration mean and fluctuation fields. J Appl Meteorol Climatol 24:1152–1166

Sawford BL (1999) Rotation of trajectories in Lagrangian stochastic models of turbulent dispersion. Boundary-Layer Meteorol 93:411–424

Schmid HP, Oke TR (1990) A model to estimate the source area contributing to turbulent exchange in the surface layer over patchy terrain. Q J Roy Meteorol Soc 116:965–988

Schmid HP (1994) Source areas for scalars and scalar fluxes. Boundary-Layer Meteorol 67:293–318

Schmid HP (1997) Experimental design for flux measurements: matching scales of observations and fluxes. Agric Forest Meteorol 87:179–200

Schuepp PH, Leclerc MY, MacPherson JI, Desjardins RL (1990) Footprint prediction of scalar fluxes from analytical solutions of the diffusion equation. Boundary-Layer Meteorol 50:355–373

Shaw RH, Patton EG (2003) Canopy element influences on resolved- and subgrid-scale energy within a large-eddy simulation. Agric Forest Meteorol 115:5–17

Sogachev A, Menzhulin G, Heimann M, Lloyd J (2002) A simple three dimensional canopy-planetray boundary layer simulation model for scalar concentrations and fluxes. Tellus 54B:784–819

Sogachev A, Lloyd J (2004) Using a one-and-a-half order closure model of atmospheric boundary layer for surface flux footprint estimation. Boundary-Layer Meteorol 112:467–502
Sogachev A, Leclerc MJ, Karipot A, Zhang G, Vesala T (2005a) Effect of clearcuts on footprints and flux measurements above a forest canopy. Agric Forest Meteorol 133:182–196
Sogachev A, Panferov O, Gravenhorst G, Vesala T (2005b) Numerical analysis of flux footprints for different landscapes. Theor Appl Clim 80:169–185
Sogachev A, Leclerc MY, Zhang G, Rannik U, Vesala T (2008) CO_2 fluxes near a forest edge: a numerical study. Ecol Appl 18:1454–1469
Sogachev A, Leclerc MY (2011) On concentration footprints for a tall tower in the presence of a nocturnal low-level jet. Agric Forest Meteorol 151:755–764
Steinfeld G, Raasch S, Markkanen T (2008) Footprints in homogeneously and heterogeneously driven boundary layers derived from a Langrangian stochastic particle model embedded into large-eddy simulation. Boundary-Layer Meteorol 129:225–248
Thomas C, Foken T (2007) Flux contribution of coherent structures and its implications for the exchange of energy and matter in a tall spruce canopy. Boundary-Layer Meteorol 123:317–337
Thomson DJ (1987) Criteria for the selection of stochastic models of particle trajectories in turbulent flows. J Fluid Mech 189:529–556
van Ulden AP (1978) Simple estimates for vertical diffusion from sources near the ground. Atmos Environ 12:2125–2129
Wang W, Rotach M (2010) Flux footprints over an undulating surface. Boundary-Layer Meteorol 136:325–340
Wichmann M, Schaller E (1986) On the determination of the closure parameters in high-order closure models. Boundary-Layer Meteorol 37:323–341
Willis GE, Deardorff JW (1976) A laboratory model of diffusion into the convective planetary boundary. Q J Roy Meteorol Soc 102:427–445
Willis GE, Deardorff JW (1978) A laboratory model of dispersion from an elevated source within a modeled convective planetary boundary. Atmos Environ 12:1305–1312
Willis GE, Deardorff JW (1981) A laboratory model of dispersion from a source in the middle of the convectively mixed layer. Atmos Environ 15:109–117
Wilson JD, Sawford BL (1996) Review of Lagrangian stochastic models for trajectories in the turbulent atmosphere. Boundary-Layer Meteorol 78:191–210
Wilson JD, Flesch TK (1997) Wind and remnant tree sway in forest openings III. A wind flow model to diagnose spatial variation. Agric Forest Meteorol 93:259–282
Zhang G, Leclerc MY, Karipot A (2010) Local flux-profile relationships of wind speed and temperature in a canopy layer in atmospheric stable conditions. Biogeosciences 7:3625–3636

Chapter 4
Footprint Studies

The corpus of scholarly works delving into aspects of footprints in the atmospheric environment has evolved from studies predicting the footprint in idealized conditions near the surface to a vast array of studies that incorporate a high degree of physical realism including over non-ideal and complex terrain in challenging conditions. Such sophistication includes modeling footprints of flux systems located on tall towers in the convective boundary layer, in the presence of atmospheric low-level jets and with minor topography; it also extends our interpretation of in-canopy flux sensors data by quantifying the footprint amidst a three-dimensional array of sources and sinks.

4.1 Footprint in the Atmospheric Boundary Layer

Spatio-temporal interpretation provided by footprint analyses can be extraordinarily useful when terrestrial biome carbon sequestration, fossil fuel, and oceanic CO_2 source/sink strengths assessments are sought. Such analyses can also be used to separate the contribution of locally generated gases from scalars transported over thousands of kilometers from several states away as has been shown to be the case in stable boundary layer cases (Corsmeier et al. 1997; Beyrich and Mengelkamp 2006; Sogachev and Leclerc 2011).

Spearheaded by the need to globally monitor rising atmospheric greenhouse gas concentrations, a welcome recent development arrived in the form of a tall tower network with locations in the USA, the Amazon, Europe, and throughout Eurasia. Tall towers have the potential to provide a significant and potentially powerful tool that can be used to obtain regional scale estimates of concentrations and fluxes of greenhouse gases. Information provided by sensors placed at these high levels from the surface can in turn be invaluable to carbon cycle research, provide

M. Y. Leclerc and T. Foken, *Footprints in Micrometeorology and Ecology*,
DOI: 10.1007/978-3-642-54545-0_4, © Springer-Verlag Berlin Heidelberg 2014

constraints on global models, and help improve the development of interactive biogeochemical models coupled to global circulation models such as the Community Climate System Model (CCSM). However, their role in constraining models at regional and continental scales from such tall towers is significantly leveraged by the application of footprint modeling framework to interpret tall tower data. A knowledge of the different contributions to the tall tower point measurements must be assessed or be they from varying land uses, fossil fuel burning before the data can be released to the international scientific community with the confidence that the data reflects the information sought for by the secondary users.

The National Oceanic and Atmospheric Administration (NOAA) Earth System Research Laboratory (ESRL) is continuously measuring CO_2 and CO at the South Carolina tower near Aiken, SC in a partnership with the U.S. Department of Energy, Office of Science and with the Savannah River National Laboratory (SRNL) and the University of Georgia. The Aiken tower (Fig. 4.1) includes three three-dimensional sonic anemometers and fast response CO_2 and H_2O sensors at each sampling height on the tower that are now also used for NOAA CO_2 and CO high-precision mixing ratio measurements. This tall tower samples, in the near field, the southeastern US within a mixed use agricultural, residential, and industrial zones.

The Aiken tower is only one of the many boundary-layer atmospheric measurements of scalars in the NOAA/ESRL, GMD Carbon Cycle Greenhouse Gases Group (CCGG) cooperative air sampling network effort. The network is an international effort which includes regular discrete samples from the NOAA ESRL/GMD baseline observatories, cooperative fixed sites, and commercial ships. Air samples are collected approximately weekly from a globally distributed network of sites. Samples are analyzed in Boulder by CCGG for CO_2, CH_4, CO, H_2, N_2O, and SF_6; and by INSTAAR for the stable isotopes of CO_2 and CH_4 and for many volatile organic compounds (VOCs) such as ethane (C_2H_6), ethylene (C_2H_4) and propane (C_3H_8). Measurement data are used to identify long-term trends, seasonal variability, and spatial distribution of carbon cycle gases.

These and other measurements have been widely used to constrain atmospheric models that derive plausible source/sink scenarios. Inverse Lagrangian footprint modeling strategies enable a diagnostic of the diurnal development of air layering in the lower troposphere, and assess the source strengths of emissions from agricultural, forest, and urban regions to a concentration measurement. The source distribution was calculated with a Lagrangian trajectory model at the European scale (Fig. 4.2), according to Vermeulen et al. (2001).

4.1.1 Tall Tower Footprints

A creative approach focusing on the interpretation of footprints from tall towers come to us via Gloor et al. (2001). In their paper, the authors used a simple,

Fig. 4.1 Tall tower (Aiken AmeriFlux site US-Akn located on the premises of the Savannah River Site, Aiken, Photograph by David Durden, Published with kind permission of)

approximate approach to estimate the concentration footprint of a passive tracer observed from a tall tower (Fig. 4.3). They used the Lagrangian HYSPLIT4A (Draxler 1997) at high resolution to calculate trajectories from the source to the tower. Gloor et al. (2001) used as tracer a tetrachloroethylene (C_2Cl_4), a dry cleaner fluid present in the atmosphere. In locations where the data was not available, they cleverly resorted to the use of population density instead. To ensure robustness in the interpretation of their data, Gloor et al. (2001) also demonstrated that their results are independent of the simplification made in their approach. However, a limitation of the study is the fact that the authors resorted to the National Centers for Environmental Prediction (NCEP) wind field, the resolution of which is coarse. The tall tower footprint contribution evaluated at the 300 m level for two distinct days at 25, 50 and at 70 % level is presented in Fig. 4.3.

In parallel to the main body of research of the previous two decades of scientific peer-review footprint literature, a group of researchers have also sought to advance our understanding of sources within a tall tower footprint and from airborne platforms: Using the original footprint studies, the group of papers by Gerbig et al. (2003a, b), Lin et al. (2004), Wang et al. (2006) examined similar issues. These studies make use of models such as the STILT model (Lin et al. 2004), a backward Lagrangian approach (Gerbig et al. 2003a, b; Lin et al. 2004). Many of these studies sought to constrain CO_2 measurements on a tall tower. Central to their quest, time integrals of footprints were calculated over a 2 days period preceding the measurement time when the surface influence was most intense. The Lin et al. (2004) study clearly shows that convection has a major impact on the decay of

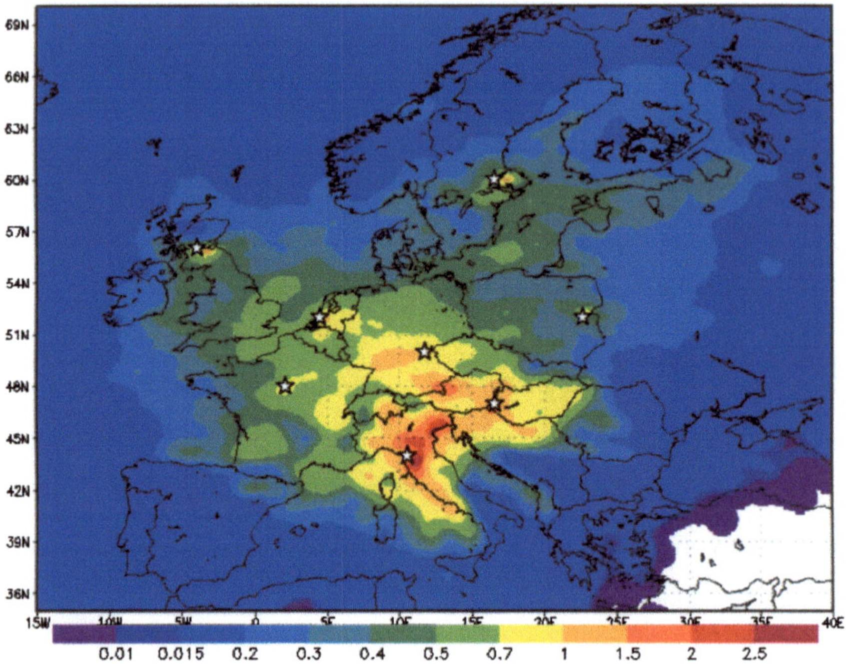

Fig. 4.2 Tall tower measurement stations and their footprint (*colours in relative units*) in the European tall tower network (*called CHIOTTO*). The *red colored areas* contribute to the network observed concentrations with at least 1 % of the maximum contribution per unit area and per unit of emission. Locations of the CHIOTTO towers: Orleans (*France*), Ochsenkopf (*Germany*), Hegyhatsal (*Hungary*), Florence (*Italy*), Cabauw (*The Netherlands*), Bialystok (*Poland*), Norunda (*Sweden*), Griffin (*UK*) (Vermeulen 2007, Published with kind permission of © Dr. Vermeulen. All Rights Reserved)

spatially integrated footprints, and that, with convection, the magnitude of that influence decays to less than 1/e of the initial value after a single day. Without convection however, the authors found the degree of influence to decay to 1/e in approximately 5 days (Gerbig et al. 2003a, b). Wang et al. (2006) also sought to decompose regional CO_2 fluxes from a tall tower over a mixed ecosystem. Their analysis showed a high sensitivity to the numerous spatial heterogeneities in surface composition within the footprint, thus showing how challenging a robust interpretation of the tall tower data proved to be.

Wang and Davis (2008) sought to assess the effect of a grass-covered clearcut to tall tower measurements when that tower is located in the middle of the grassy patch (Fig. 4.4). Using the Large-Eddy Simulation, they modeled the footprint of the grass patch located directly beneath the tower in an effort to identify the dependence of the footprint weight on clearcut size, atmospheric stability, measurement height, and entrainment flux. The authors also used an analytical solution to the diffusion equation to model individual source weight elements as a function

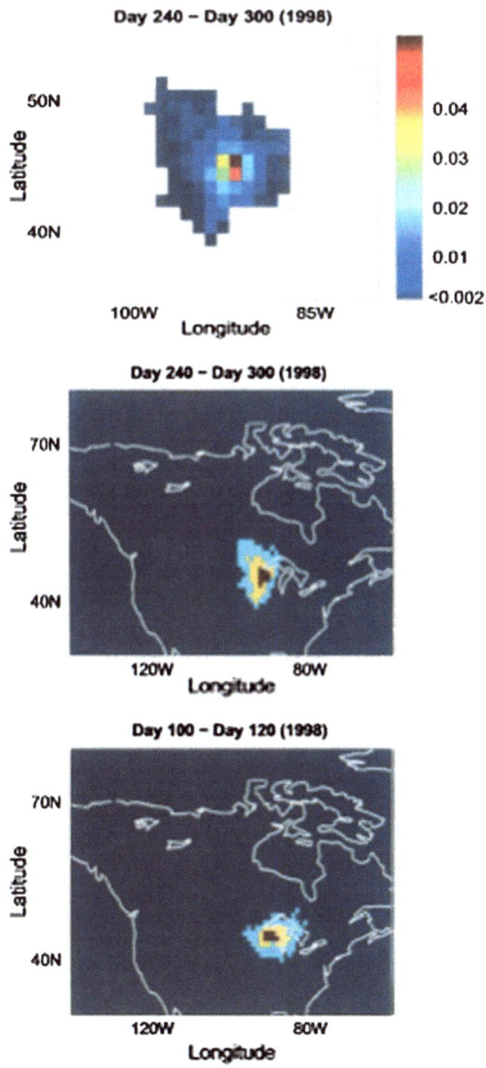

Fig. 4.3 *Top* Footprint estimate for day 240–300 in 1998 and area wherefrom fluxes contribute 25, 50 and 75 % to the correlation at the tower for (*middle*) day 240–300 in 1998 and (*bottom*) day 100–120 1998 (Gloor et al. 2001, Published with kind permission of © American Geophysical Union (Wiley), 2003. All Rights Reserved)

of height on the tall tower. They found that the contribution of the grassy patch lying just beneath the tower decreases with wind speed, atmospheric stability, increasing measurement level, and with decreasing boundary layer height. They found it to also be inversely proportional to clearcut size. The authors' results suggest that analytical footprint models cannot be used to accurately evaluate the influence of the clearcut on eddy-covariance flux measurements at the tall tower.

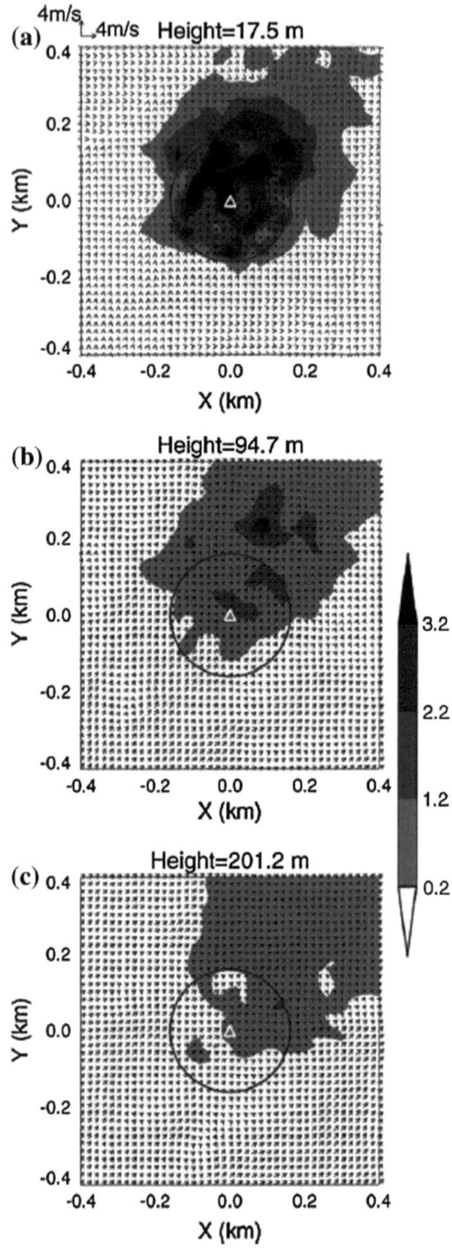

Fig. 4.4 The mean horizontal wind vectors and normalized mixing ratio of the tracer released from the surface clearcut (*circle*) at **a** 17.5 m **b** 94.7 m, and **c** 201.2 m above the ground in the tower area. The *triangle shows* the location of the tower (Wang and Davis 2008, Published with kind permission of © Elsevier, 2008. All Rights Reserved)

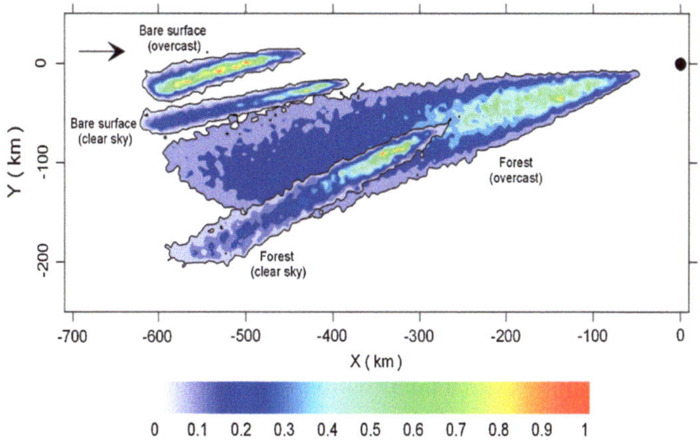

Fig. 4.5 Concentration footprint for a mixing ratio measured at 05:00 AM local time at 500 m level above a bare surface and a forest in both overcast and in clear skies. The receptor point (*concentration measurement point*) on the tall tower is indicated by a black circle (Sogachev and Leclerc 2011, Published with kind permission of © Elsevier, 2011. All Rights Reserved)

Addressing a concept fraught with theoretical difficulties, Sogachev and Leclerc (2011) sought to determine the origin of greenhouse gases measured on tall towers in the stable boundary layer. They determined the dynamic temporal and spatial evolution of sources and sinks to concentration measurements made at different times throughout the night at 25, 100 and at 500 m levels on a tall tower (Figs. 4.5 and 4.6). In nighttime conditions, the study's model tracks fluid particles traveling backward in time, mostly horizontally during the evening/night hours, to earlier source/sinks exchange hours characterized by an upwind regime of unstable conditions and greatest mixing. Since the results hinge so heavily on the state of the atmosphere, the above study discussed both clear and overcast sky conditions as fluid elements are advected over different terrain throughout different atmospheric conditions. The model uses a backward Lagrangian stochastic model (Legg and Raupach 1982; Ley and Thomson 1983) with input variables obtained from the higher-order closure model SCADIS. Their results simulate the source contribution measured at different times throught the night on a tall tower both in the presence of and in the absence of a jet. The study depicts how the presence of the jet, as shown earlier in Corsmeier et al. (1997), Mathieu et al. (2005) and Karipot et al. (2006), impedes the vertical transfer of scalars above the jet with the surface. From the carbon cycle community modeling standpoint, this is a result of significance.

The resulting tracer concentration profile on the tall tower then also results from the contribution of time-dependent and location-dependent sources whose upwind position is largely a function of both time and measurements level. For instance, the Ekman layer will make the contributing sources shift throughout the night. Both directional shear with height and directional wind change over time influence the distance over which the scalar travels and its direction. This directional change in source contribution, has received scant attention except for the work by Markkanen et al. (2009). When this effect is added to the ever changing source location characteristics of surface roughness and other surface properties, it is easy to appreciate the complexity of the interpretation.

The shape and the spatial extent of footprints for both forest and open place in the presence of overcast skies are similar. With respect to the spatial extent (from small to large), the spread (from small to large) and the location of the area with maximal influence regarding the sensor (from nearby to distant), they could be ranged as follows: forest (clear sky), forest (overcast) and bare surface (overcast), respectively. For these surfaces, footprints have, in a first approximation, a spatial extent of the area with maximal contribution to a concentration signal from 70 to 120 km away. The shape of the footprint for the measurement point located above the bare surface under clear sky conditions is more complex. The simulation suggests that none of the material emitted from as far as 50 km upwind gets to the hypothetical sensor. The footprint is spatially spread within the area of a more remarkable contribution situated 10–15° to the left concerning the wind direction observed at the sensor and at distance as far as 100–300 km away.

That study also showed (Fig. 4.6) that the most important sources contributing to the tall tower mixing ratio are those which are formed during the day in regime of convection when the particles are exchanged rapidly throughout the boundary layer. In clear nights, the footprint is narrower than in overcast days, a reflection of the fact that clear nights are usually more stable leading to a hampering of the lateral and vertical dispersion.

4.1.2 The Influence of Coriolis Forces on Footprint

It is in the atmospheric boundary layer that Coriolis forces are most important. With this in mind, Markkanen et al. (2009) examined the influence of the Coriolis force on footprints and compared LES results with and without it. Given that the dispersion in weakly to moderately stably stratified boundary layers is impacted by the Ekman layer, it can be expected that the concentration and flux footprints veer with height. This was demonstrated recently by both Steinfeld et al. (2008) and by Markkanen et al. (2009). Steinfeld et al. (2008) showed a combination of a nocturnal jet, a strong turning of the wind with height and a triple-layer structure of the potential temperature profile, all indications of a stably stratified boundary layer. Figure 4.7 suggests that the footprint model should be able to reproduce a height-dependent wind direction. This wind turning with height is currently not accounted for in most footprint Lagrangian stochastic simulations or in analytical solutions to the advection-diffusion equation, except in Sogachev and Leclerc (2011).

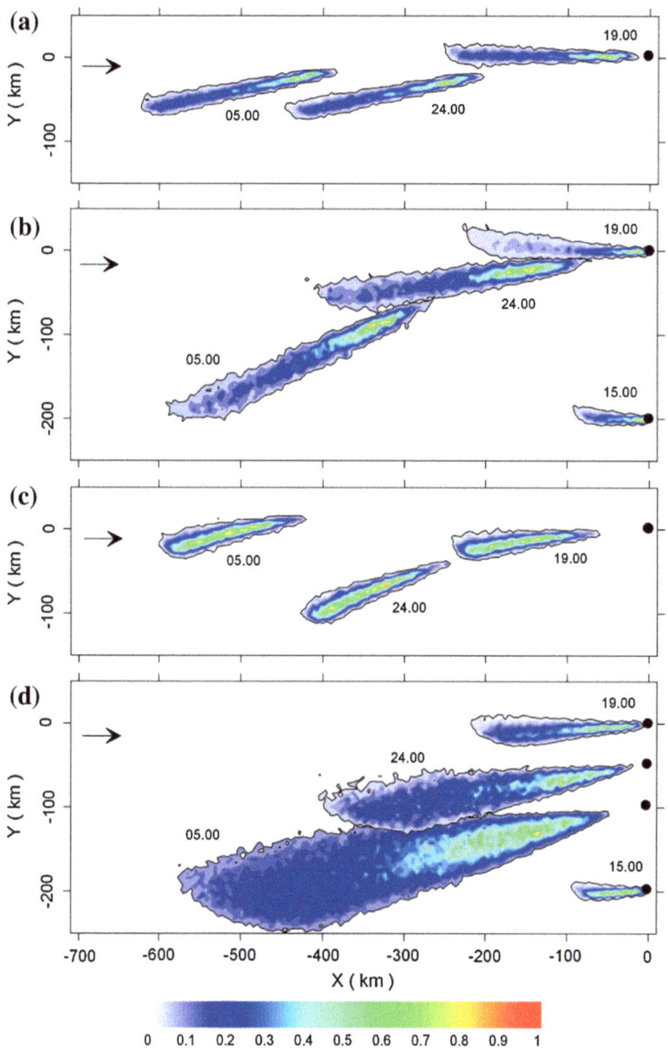

Fig. 4.6 Concentration footprint of a mixing ratio from a tall tower shown at 3:00 PM local time; 7:00 PM local time; 24:00 local time; and 05:00 AM local time before sunrise (*time of day shown directly in the figures*) at 500 m level above **a** above an open site and **b** above a forest canopy in clear sky conditions and above **c** above an open site and **d** above a forest in the presence of cloudy conditions. The concentration footprint is normalized by its maximal value. The receptor point is indicated by a *black circle*. Footprints for times 05:00 AM local time, 3:00 PM local time and also for 24:00 local time in (**d**) and at 15:00 local time in (**b**) are shifted for clarity (Sogachev and Leclerc 2011, Published with kind permission of © Elsevier, 2011. All Rights Reserved)

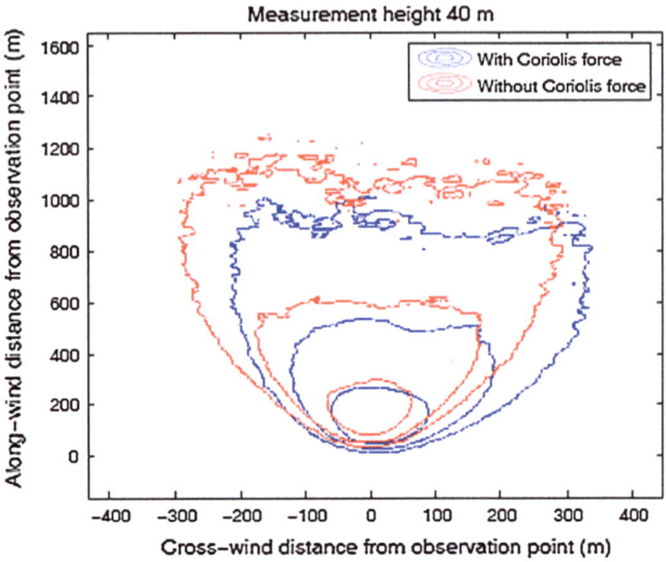

Fig. 4.7 Footprint calculations obtained using the Large-Eddy Simulation of Steinfeld et al. (2008) with and without the effect of the Coriolis force at 40 m level (Markkanen, personal communication, Published with kind permission of © Dr. Markkanen, 2012. All Rights Reserved)

4.1.3 Flux Footprints in the Convective Boundary Layer

The first study examining the behavior of footprints in the convective boundary layer is attributed to Leclerc et al. (1997) in a study using a combination of Large-Eddy Simulation, Lagrangian simulation and a tracer study used to simulate the behavior of footprints in the atmospheric surface layer. To give credence to their results, they first compared footprint prediction results of their Large-Eddy Simulation against tracer flux data by Finn et al. (1996) and against those of a Lagrangian simulation. Their study demonstrated a very good general agreement before, at, and past the footprint peak between all three methods when available measurements and Lagrangian simulations in the convective surface layer were used. Outside the atmospheric surface layer, their results showed a gradual yet marked decoupling of flux footprints with increasing distance from the surface, a result of high relevance for both airborne measurements and tall tower measurements.

Figure 4.8 illustrates an LES output depicting the amount of fetch required for a flux measurement at 200, 250, and 300 m level in a mixed layer with an Obukhov length of −55 m either from a hypothetical airborne or in situ tower platform. The 350 m level shows that the most sensitive source contribution occurs at approximately 1,700 m. Results further suggest the cumulative flux to converge extremely slowly with only approximately 10 % of the total flux recovered 4 km away from the flux tower.

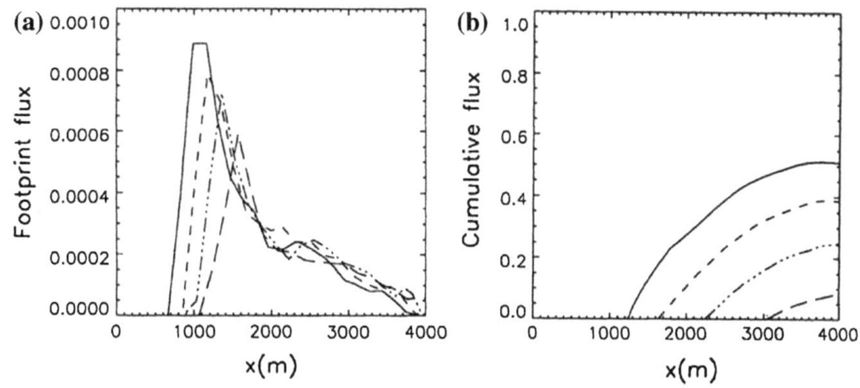

Fig. 4.8 Normalized footprints (**a**) and cumulative fluxes (**b**) from the Large-Eddy simulation for observation level of 250 (*solid line*), 300 (*dashed line*), 350 (*dashed-dotted line*) and 400 m (*long dasehd line*) with L = −55 m. The height of the boundary layer is 505 m. (Leclerc et al. 1997, Published with kind permission of © American Geophysical Union (Wiley), 1997. All Rights Reserved)

Fig. 4.9 Dimensionless flux footprint versus the dimensionless downwind distance with $X = w{-}x/Uz_i$, for h in figure read z_i, for *crosswind line* source at the surface (Weil and Horst 1992, Published with kind permission of © Hemisphere Publishing, 1992. All Rights Reserved)

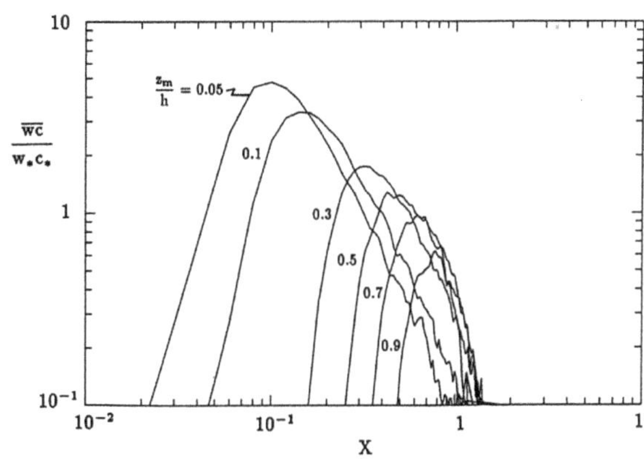

Weil and Horst (1992) evaluated results from a Lagrangian simulation based on Thomson (1987) against laboratory simulations of dispersion in the convective boundary layer (Weil 1989) and showed that the Lagrangian footprint model prediction agrees well with the Willis and Deardorff (1976) laboratory data for a near-surface release. The authors point out that the dimensionless footprint has a maximum that decreases monotonically with height as shown in Fig. 4.9.

Weil and Horst (1992) found that a key feature of the footprint is that the footprint has a horizontal scale of the order of Uz_i/w_* where U is the mean wind speed, z_i is the convective boundary layer depth, and w_* is the convective velocity scale and is confined mainly to the dimensionless distance X < 1 or 2 even when the measurement height is large. The authors suggest that the vigorous mixing

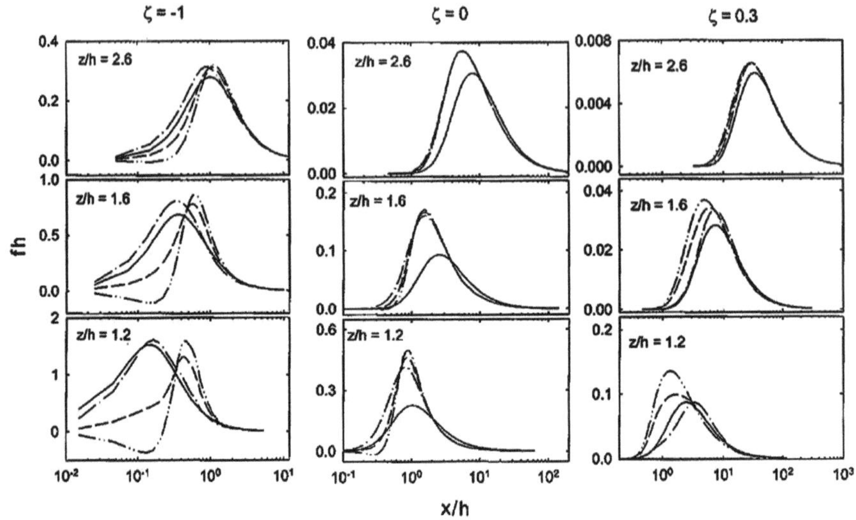

Fig. 4.10 Normalized flux footprints at two heights within the roughness sublayer ($z/h = 1.2$ and 1.6) and at one height in the inertial sublayer ($z/h = 2.6$): *Dashed line*, $F_e/F_T = 0.8$ (*configuration A*); *dash–dot–dot line*, $F_e/F_T = 1.2$ (*configuration B*); *dash–dot line*, $F_e/F_T = 0.2$ (*configuration C*); *solid line*, calculation using $d/h = 0.6$ and $z_0/h = 0.1$ (Lee 2003)

caused by the large eddies and making the contribution of nearby sources more important, is the prime reason responsible for the compact nature of the footprint.

Judicious advice is provided to us by authors Weil and Horst (1992) who point out the need to include the longitudinal velocity fluctuations in light winds. Their study also recommends that sources and sinks near the top of the convective boundary layer be accounted so that improved convective boundary layer footprint estimates can be achieved.

4.1.4 Footprint in the Roughness Sub-Layer of Plant Canopies

Footprint of fluxes in the roughness sub-layer over canopies with elevated sources were examined by Lee (2003) who used Raupach's (1989) localized near-field (LNF) theory combined with parameterization of canopy turbulence to examine source configuration and atmospheric stability impact the footprint. Figure 4.10 illustrates results using the Lee approach in the roughness sub-layer for contrasting stabilities (one stable, one neutral and one unstable case for a vertical range between 1.2, 1.6, and 2.6 the canopy height h. It can be seen that the impact of the proximity of the sources is modest and is seen in unstable conditions near treetop.

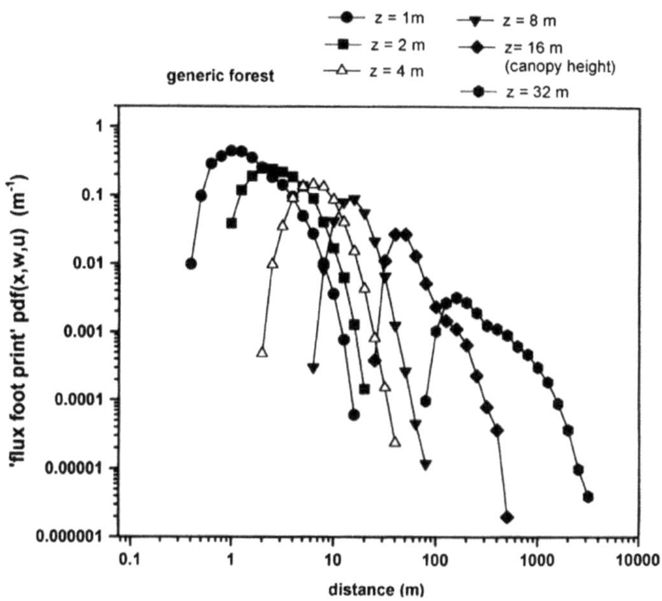

Fig. 4.11 The horizontal distribution of flux footprint probability density functions at an arbitrary level within and above a generic forest by Baldocchi (1997)

4.2 In-Canopy Footprints

Characterizing the footprint behavior inside vegetation canopies constitutes an important step forward credited to Baldocchi (1997). That study was the first to describe the footprint behaviour inside a forest canopy amidst sources and sinks in neutral conditions. The application of Lagrangian simulation to in-canopy flux footprints offers a more natural approach to describe diffusion in strongly inhomogeneous, shear and anisotropic turbulence. Baldocchi (1997) used the Lagrangian approach formulated by Flesch and Wilson (1992) following the Thomson (1987) criteria with parameterized turbulence vertical profiles inside the canopy and similarity relationships above the tree layer. Section 2.4.2 discusses the Lagrangian simulation in greater detail. Figure 4.11 illustrates the behavior of the footprint within a model canopy (Raupach 1988) as a function of height from the ground within the forest up to approximately twice the canopy height.

Figure 4.12 shows that the location of the footprint with height decreases with increasing vertical distance from the surface source. It can also be seen that the footprint is contracted within the canopy when compared against their counterparts in the region above the canopy. This is attributed to the high shear characteristic of canopy flows.

The influence of canopy density on the flux footprint within the canopy is also examined using their model canopy along with a very dense canopy such as a

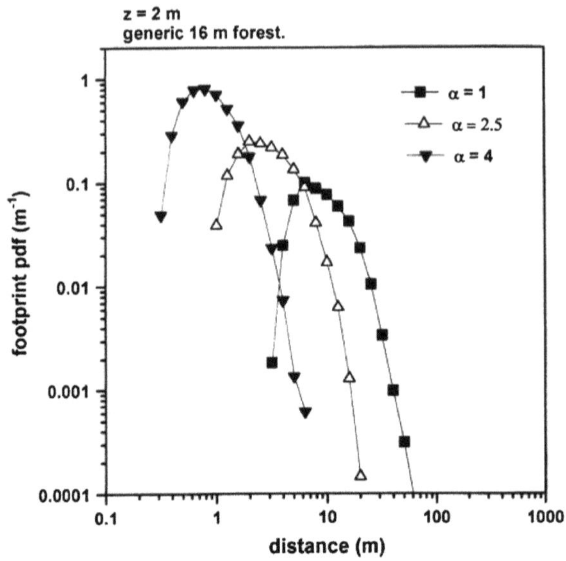

Fig. 4.12 The impact of the horizontal wind speed attenuation coefficient α on the horizontal distribution of the flux footprint probability density function assessed at 2 m above the floor of a generic forest. The canopy height was 16 m and $\sigma_w(0)$ was 0.25 u_* (Baldocchi 1997)

mature temperate broad-leaved or a tropical forest characterized by a high profile parameter α (Table 2.7) of approximately four (Baldocchi 1997). Due to the resulting higher shear of the dense canopy, results exhibit a more compact footprint.

The modeled flux footprint estimated at a hypothetical sensor location near the ground suggests that the greatest contributions result from sources in close proximity to the sensor location within the first 8 m with a peak of about 80 cm. Baldocchi (1997) points out that these results support anecdotal evidence based on field observations.

The sensitivity of the footprint function to various Lagrangian timescale formulations was also examined in the same study and shown to have only a rather modest influence on the results. Given the importance of thermal stability on the behavior of turbulence statistics inside the canopy layer (Shaw et al. 1988; Amiro 1990; Leclerc et al. 1990, 1991; Greens et al. 1995; Kruijt et al. 2000; Launiainen et al. 2007; Zhang et al. 2011), its impact on the footprint has to be considered.

The first Large-Eddy Simulation of footprints inside plant canopies is attributed to Prabha et al. (2008). This LES study of canopy footprints used a tree canopy with three vertically distributed sources with unit source strength to model the flux footprint in neutral conditions. Their choice of vertically distributed sources and their respective source strength is intended to reproduce the vertical distribution of sources and sinks inside the tree layer. The reader is referred to Shaw and Patton (2003) for details of the LES version.

Figure 4.13 shows that LES simulations by Prabha et al. (2008) suggest that, for sources located at about a third of the canopy height and in the upper canopy, when flux footprints behave differently at several levels in the canopy environment, a higher peak for flux footprint for elevated sources inside the canopy is noted,

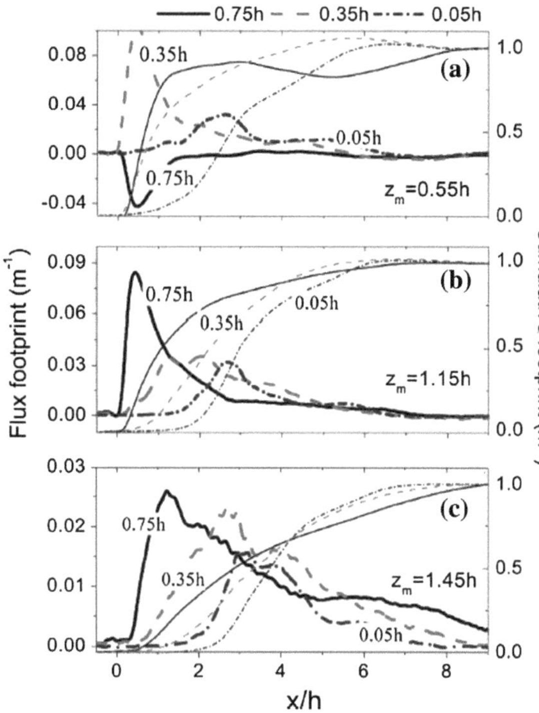

Fig. 4.13 LES flux footprint (*thick lines*) and cumulative (*thin lines*) flux corresponding to sources at 0.05, 0.35 and 0.75 h for measurement heights **a** 0.55 **b** 1.15 and **c** 1.45 h. (Prabha et al. 2008, Published with kind permission of © American Meteorological Society, 2008. All Rights Reserved)

in general agreement with results of both Rannik et al. (2003) and Lee (2004). The downwind shift of the upwind locus of maximum flux sensitivity can be seen with increasing level within the tree layer leading to a larger footprint envelope and a greater fetch distance. The authors also presented the behavior of concentration footprint from a mid-canopy source.

Prabha et al. (2008) also examined the behavior of the centroid of the plume for sources released within the canopy layer, along with the flux and concentration footprints modeled within and above the model forest layer. Figure 4.14 depicts such results from a source releasing material in mid-canopy.

Vertical profiles of wind speed, standard deviations of velocity components, Reynolds stress and Lagrangian timescale inside and immediately above the canopy are provided to implement random-walk formulations based on the Langevin equation. The Lagrangian simulation model releases imaginary fluid elements at each of the three (0.75, 0.35 and 0.05 h) imaginary sources inside the tree layer (Fig. 4.15).

Prabha et al. (2008) initialized the Lagrangian simulation using flow statistics obtained from the LES. They used two approaches to derive the time scale (τ) needed in the LS. The study points out that the Lagrangian timescale is a critical variable required to correctly dictate the dispersion inside the canopy layer.

Figure 4.16 shows cumulative flux footprints at three measurement levels, with one measurement height at mid-canopy and the two others either right above it or

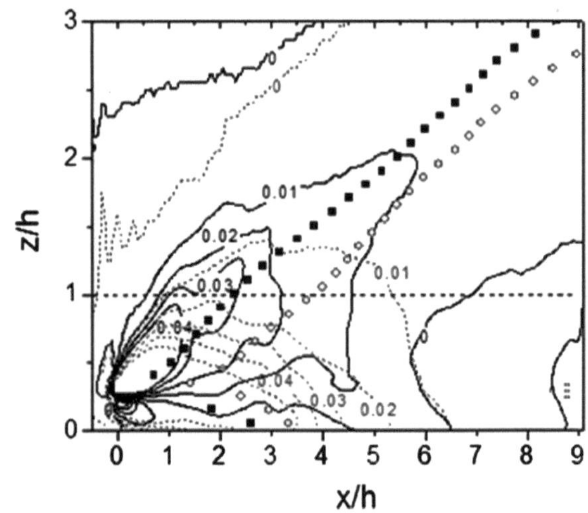

Fig. 4.14 Distribution of concentration (*solid contours*) and flux footprint (*dashed contour*) in the xz plane from LES for a mid-canopy source. Centroids of scalar-concentration (*black squares*) and flux footprint (*open circles*) found from the stream wise distribution direction is also presented (Prabha et al. 2008, Published with kind permission of © American Meteorological Society, 2008. All Rights Reserved)

at an often seen z_m ($z_m = 1.45\ h$); these footprints originate from sources placed at 0.75, 0.35 and 0.05*h* i.e. corresponding to the canopy depths used in the LES study by the same authors. The footprint of the elevated sources (0.75 and 0.35 *h* releases) is contracted (less than 1 *h* if $z_s < z_m < h$) with a peak value higher than that of the surface source. Their results agree with those of Rannik et al. (2003) and Lee (2004) noting a higher peak for flux footprints associated with elevated sources inside the canopy and that the flux is weighted more heavily by contributions from the upper canopy. All three sources characterize a downwind shift of the flux footprint peak with increasing measurement height. Of interest is the presence of a negative flux footprint below the elevated source.

Above treetop, at $z_m = 1.45\ h$, the maximum contribution comes from the 0.75 *h* source and peak flux is found at $x = 1\ h$. The peak flux footprint of mid canopy and surface sources lies within a downwind distance of 2.5 and 3 *h*. The cumulative flux of all three sources becomes virtually identical at 4 *h*. The cumulative flux of surface and mid-canopy sources becomes higher than that of 0.75 *h* level source at 5.8 and 4 *h* respectively. This happens due to the broadening of the footprint envelope. Practical implications of such a behavior to flux measurements above the canopy can be interpreted by inverting the flux footprints presented here. The cumulative flux footprint depicts contributions from below-canopy sources ranging from the immediate upwind region right up to far-field sources. The flux from surface and mid-canopy sources at locations close to the measurement tower ($x < 3\ h$) contributes to less than 20 % while sources at far away locations (>4 *h*) contribute to 50 % of the flux at the measurement height of 1.45 *h*.

In the case of a surface source, a downwind distance of 2.5 *h* separates footprint peaks if the measurement height is chosen between the canopy top and $z_m = 2\ h$. While the peak location is nearly constant at 3 *h* when the measurement level is between 0.4 and 1 *h*, below 0.4 *h*, the footprint peaks are separated by a

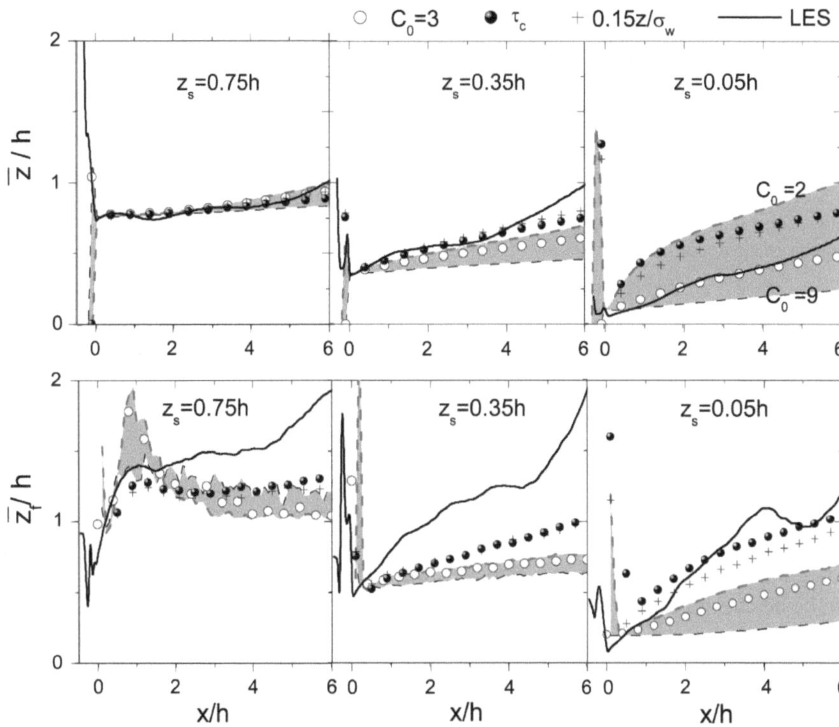

Fig. 4.15 Comparison of LES (*thick solid line*) and LS plume heights from the (*top*) scalar-concentration and (*bottom*) flux footprint associated with (*left to right*) three sources $z_s = 0.75$, 0.35, and 0.05 *h*. Results from LS with three different time scales, $2\sigma_w^2/(C_0\varepsilon)$, where $C_0 = 3$ (open circles), $\tau_c = K_{c\_0.05h}/\sigma_w^2(z)$(*filled black–gray circles*), and $0.15z/\sigma_w$ (+ signs), are presented. Shaded areas encompass LS results bounded by $C_0 = 1$ (thick dashed line) and $C_0 = 9$ (*thin dashed line*) (Prabha et al. 2008, Published with kind permission of © American Meteorological Society, 2008. All Rights Reserved)

distance of 3 *h* as the released tracer from 0.05 *h* resides in the lower part of the canopy due to transient horizontal motions before being carried upward by weak vertical motions. As a result, the flux footprint peaks at locations below the canopy crown are broader than those at higher elevations.

Some of the tracer released at in the upper canopy is transferred below the source level and contributes to a negative flux footprint. Wind tunnel experiments of Legg et al. (1986) showed negative flux below the elevated line source. Negative flux footprints below elevated sources have also been observed in the Lagrangian simulation of Baldocchi (1997) and of Markkananen et al. (2003). Finnigan (2004) argues that negative flux footprints in the models are an artifact of reducing the complex source-sink distribution of the canopy with a single layer. In the wind tunnel diffusion study from an elevated heat source (Coppin et al. 1986) at 0.85 *h*, almost all the heat was transported below the source level by sweeps.

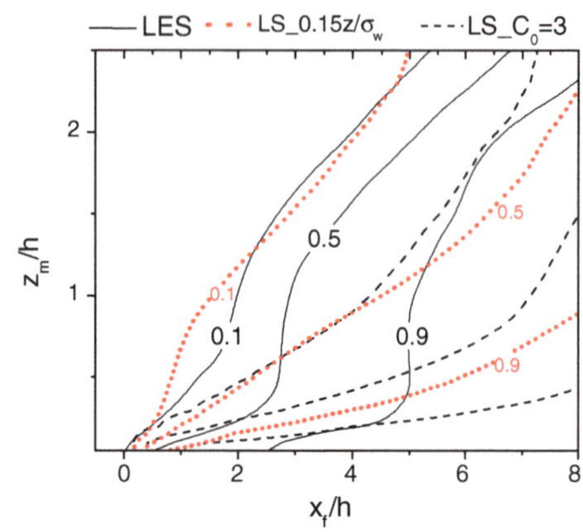

Fig. 4.16 Distribution of fetch distances for 10, 50 and 90 % of flux for different measurement heights for the surface source. *Solid lines* correspond to data from LES and *dashed lines* for LS result with $C_0 = 3$ and *dotted lines* with a time scale 0.15 z/σ_w (Prabha et al. unpublished)

4.3 Flux Footprint in Canopy Over Hills

Finnigan (2004) examined the footprint concept when applied to model forest canopies over hills. Theoretical considerations show that the concentration footprint can be described by the Green function of the conservation equations of a scalar but this is not necessarily so for the flux footprint. This solution is applicable to homogeneous shear flows despite the fact that the application to canopy-type flows represents a special case. By no means, it implies that this solution can be extended to other flows. Finnigan (2004) also points out that homogeneous canopy flows represent an artifact of reducing a canopy source-sink distribution to a single layer.

Incanopy flows located over a ridge, model studies indicate that truly anomalous behavior can be expected in the flux footprint to the point that the application of the footprint concept to canopy-flow atop a ridge is not currently advisable. Further demonstrations based on experimental data was shown by Katul et al. (2006).

4.4 Influence of Contrasting Adjoining Surfaces on Footprints

4.4.1 Role of Contrasting Thermal Land Surfaces on Fluxes and Footprints

Markkanen et al. (2010) examined footprints arising from a heated surface with heat advected to an adjoining surface of contrasting temperature. Using the inverse Lagrangian simulation LPDM-B and the Large-Eddy Simulation PALM (Raasch and Schröter 2001), they modeled the three-dimensional spread of the plume. Their results suggest that the LPDM-B model exhibits a larger plume spread in the transverse component of the flow than the LES (Fig. 4.17).

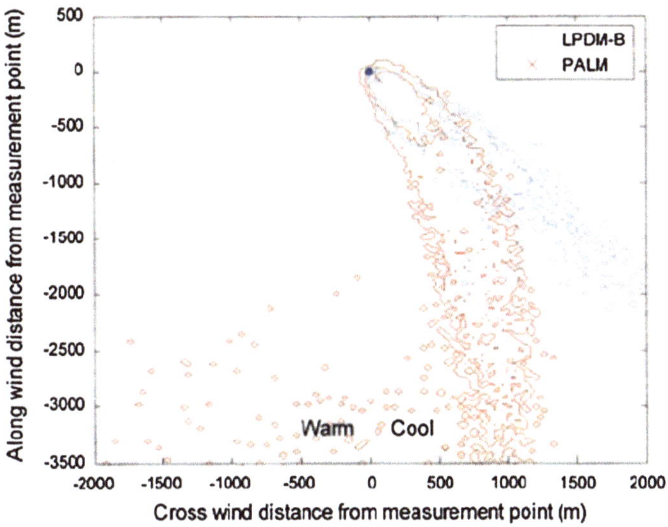

Fig. 4.17 Concentration footprints from the LPDM-B (*blue contour*) and PALM (*red contour*) at 30 m. The step change of surface properties (*blue line*) is 50 m east of the measurement position. Contours indicate the smallest areas contributing 50 and 80 % of the total concentration signal (i.e. contribution from the whole domain area). *Crosses indicate* the respective location of the maximum of the footprint function and the dot indicates the measurement point (Markkanen et al. 2010, Published with kind permission of © Borntraeger Verlagsgesellschaft, 2010. All Rights Reserved)

The thermal contrast between the two adjoining surfaces induce a circulation whose strength is proportional, all other variables being constant, to the contrast in their respective thermal strength. Markkanen et al. (2010) found the LES (PALM) model in which a flux footprint model was embedded to be superior to that of the LPDM-B for dissimilar adjoining surfaces giving strong organized circulations (Fig. 4.17). This result naturally arises since the LES faithfully reproduces the two opposite wind directions at different levels while the inverse Lagrangian method reflects only the local wind direction at one height to infer the footprint flux and concentration of a circulation, the flow is in opposite direction at different levels.

Markkanen et al. (2010) also found that the Lagrangian footprint analysis works well in this physical system as long as the circulation remains weak. While the authors point out that Inagaki et al. (2006) found that the mesoscale flux can result in an underestimation of several tens of percent of the areal averaged flux, Markkanen et al. (2010) did not include the mesoscale flux in the paper either. This subject constitutes an interesting avenue to pursue in subsequent studies.

4.4.2 Role of Clearcuts on Forest Fluxes/Footprints

Another application of the role of spatial inhomogeneities in upwind surfaces on fluxes and flux footprints was introduced to examine fluxes for large swaths of

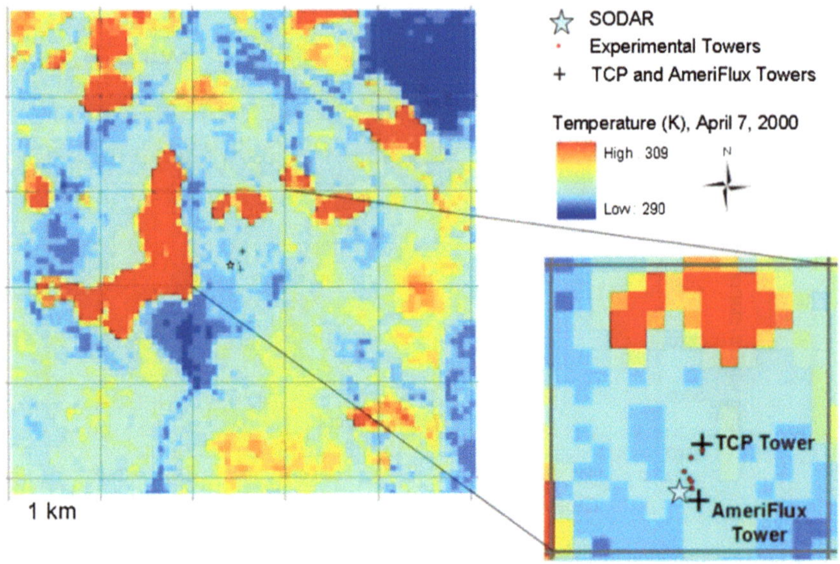

Fig. 4.18 Remote sensing observation of surface temperature by Landsat 7 in April 7th, 2000 at 10:55 AM around the Florida AmeriFlux site, where the sodar was later deployed (May, 2000). An arc-shaped logged patch can be seen to the west side of the sodar location. (Zhang et al. 2011)

forest stands removed in forest plantation with the study of Leclerc et al. (2003b) augmented by that of Zhang et al. (2011). In the first study, the authors validated two flux footprint models using SF_6 as a passive tracer in the presence of a clearcut and without. This was done using a wind-direction analysis to select periods with clearcuts versus measurements of fluxes over a uniform canopy. Figure 4.18 highlights the contrast in surface temperatures exhibited between the clearing and the forest can be seen using the satellite information centered around the flux tower (Fig. 4.18) show how thermal effects dominate the flow of varying proportions throughout the day: these effects can be strongly felt throughout the range of the sodar (Figs. 4.19 and 4.20) and it can also be seen that these impact the magnitude and direction of the flow. Figure 4.19 conclusively demonstrate (Zhang et al. 2011) that the flow properties when the wind is from the direction of the clearcut is by no means attributed to synoptic conditions. Given the preponderance of thermal features arising from land-use characteristics, is it any wonder that their impact could overflow their immediate region and be advected into the forest downwind where flux measurements were made?

The methodology and model robustness had been previously tested over a 1 m tall sagebrush canopy (Finn et al. 1996) and then over a in a peach orchard i.e. a surface characterized by a canopy of intermediate roughness (Leclerc et al. 2003a). The Finn et al. (1996) study demonstrated the ability of the model to accurately reproduce footprint behavior well outside the roughness sublayer using several

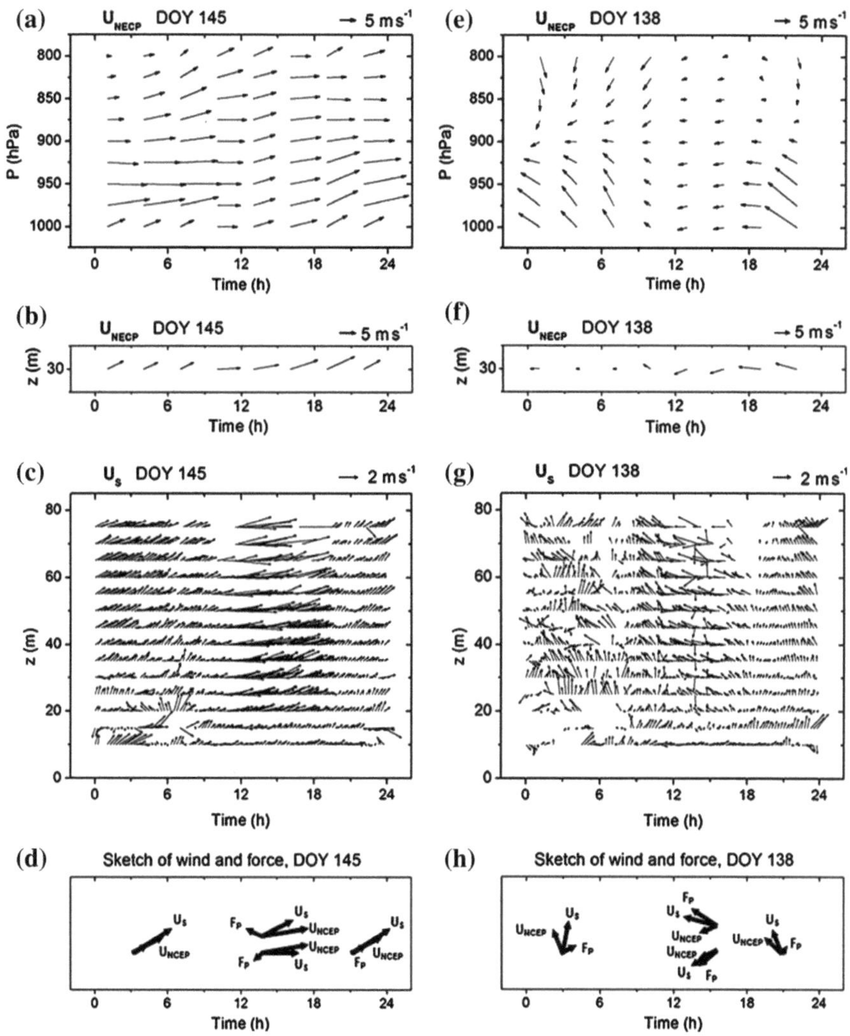

Fig. 4.19 Comparison between synoptic wind vector from NCEP reanalysis and local wind vector from sodar (U_s) on DOY 145 with westerly wind (*left figures*) and on DOY 138 with easterly wind (*right figures*). NCEP and U_S profiles are shown in **b–c** and **f–g**. A sketch of NCEP, U_S, and pressure gradient force (F_P) due to the clearcut, representing typical situations in plots **b–c** and **f–g**, is shown in **d** and **h** (Zhang et al. 2011)

modeling strategies. More difficult conditions were imposed on the models when a tracer experiment was used in a homogeneous peach orchard (Figs. 4.21 and 4.22).

While the results showing model strength and robustness bode well, the notion of applying the model within the roughness sub-layer had not been until the validation study was applied to a tall pine canopy was attempted Leclerc et al.(2003b) as shown in Figs. 4.21 and 4.22.

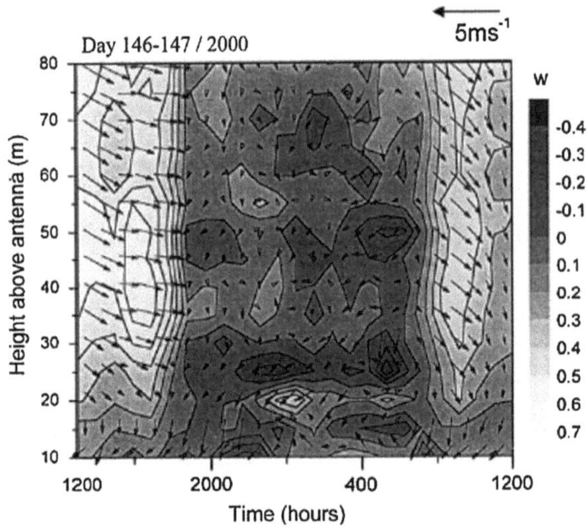

Fig. 4.20 Hourly averaged horizontal and vertical velocities as a function of height from mini-sodar data (*vertical velocities are given in color code and horizontal velocities are given in terms of arrows; the length of the arrows gives the magnitude and the orientation of the arrows give the direction*). The data presented is from 12:00 h on day 146 to 12:00 h on day 147, 2000 (Leclerc et al. 2003a, Published with kind permission of © Elsevier, 2003. All Rights Reserved)

In the pine plantation, the presence of the large clearcut outside the footprint area was found to alter dramatically the structure of the exchange above the adjoining forest canopy, with major modifications to the flux. When the wind passed over the clearcut located well outside the footprint area, the attribution of fluxes to the neighboring surfaces were off by up to 300 % when compared against footprint fluxes predicted in the absence of the large bare swath as shown in Fig. 4.23. This fact would have gone underreported by experimentalists who would have kept believing that the fluxes measured above a forest canopy would be a reflection of the neighboring surface most of which contained within a radius smaller than 500 m in daytime conditions. This was detected thanks to a rare and fortuitous combination of the simultaneous use of a sodar on site and measurements of high-frequency tracer eddy flux, both typically absent in most measurement field campaigns or at routine flux sites. In an analysis done jointly with the sodar data, a wind directional analysis reveals a particular modification of the flow in terms of mean wind speed and direction resulting from the newly cleared land: the presence of organized systematic circulations arising from the contrast in surface temperature between the forest and the clearing itself was detected. The use of the MODIS satellite imageryproved to be an invaluable tool in the appreciation of the importance of a landscape-wide perspective on local forest flux measurements.

Large-scale inhomogeneities induced by logging or by modifications to the energy balance of the neighboring surface, despite their location outside the footprint area, play an important role in the modulation of the surface-atmosphere exchange modifying fluxes. The large scale surface inhomogeneities generated as a result of recent logging activities in the area several hundred of meters away from the site appear to lead to much higher experimental tracer fluxes during NNW-NW winds. This represents an unaccounted but significant forcing not represented in either the Lagrangian simulation or in the analytical solution.

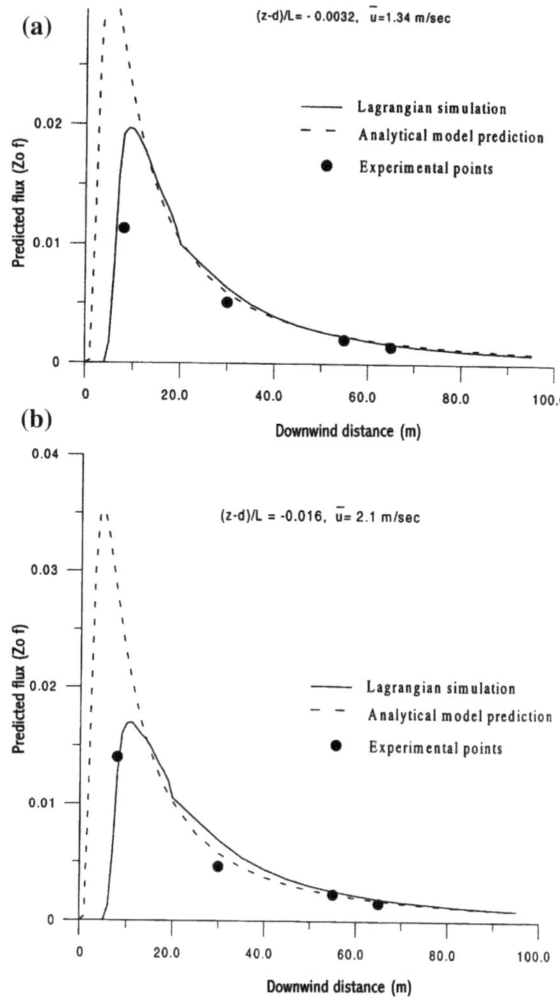

Fig. 4.21 Validation of fractional flux density above the orchard using the Lagrangian simulation (*solid line*) and the analytical solution to the diffusion-equation (*dashed line*) for near neutral conditions **a** $(z - d)/L = -0.0032$, run no. 12, 22 September 1998 **b** $(z - d)/L = -0.016$, run no. 10, 22 September 1998. Closed circles represent the experimental data points at each tower (Leclerc et al. 2003b, Published with kind permission of © Elsevier, 2003. All Rights Reserved)

Until the flow dynamics of this complex three-dimensional surface-atmosphere interaction are adequately understood, this study underscores the importance of judicious site selection and the need to take into account surrounding landscape properties, not only the region delineated by the footprint envelope, but also outside the footprint envelope.

Lessons learned from these experiments teach us that the effects of surface inhomogeneities associated with topography, varying vegetation, and management practices should be mapped before initiating a flux program. In addition, remote-sensing maps are also recommended using remote sensing data such as that of the MODIS satellite so that thermal gradients can be identified and included in the interpretation of flux measurements. Thermal maps should be obtained routinely for flux systems placed in forest plantations and in any region where land-use

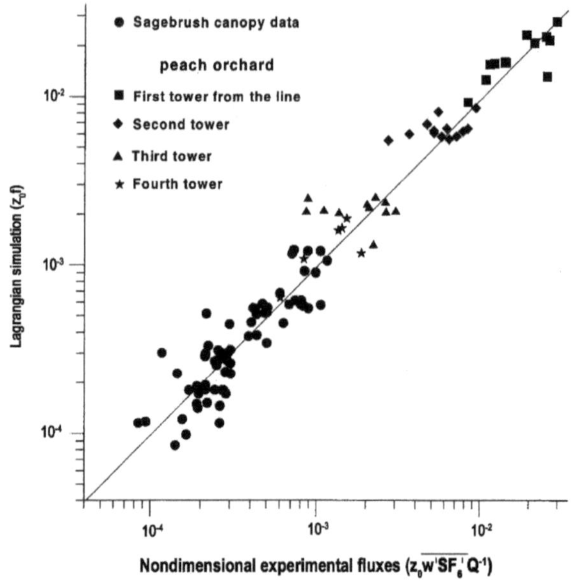

Fig. 4.22 Comparison of fractional flux density between the Lagrangian simulation above the orchard and above the sagebrush canopies and the tracer fluxes (Leclerc et al. 2003b, Published with kind permission of © Elsevier, 2003. All Rights Reserved)

changes undergo changes within the timescale of the field monitoring program. In addition, the presence of an acoustic sodar at a flux site can provide important clues which otherwise could go unnoticed. The Leclerc et al. (2003a) tracer study over the pine forest was enhanced by the use of using an on-site acoustic Doppler sodar which, against all expectations, later provided an early diagnostic of anomalies in the upwind field. The overlap and interaction between mesoscale flow and localized flow features is complex and highlights the need to further examine their interaction and subsequent impact at a particular flux site.

4.4.3 Footprints in the Presence of a Transition from the Forest Leading Edge

Transitions at a forest edge are scenarios often encountered by the experimentalist leading to questions such as whether the flux system is placed in the region of influence of the transition, be it of a clearing-forest edge, a lake-forest transition, or a crop-forest mosaic. These transitions give rise to sharp and sudden horizontal pressure gradients accompanied by changes in thermal and mechanical turbulence with a locus of flow re-attachment downwind uncertain with corresponding uncertain atmospheric variables.

Important errors in the evaluation of surface-atmosphere exchange are introduced by the use of single-point measurements anywhere the assumption of horizontal flow homogeneity is violated. This oft-encountered field scenario is therefore more than an academic exercise, it is an ubiquitous field situation: Since

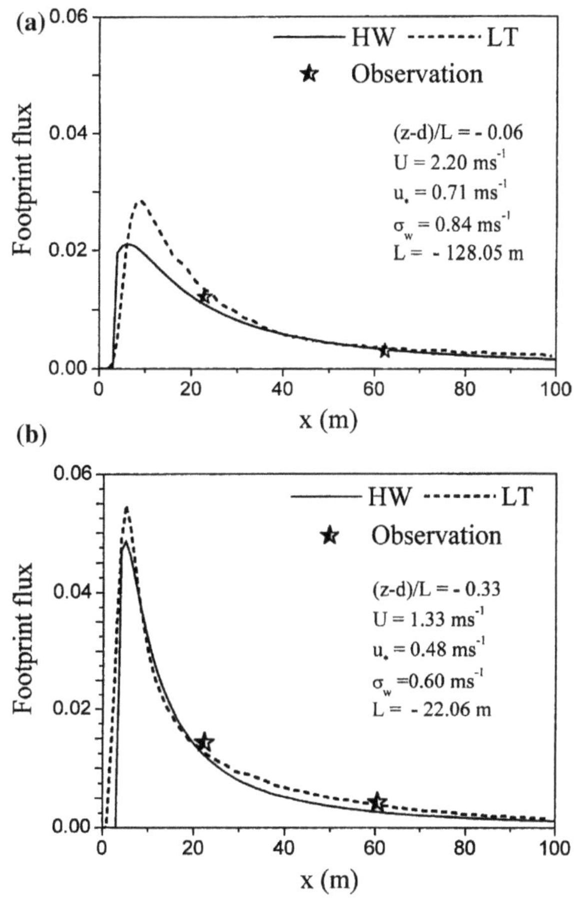

Fig. 4.23 **a** Comparison of the Lagrangian footprint simulation (- - -) and the analytical solution (—) with observed footprint flux during NNE–NE winds and mildly unstable conditions $(z - d)/L = -0.06$. **b** Same as in (**a**), but for unstable conditions $(z - d)/L = -0.33$. (Leclerc et al. 2003a, Published with kind permission of © Elsevier, 2003. All Rights Reserved)

most field research sites are located in a mosaic of mixed land-use i.e. with grasslands adjoining tree stands, managed and natural plantations with natural and manmade clearings, land sectors are frequently discarded because of the presence of such transitions. The question then becomes for the experimentalist 'Where downwind of the surface roughness step-change can a tower flux system be placed?' In such cases, the footprint function in the region neighboring the discontinuity is of considerable practical significance.

In contrast with the concentrated, time-consuming, and labor intensive logistical efforts required in a field deployment to quantify the in-depth spatial variation of physical processes and related footprint functions at these spatial discontinuities in the flow field, modeling is both cost-efficient and quantifies the impact of these inhomogeneous surface properties inherent to the site on field deployment planning.

Sogachev et al. (2005) used the higher-order closure model to quantify the effect of clearcuts on flux footprints downwind above a forest canopy. That study investigated the CO_2 flux footprint magnitude as a function of clearcut widths

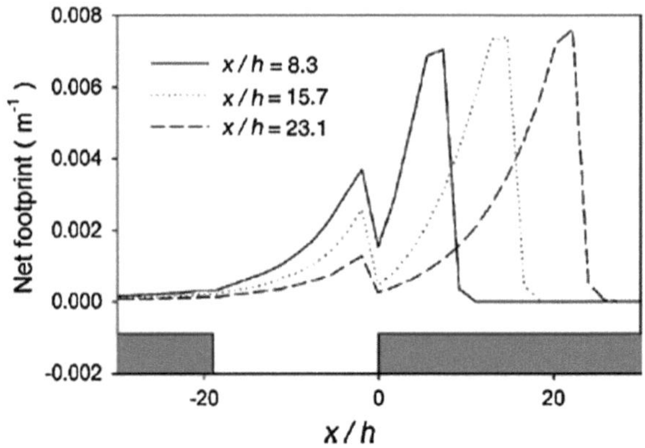

Fig. 4.24 Examples of net footprints (*joint contribution of sources located within the canopy layer and on the soil surface are considered*) derived by the model for a 17 *h*- wide clearcut for sensors located at various normalized distances, x/h downwind of the forest edge at $z_m = 1.4\ h$ after Sogachev et al. (2005), Published with kind permission of © Elsevier, 2005. All Rights Reserved

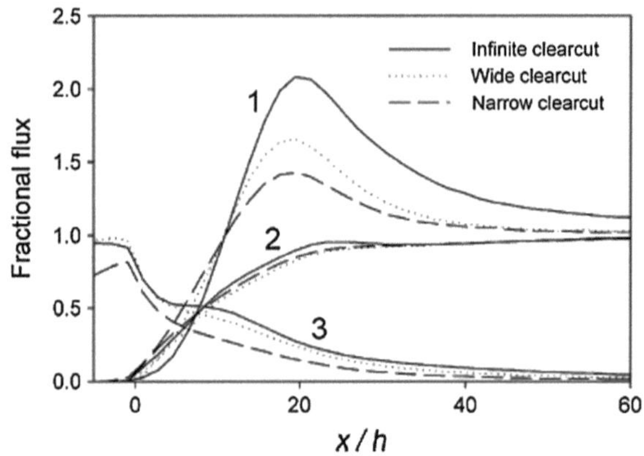

Fig. 4.25 Variation of the fractional flux functions at a height of 1.4 *h* with normalized distance, x/h downwind of the forest edge, derived by footprint modeling for sources on forest floor, inside a tree layer and on the clearcut. These functions describe the contribution of corresponding sources to a measured signal at an arbitrary location downwind of the clearcut-forest edge, after Sogachev et al. (2005), Published with kind permission of © Elsevier, 2005. All Rights Reserved

using flow statistics measured at a managed pine plantation (Leclerc et al. 2003b). Figure 4.24 depicts the net footprint for a swath width 27 times the forest height and its behavior downwind of the transition zone. In a numerical experiment, Sogachev et al. (2005) used experimental flux and turbulence data downwind of a large upwind logged area. Interestingly, despite the fact that modeled logged areas were many hundreds of meters away from the modeled flux tower, the study found scalar fluxes to be sensitive to clearcut widths. Footprint results are supported here using examples of footprints derived by the model to describe the joint contribution of sources located within the canopy layer and on the soil surface (net footprints) along a forest gap 17 *h* wide. The results suggest that the contribution of logged swaths of land to the flux signal downwind of a forest edge peaks within the first 20 canopy heights *h* to then gradually vanish for towers located at a downwind distance of approximately 30 *h*; from that point onward, results then suggest that the flow is then in equilibrium with the underlying surface and that the surface can then be treated as horizontally homogeneous. It should be pointed out at that the Sogachev et al. (2005) study uses neutral conditions and there is no doubt that both scalar and flow adjustment downwind would be vastly different in the presence of buoyancy forces. It is however, almost certainly a significant agent in the modulation and rate of adjustment of the flow/scalar the distance from the transition zone.

The knowledge of the footprint itself considerably improves our ability to deconstruct a flux signal into its different source signatures. Sogachev et al. (2005) pointed out that, for the purpose of selecting the optimum location of flux towers, the information provided by the footprint function is more convenient when presented using the fractional flux function describing the contribution of given source into a signal at that imaginary flux tower. Here, these fractional flux functions at measurement level $z = 1.4\ h$ are presented for varying dimensions of modeled clearcuts. The behavior of these functions depends on the flow structure in the clearcut-forest transition zone, which in turn is defined by the canopy structure. The flow acceleration in the lower canopy and above, the flow deceleration in the upper canopy region together with the vertical air motions, all occurred in this zone resulting in a complicated distribution of the scalar field and vertical fluxes. Using several variables such as ground flux information coupled with soil-canopy flux partitioning, Net fluxes for any given level downwind of the forest edge can be derived.

The study by Sogachev et al. (2005) shows how flow distortion created by the clearing-forest transition leads to the formation of complex flow motions both inside and above the canopy layer downwind and how these motions modulates the scalar distribution throughout vegetated canopies. Figure 4.25 illustrates how a simultaneous interaction of sources located on the surface and in the canopy layer can produce a net flux enhancement over different fetches, the amplitude and distribution of which is a function of the ratio of source strengths of the surface to that of the canopy layer.

For towers located in complex terrain, the approach based on two and three-dimensional flow models capable of taking into account the heterogeneity in surface properties is strongly recommended for footprint estimation. The interpretation of eddy covariance flux measurements over Lake Valkea-Kotinen in the framework of Helsinki Environment Research Centre (HERC) project (Smolander and Stenberg 2005; Vesala et al. 2006) is a practical example confirming the adequacy and usefulness of this approach.

Klaassen and Sogachev (2006) used the higher-order closure model SCADIS to characterize the fetch dependence on the integrated flux footprint at a height of 1.35 h_c and past a bog-forest edge. That study compared this higher-order closure

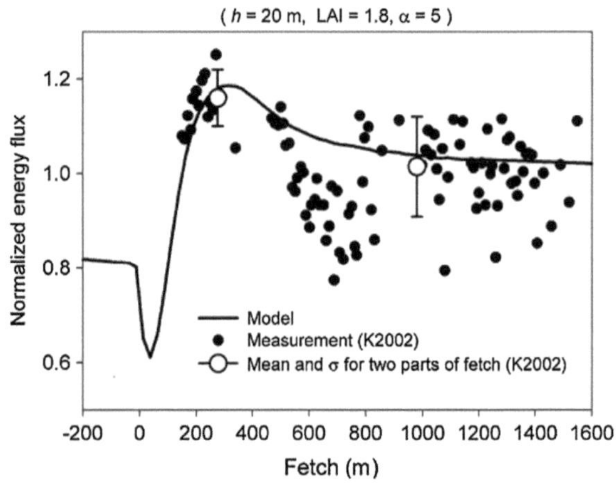

Fig. 4.26 Normalized energy flux at 1.35 h_c at the forest edge versus fetch downwind. Comparison of measurements (Klaassen et al. 2002, K2002) and the SCADIS model according to Klaassen and Sogachev (2006)

model against flux measurements (Klaassen et al. 2002). The results are roughly in the same order of magnitude, tending to overshoot at approximately at a downwind distance of 30 h_c and modeling correctly the fluxes between 40 and 80 h_c. Klaassen and Sogatchev (2006) found an enhancement of upwind surface flux by a factor of 2–3 in their case study (Fig. 4.26). They attributed the presence of vertical advection arising from the step change at the forest edge and recommended that field observations downwind of roughness transition. The authors have further cautioned us that footprint models should take the actual turbulence field into consideration when atmospheric flux measurements data downwind of a leading edge are analyzed. In their study, the integrated footprint as a measure of atmospheric flux enhancement exceeds the value 1.1. Enhanced scalar fluxes downwind to 25 h_c while 15 h_c was required for a momentum adjustment over the forest canopy.

Belcher et al. (2008) also examined the dynamical processes that control flow and turbulence above and within tall forest canopies in complex terrain. They examined the adjustment of the flow to a forest edge (Fig. 4.27). The mean flow within the canopy adjusts to the forest edge over a few multiples of the canopy drag length scale, roughly 3 L_c, where L_c is inversely proportional to the leaf area of the forest. In practice this adjustment length varies from about 10 m in very dense plantations, to about 100 m in more sparse woodland. Over this adjustment region, air is systematically advected out of the top of the forest canopy. Turbulence in the flow within the forest adjusts following the adjustment of the mean flow and we have developed scaling arguments that suggest that this occurs over a length that again scales on L_c. In this way, the adjustment of the mixing and transport to a forest edge may require a fetch of between 20 and 200 m before the turbulence resembles the mixing layer eddy structure of homogeneous canopies.

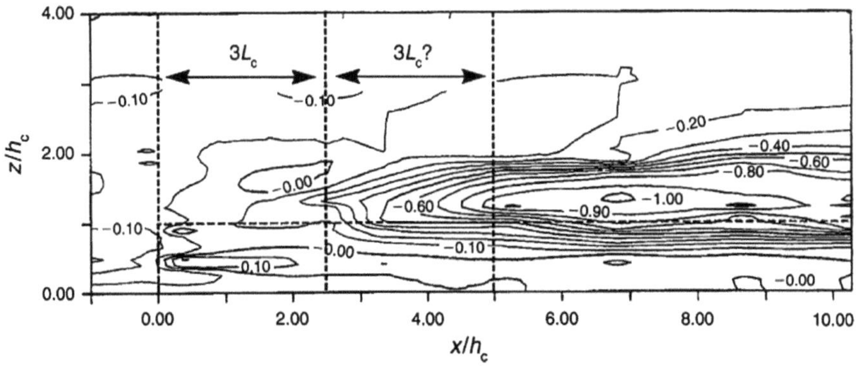

Fig. 4.27 Development of the turbulence stress downwind of a forest edge calculated by Morse et al. (2002) from their wind tunnel measurements. The *vertical dashed lines* indicate distances $3L_c$ and $6L_c$ downwind of the forest edge. Notice how the turbulent stress changes very little in the adjustment region of the mean flow, of length $3L_c$ (*added to the figure by* Belcher et al. 2008), but does then adjust within a second region, which is also of length $3L_c$

4.5 Flux Footprints Over Complex Topography in Forests

Sogachev et al. (2004) used a second-order flow model using a terrain-following coordinate system, suitable for simulation of flow over hilly terrain covered by forests. This model, SCADIS, was able to reproduce flow features arising from pressure gradient forces. The model was applied to constant flux boundary conditions at the surface and a pre-defined canopy exchange rate to simulate concentration and flux fields, and the relationship between fluxes and spatially distributed sources and sinks. They found that fluxes were most 'disturbed' in comparison with values at the same level but far upwind from the ridge, at the leeward side of the crest, near the leeward slope of the ridge in the wake region and at the upwind foot of the ridge. They also found that the footprint function is strongly dependent on the location of the flux measurements, with a relatively higher contribution from distant sources for flux measurements near the top of the ridge. The footprint function is characteristic of the reversed flow in the recirculation zone on the leeward side, recirculation which actually occurs close to the surface at the leeward foot of the ridge. This result was also supported theoretically by Belcher et al. (2008). The fluxes at a real measurement site are in gently hilly terrain, something frequently encountered for those making flux measurements. In these conditions, fluxes are influenced by upwind heterogeneity of the vegetation and by the characteristics of the local topography.

Thus, it can be said unequivocally that the position of the tower in relation to the terrain characteristics is extremely important as seen in the Fig. 4.28, in contrast with a site located over a flat homogeneous canopy. The reader is reminded of the importance of a pre-existing local climatological study to assess the relative importance of local relief features to a point flux measurement.

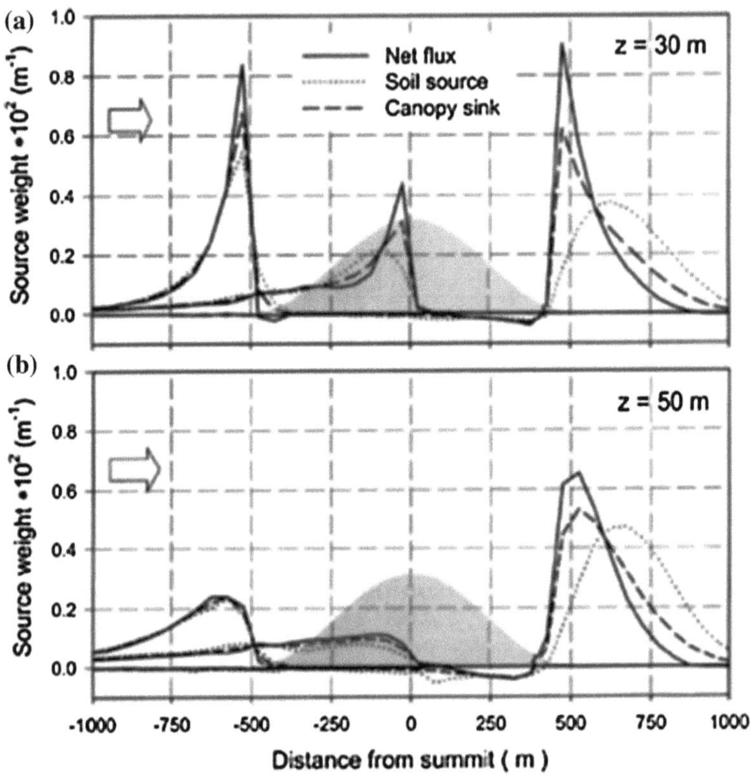

Fig. 4.28 Source weight functions for fluxes at different locations over a ridge: upwind foot (*distance* –500 m) and leeward foot (*distance* 500 m) and at the ridge crest (*distance* 0 m). Two heights over the local surface were considered: **a** 30 m and **b** 50 m. The ground respiration was 4 mmol m^{-2} s^{-1} and photosynthetic CO_2 exchange was -8 mmol m^{-2} s^{-1}. Source weight functions were normalized by the vertical fluxes at the given points of interest. The height of the model ridge was 100 m with a width of 1,000 m. The topography variations are shown by grey area. Arrows show the direction of the airflow. (Sogachev et al. 2004, Published with kind permission of © Elsevier, 2004. All Rights Reserved)

The importance of topography is greatest for sources close to ground and influences the source function. This is something that has implications for respiration measurements. In the presence of katabatic flow in a stably stratified boundary layer, the sensitivity of the location of the tower is further compounded. Considering topographical variations, the optimal location of a flux measurement tower in hilly terrain depends also on the measurement level.

Belcher et al. (2008) also found that even hills of low slope change the flow within and above the forest substantially and documented the presence of a region of reversed flow within the canopy in the lee of even very gentle topography, when the slope of the hill is so small that, in their own words, 'in the absence of the canopy, the streamlines just follow the hill surface with no separation'. This reversed flow is forced by pressure gradients that accelerate air towards the crest on the both the upwind and downwind slopes. Air is then ejected from the forest canopy just downwind of the crest. Wind tunnel studies have now confirmed the existence of this reversed flow, and carefully gathered field data documenting this physical process would be helpful.

These results in neutral flow over hills have important implications for the mixing and transport of scalar. Modeling studies suggest that the flow up the slopes of the hill carries the scalar which is then ejected from the top of the canopy just downwind of the crest. This leads to marked variations in the flux of scalar above the top of the forest by a factor of 10 or more. Belcher et al. (2008) examined the nocturnal flow over hills. As the air cools and a stable temperature profile develops, turbulence in the canopy collapses as the flow above remains turbulent. In neutral and unstable conditions, turbulent mixing ensures that both the flows above and within the forest are dynamically coupled. The collapse of canopy turbulence leads to a decoupling so that both flows evolve independently. The cooler air within the canopy drains down the slopes, forced by a pressure gradient. In contrast with the hydrodynamic pressure gradient associated with the flow over the hill, the cool air pushes the air towards the crest. Hence drainage currents form in the canopy when the hydrostatic pressure gradient exceeds the aerodynamic pressure gradient. Perhaps surprisingly, the condition for this to happen depends not on the slope of the hill, but rather uniquely on the length of the slope resulting in drainage currents forming. This is the case even for small slopes.

Dimensional analysis of unsteady gravity currents (e.g. Hatcher et al. 2000) suggests that the slope may play a role in the time taken to reach the steady state. The studies by Belcher et al. (2008) and Sogachev et al. (2008) offer some insight in the interpretation of flux tower measurements in such complex terrain. Belcher et al. (2008) state that it is their intent to provide simple scaling laws that can be used to provide first estimates of the impact of complex terrain on measurements at any particular flux tower; they also stress that this information can help improve high-resolution mesoscale numerical models that can be configured specifically for individual tower sites. The combination of such high-resolution numerical models with the measurements taken at the flux towers can then be combined using inverse modeling techniques to obtain optimal estimates of the net ecosystem exchange on the scale of the landscape. These advances, however ambitious, are deemed necessary to quantify the global exchange between the terrestrial biosphere and the atmosphere.

The study by Katul et al.(2006) indicates the dominant role of advection in the scalar mass balance for mass transport on a forested hill. The above study describes the flow over hills using a first-order closure model of a gentle cosine-shaped hill aimed at studying the impact of the terrain on scalar exchange in a hypothetical forest canopy in neutral conditions, Fig. 4.29 shows how the concentration field and the fluxes are insensitive to the lower boundary conditions, an insightful result. In addition, while the modeled above ground sources and sinks are much smaller than H/L (hill slope with H: hill height, L: length scale), the proportional variability in the eddy flux, for instance, is at least an order of

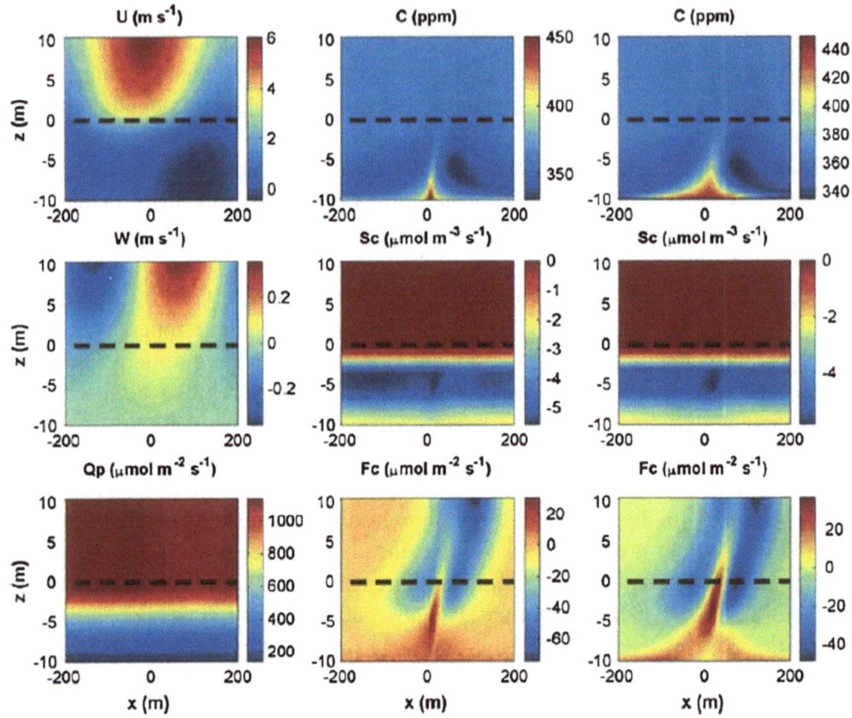

Fig. 4.29 Spatial variation of the forcing (*left panels*) and response variables for flux (*middle panels*) and concentration (*right panels*) boundary conditions. The forcing variables include the mean wind field $(\overline{u}, \overline{w})$ and incident photosynthetically active radiation (*Q*p) and the response variables include mean CO_2 concentration (c), bove ground CO_2 sources and sinks (S_c), and the turbulent CO_2 fluxes $\left(F_c = \overline{w'c'}\right)$. For reference, the canopy *top* is shown as a dashed horizontal line. The model calculations are for steady-state neutral flows (Katul et al. 2006)

magnitude larger than H/L. The value of the horizontal gradient in the concentration across the hill, especially near the hilltop and within the recirculating region suggest that the advection terms must be included in a flux measurement program located on gentle terrain.

This is attributed to air carried out of the canopy volume by the re-circulating region at the lee side of the hill. While the authors of the above study have not explicitly calculated the footprint over the hilly terrain, these conclusions point to the fact that topographical features must be accounted for when inferring fluxes from tower measurements in these conditions. The imbalance between $u\partial c/\partial x$ and $w\partial c/\partial z$ is sufficiently large to decouple the local canopy photosynthesis from the local turbulent flux. Hence, linking tower-based eddy-covariance measurements to local biological sources and sinks on hilly terrain, to be correct, must include both horizontal and vertical advective terms.

While the direct effects of topographic forcing on photosynthesis are small, the errors involved in ignoring one or both advection terms when inferring net ecosystem productivity from flux measurements on a single tower are sufficient to invalidate the estimates of net canopy exchange using eddy covariance measurements. It should be emphasized that the example of a two-dimensional hill maximizes the flow perturbation and, therefore these errors, but even in gentle three-dimensional terrain, errors of order 100 % can be anticipated if advection is ignored.

4.6 Emissions of Odor and Reactive Trace Gas Fluxes Using the Flux Footprint Method

An innovative and potentially transformative application of the flux footprint method to the field of waste management arose with the work of Sarkar and Hobbs (2003). In fact, issues related to odour complaints from a local community around industrial sites dealing with solid wastes are offending, frequent and multi-faceted in their air quality and socio-economic ramifications. That study extended the scope of applications of the flux footprint method by applying it to determine the emissions of odor from landfill. This is of significance since most methods have resorted to standard micrometeorological techniques which measure the flux but do not provide any insight on the location of the offending sources. Furthermore, these techniques do not satisfy the problems of the amount of odorant loss to the atmosphere in heterogeneous conditions. The footprint model used is that of the analytical solution to the advection—diffusion equation.and provides a location of the emissions which was much needed. The footprint method represents a step forward in the field of gaseous emissions from landfill and solid waste sites where a large source of tpically indefinite geometery and characterized with spatially inhomogeneous surface is the norm. The authors used the SAM model by Schmid (1994) and tested their footprint methods against the Linvall hood measurements but are also point out in the same breath that a reliable use of the Lindvall hood method hinges on the biases, variability associated with the emission source under consideration.

Karahabata et al. (1999) investigated the footprint size and source distribution within the footprint and with respect to the observation point over a boreal forest canopy. The source distribution was very uneven and the authors reproduced the variability by looking at the streamwise component and the lateral component of the flow. They found that, in daytime unstable conditions and steady wind direction, flux measurements varied little but in turbulent gusty conditions, the variability doubled.

Karahabata et al. (1999) raise the important question, based on their data, of whether the measured flux is a product of the changing emissions or whether it is a function of the changing footprint. Given the variability within the footprint which depends on the heterogeneity of the forest and tree types and the sampling protocol, the difference between modelled emissions and measured emissions can be can be minimized by identifying the footprint area to tag the emission rate to that particular area, thus decreasing the discrepancy between measured and modeled results. Such an approach is also helpful in improving ecophysiological emission models so that they can be used with more confidence as forcing modules within climate models.

A footprint model containing a biochemistry module and coupled to a simple turbulence model was used in an original application of footprint research by authors Strong et al. (2004). For reactive trace gases the possible reaction time must be included into a footprint model. The above study noted the need to add the reactive hydrocarbon flux into the footprint integral of a Lagrangian footprint model. The new footprint expression to describe the reaction rate from the source area to the measuring point thus can be rewritten as additional function which:

$$F(x, z_m) = \int_{-\infty}^{x_m} \int_{0}^{z_m} \psi(x_m - x, z_m) Q(x, z) f(x_m - x, z_m) dz dx, \qquad (4.1)$$

where Q is the source area function and f the footprint function. The function ψ sums up the concentrations χ of each particle p arriving the measuring point

$$\psi(x_m - x, z_m) = \frac{\sum_{p=1}^{N} \chi_p}{N} \qquad (4.2)$$

Leaf senescence and fall alter the structure of the canopy which in turn alters both the flow and the source distribution of these biogenic sources within a canopy. With the footprint expression derived in (4.1) and (4.2) tailored to take into account the presence of these reactive sources, Strong et al. (2004) documented the influence of foliage senescence and abscission on the effective source distribution of isoprene for a mixed deciduous forest canopy. The resulting footprint was found to expand in the presence of a sparse canopy, likely a result of decreased shear and the turbulence transport being greater in those conditions in the lower canopy layer. For a canopy that is defoliated at 50 % of its original full leafed original canopy leaf area, the footprint spatial extent increases by 46 %. The results below show also the impotance of the location within the canopy of those respective sources (Fig. 4.30).

Another footprint study involving hydrocarbon flux footprints was done by Rinne et al. (2007) at the Hyytiäla boreal forest site. That work used the Lagrangian model by Markkanen et al. (2003) and also included a reaction term, which calculates the time dependent reaction of the submitted particles. Because different reactions have a different reaction time, the footprint of the different trase gases is unique: for short reaction time, a very short footprint was found while very long reaction time had a footprint similar to inert gases (Fig 4.31).

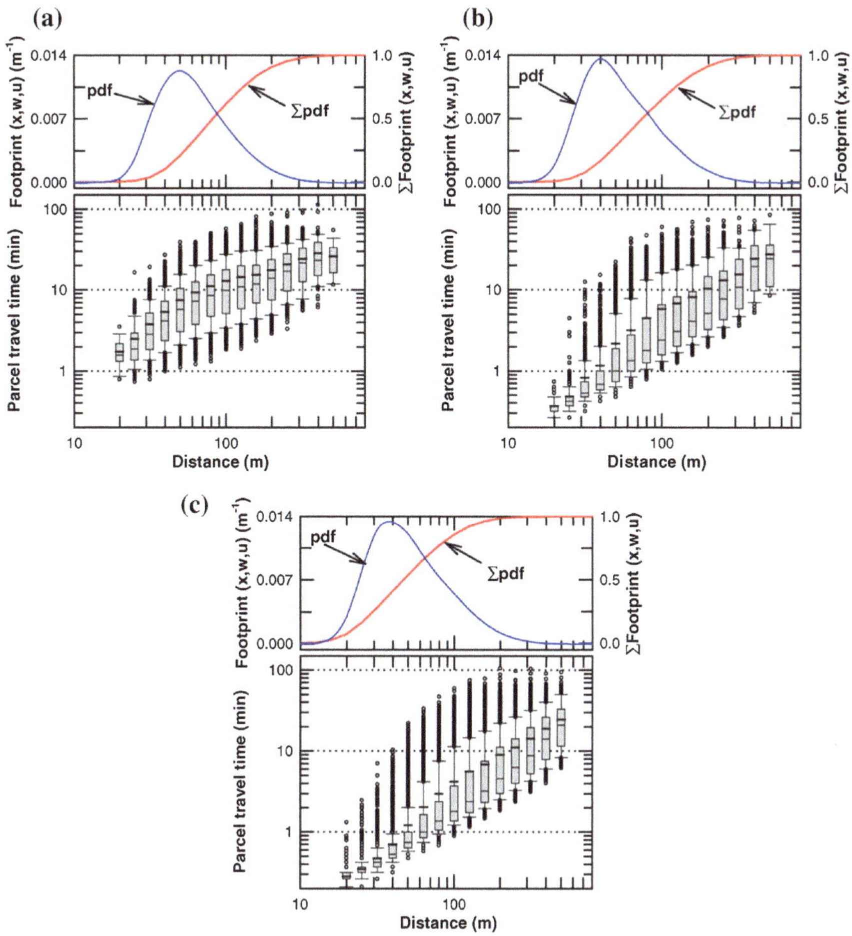

Fig. 4.30 **a** Flux footprint probability density function (*pdf*) and cumulative flux footprint pdf for the 33 m level above the canopy, release at $z/h_c = 0.25$, and fully foliated conditions. **b** The same as (**a**), except for release at $z/h_c = 0.75$. **c** The same as (**a**), except for release at $z/hc = 0.90$. For the box plots, the *bold vertical line* is the mean, the *thin vertical line* is the median, the *shaded box* shows the inner-quartile range, the error bars denote the 10th and 90th percentiles, and outlier data are shown as circles (Strong et al. 2004, Published with kind permission of © Elsevier, 2004. All Rights Reserved)

4.7 Footprints in Urban Areas

Experimental and modeling studies in urban areas remain a challenge for micrometeorologists. This is an area that creates potentially exciting and useful contributions: Urban areas are extremely heterogeneous. In addition, proper height-dependent scaling can be tricky. The determination of the zero-plane

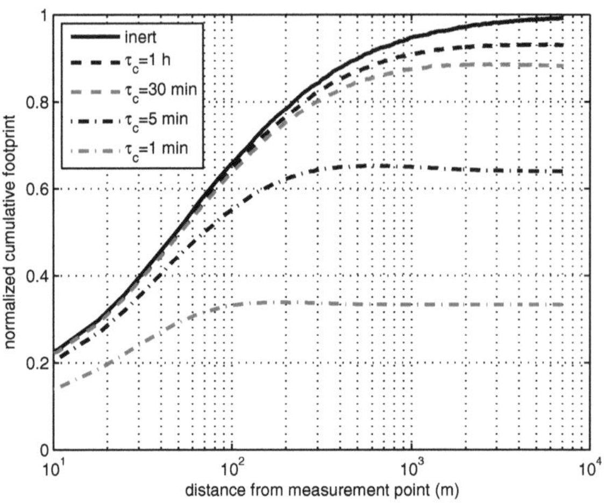

Fig. 4.31 Effect of chemical degradation on cumulative footprints at a height of 22 m for components with different chemical life time within the canopy and for a source height of 11.2 m (Rinne et al. 2007, Published with kind permission of © Copernicus Publications, distributed under the Creative Commons Attribution 3.0 License, 2007. All Rights Reserved)

displacement—which is important for the application of any footprint model—is problem of high complexity (see Sect. 2.2.2). Even footprint techniques were applied to find the zero-plane displacement (Kanda et al. 2002). Nevertheless, flux measurements are a great contribution to most urban meteorological studies (Grimmond 2006; Feigenwinter et al. 2012). This is not the case with footprint models. Previous studies (Schmid et al. 1991; Grimmond and Oke 1999; Kanda et al. 2002; Moriwaki and Kanda 2004) used the analytical model by Schmid (1994, 1997). These studies give only a first guess about the footprint because of the simplifying assumptions of the model. In a recent study by Goldbach and Kuttler (2013) the model was again used for an urban and suburban flux site in the town of Oberhausen, Germany.

More recent studies used higher order closure or Lagrangian footprint models. The SCADIS model (Sogachev and Lloyd 2004) was applied for an urban meteorology study in the city of Helsinki. Järvi et al. (2009) investigated the footprint in two road sectors with a wind direction perpendicular to the road. That study included in their calculation surface types like road, parking area, soil, trees and buildings with different height. The measurements were done at 31 m over ground. Vesala et al. (2008) determined the footprints for sources in the urban canopy level (approximately height of the zero-plane displacement), which is much smaller than the footprint with sources from the ground level (Fig. 4.32). The sources from the ground level are often under the influence of the channel effect due to the street canyons than sources at higher levels above the city buildings. Such studies are of great consequencein relation with air pollution measurements seeking to identify the sources origin.

Another recent study with the footprint model by Kormann and Meixner (2001) was done on the roof of King's College London north of the Thames river (Kotthaus and Grimmond 2012). In a pre-study from April 18 to July 15, 2009, the turbulent fluxes of momentum and sensible heat were measured. The footprint was

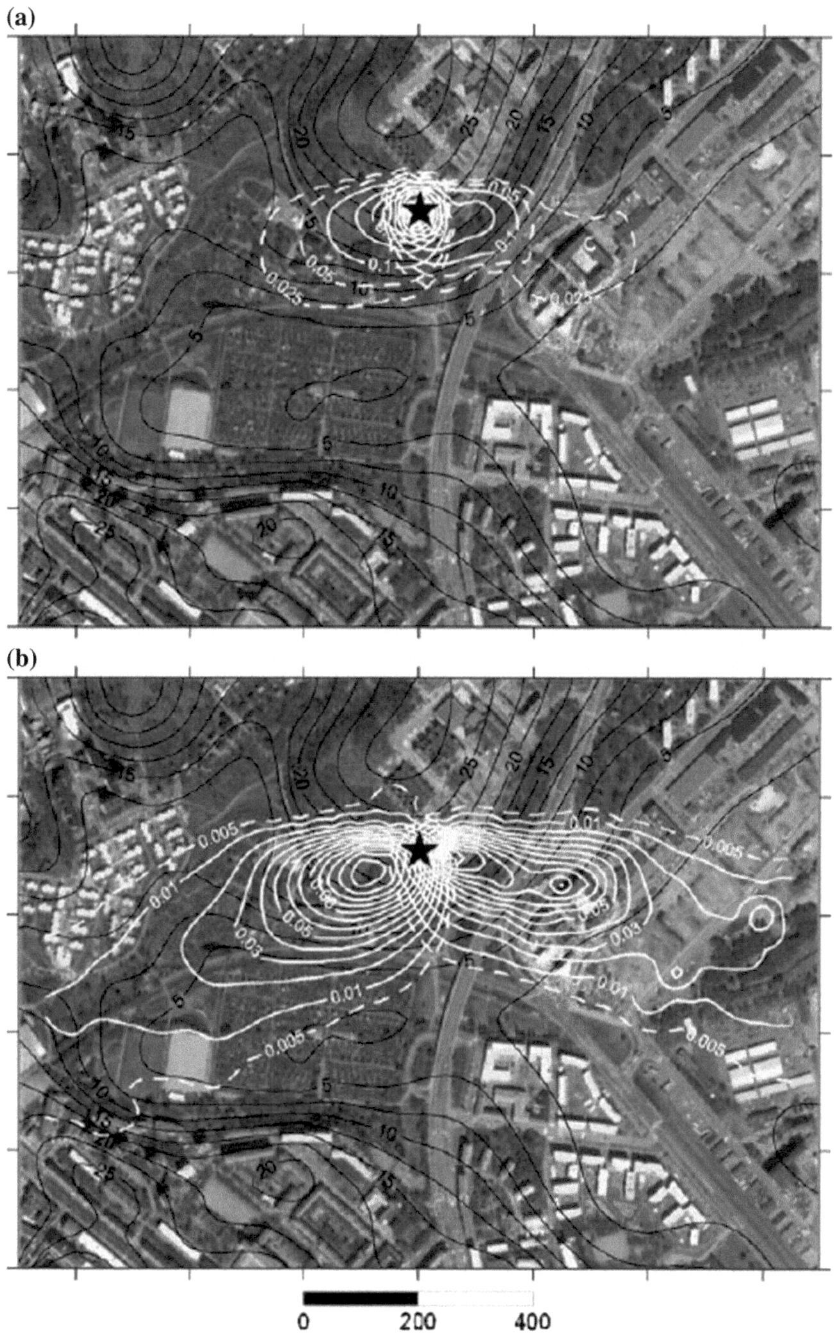

Fig. 4.32 Topography of the measurement site is drawn using *black contours* (Vesala et al. 2008). White contours give footprints for **a** canopy (*upper*) and **b** soil (*lower*) sources: scale (10^{-4} m^{-2}), Published with kind permission of

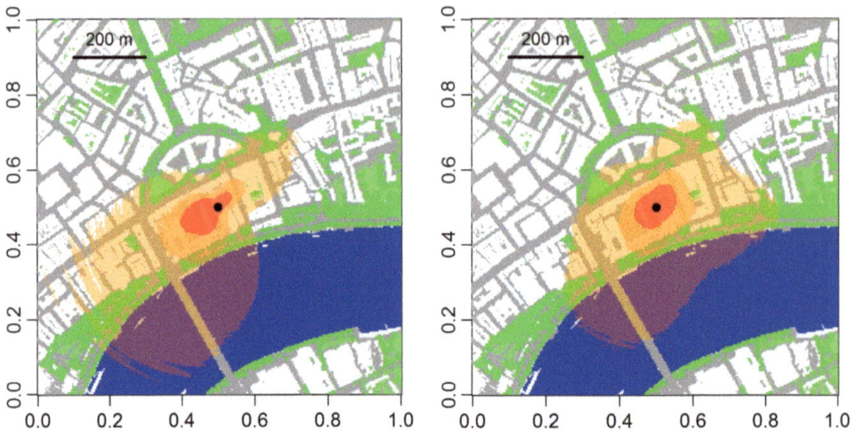

Fig. 4.33 Footprint climatology of the eddy-covariance tower at 39 m a.g.l. for 28 April—15 July 2009; on the left side neutral ($0 > z/L > -0.0625$), and on the *right side* unstable conditions ($z/L < -0.0625$); the *purple colors* indicate a source contribution of 80, 50 and 30 % (*from light to dark*); *orange* street canyon/courtyard, *red* rooftop, *yellow* river front, *green* vegetation (Pauscher, Kotthaus, Grimmond, personal communication, 2012, Published with kind permission of © Mr. Pauscher, Dr. Kotthaus, Prof. Dr. Grimmond, 2012. All Rights Reserved)

analyzed using the Rannik-Göckede-approach (Rannik et al. 2000; Göckede et al. 2006, Pauscher, Kotthaus, Grimmond: personal communication, 2012) for an eddy-covariance system at 48 m a.s.l. The displacement height was determined with the morphological approach (see Sect. 2.2.2 and MacDonald et al. 1998) and was found to be 17 m for the North sector (town) and 8.4 m for the South sector (Thames river). The calculated footprints are shown in Fig. 4.33 over this nearly 3 month period in neutral and unstable conditions.

References

Amiro BD (1990) Comparison of turbulence statistics within three boreal forest canopies. Bound-Layer Meteorol 51:99–121

Baldocchi D (1997) Flux footprints within and over forest canopies. Bound-Layer Meteorol 85:273–292

Belcher SE, Finnigan JJ, Harman IN (2008) Flows through forest canopies in complex terrain. Ecol Appl 18:1436–1453

Beyrich F, Mengelkamp H-T (2006) Evaporation over a heterogeneous land surface: EVA_GRIPS and the LITFASS-2003 experiment—an overview. Bound-Layer Meteorol 121:5–32

Coppin PA, Raupach MR, Legg BJ (1986) Experiments on scalar dispersion within a model plant canopy Part II: an elevated plane source. Bound-Layer Meteorol 35:167–191

Corsmeier U, Kalthoff N, Kolle O, Kotzian M, Fiedler F (1997) Ozone concentration jump in the stable nocturnal boundary layer during a LLJ-event. Atmos Environ 31:1977–1989

Draxler RR (1997) Hybrid Single Particle Lagrangian Integrated Trajectory (HYSPLIT4) model. NOAA Air Resources Laboratory, Silverspring MD. http://www.arl.noaa.gov/ready/hysplit4.html
Feigenwinter C, Vogt R, Christen A (2012) Eddy covariance measurements over urban areas. In: Aubinet M et al (eds) Eddy covariance: a practical guide to measurement and data analysis. Springer, New York, pp 377–397
Finn D, Lamb B, Leclerc MY, Horst TW (1996) Experimental evaluation of analytical and Lagrangian surface-layer flux footprint models. Bound-Layer Meteorol 80:283–308
Finnigan J (2004) The footprint concept in complex terrain. Agric Forest Meteorol 127:117–129
Flesch TK, Wilson JD (1992) A two-dimensional trajectory-simulation model for non-Gaussian inhomogeneous turbulence within plant canopies. Bound-Layer Meteorol 61:349–374
Gerbig C, Lin JC, Wofsy SC, Daube BC, Andrews AE, Stephens BB, Bakwin PS, Grainger CA (2003a) Toward constraining regional-scale fluxes of CO2 with atmospheric observations over a continent: 2. Analysis of COBRA data using a receptor-oriented framework. J Geophys Res 108:4757
Gerbig C, Lin JC, Wofsy SC, Daube BC, Andrews AE, Stephens BB, Bakwin PS, Grainger CA (2003b) Toward constraining regional-scale fluxes of CO2 with atmospheric observations over a continent: 1. Observed spatial variability from airborne platforms. J Geophys Res 108:4756
Gloor M, Bakwin P, Hurst D, Lock L, Draxler R and Tans P (2001) What is the concentration footprint of a tall tower?. J Geophys Res 106(D16):17,831–817,840
Göckede M, Markkanen T, Hasager CB, Foken T (2006) Update of a footprint-based approach for the characterisation of complex measuring sites. Bound-Layer Meteorol 118:635–655
Goldbach A, Kuttler W (2013) Quantification of turbulent heat fluxes for adaptation strategies within urban planning. Int J Climatol 33:143–159
Greens SR, Grace J, Hutchings NJ (1995) Observations of turbulent air flow in three stands of widely spaced Sitka spruce. Agric Forest Meteorol 74:205–225
Grimmond CSB, Oke TR (1999) Aerodynamic properties of urban areas derived from analysis of surface form. J Appl Meteorol 38:1262–1292
Grimmond CSB (2006) Progress in measuring and observing the urban atmosphere. Theor Appl Climat 84:3–22
Hatcher L, Hogg AJ, Woods AW (2000) The effects of drag on turbulent gravity currents. J Fluid Mech 416:297–314
Inagaki A, Letzel MO, Raasch S, Kanda M (2006) Impact of surface heterogeneity on energy balance: a study using LES. J Meteor Soc Japan 84:187–198
Järvi L, Rannik Ü, Mammarella I, Sogachev A, Aalto PP, Keronen P, Siivola E, Kulmala M, Vesala T (2009) Annual particle flux observations over a heterogeneous urban area. Atmos Chem Phys 9:7847–7856
Kaharabata SK, Schuepp PH, Fuentes JD (1999) Source footprint considerations in the determination of volatile organic compound fluxes from forest canopies. J Appl Meteorol 38:878–884
Kanda M, Moriwaki R, Roth M, Oke T (2002) Area-averaged sensible heat flux and a new method to determine zero-plane displacement length over an urban surface using scintillometry. Bound-Layer Meteorol 105:177–193
Karipot A, Leclerc MY, Zhang G, Martin T, Starr D, Hollinger D, McCaughey H, Hendrey GM (2006) Nocturnal CO_2 exchange over tall forest canopy associated with intermittent low-level jet activity. Theor Appl Climat 85:243–248
Katul GG, Finnigan JJ, Poggi D, Leuning R, Belcher SE (2006) The influence of hilly terrain on canopy-atmosphere carbon dioxide exchange. Bound-Layer Meteorol 118:189–216
Klaassen W, van Breugel PB, Moors EJ, Nieveen JP (2002) Increased heat fluxes near a forest edge. Theor Appl Climat 72:231–243
Klaassen W, Sogatchev A (2006) Flux footprint simulation downwind of a forest edge. Bound-Layer Meteorol 121:459–473

Kormann R, Meixner FX (2001) An analytical footprint model for non-neutral stratification. Bound-Layer Meteorol 99:207–224
Kotthaus S, Grimmond CSB (2012) Identification of micro-scale anthropogenic CO_2, heat and moisture sources—Processing eddy covariance fluxes for a dense urban environment. Atmos Environ 57:301–316
Kruijt B, Malhi Y, Lloyd J, Norbre AD, Miranda AC, Pereira MGP, Culf A, Grace J (2000) Turbulence statistics above and within two amazon rain forest canopies. Bound-Layer Meteorol 94:297–331
Launiainen S et al (2007) Vertical variability and effect of stability on turbulence characteristics down to the floor of a pine forest. Tellus 59B:919–936
Leclerc MY, Beissner KC, Shaw RH, den Hartog G, Neumann HH (1990) The influence of atmospheric stability on the budgets of the Reynolds stress and turbulent kinetic energy within and above a deciduous forest. J Appl Meteorol 29:916–933
Leclerc MY, Beissner KC, Shaw RH, den Hartog G, Neumann HH (1991) The influence of buoyancy on third-order turbulent velocity statistics within d deciduous forest. Bound-Layer Meteorol 55:109–123
Leclerc MY, Shen S, Lamb B (1997) Observations and Large-Eddy simulation modeling of footprints in the lower convective boundary layer. J Geophys Res 102(D8):9323–9334
Leclerc MY, Karipot A, Prabha T, Allwine G, Lamb B, Gholz HL (2003a) Impact of non-local advection on flux footprints over a tall forest canopy: a tracer flux experiment (Special issue: Advances in micrometeorology: Tribute to G. W. Thurtell). Agric Forest Meteorol 115:19–30
Leclerc MY, Meskhidze N, Finn D (2003b) Comparison between measured tracer fluxes and footprint modeling predictions over a homogeneous canopy of intermediate roughness. Agric Forest Meteorol 117:145–158
Lee X (2003) Fetch and footprint of turbulent fluxes over vegetative stands with elevated sources. Bound-Layer Meteorol 107:561–579
Lee X (2004) A model for scalar advection inside canopies and application to footprint investigation. Agric Forest Meteorol 127:131–141
Legg BJ, Raupach MR (1982) Markov-chain simulation of particle dispersion in inhomogeneous flows: the mean drift velocity induced by a gradient in Eulerian velocity variance. Bound-Layer Meteorol 24:3–13
Legg BJ, Raupach MR, Coppin PA (1986) Experiments on scalar dispersion within a model plant canopy, Part III: an elevated line source. Bound-Layer Meteorol 35:277–302
Ley AJ, Thomson DJ (1983) A random walk model of dispersion in the diabatic surface layer. Quart J Roy Meteorol Soc 109:867–880
Lin JC, Gerbig C, Wofsy SC, Andrews AE, Daube BC, Grainger CA, Stephens BB, Bakwin PS, Hollinger DY (2004) Measuring fluxes of trace gases at regional scales by Lagrangian observations: application to the CO_2 budget and rectification airborne (COBRA) study. J Geophys Res 109:D15304
MacDonald RW, Griffiths RF, Hall DJ (1998) An improved method for the estimation of surface roughness of obstacle arrays. Atmos Environ 32:1857–1864
Markkanen T, Rannik Ü, Marcolla B, Cescatti A, Vesala T (2003) Footprints and fetches for fluxes over forest canopies with varying structure and density. Bound-Layer Meteorol 106:437–459
Markkanen T, Steinfeld G, Kljun N, Raasch S, Foken T (2009) Comparison of conventional Lagrangian stochastic footprint models against LES driven footprint estimates. Atmos Chem Phys 9:5575–5586
Markkanen T, Steinfeld G, Kljun N, Raasch S, Foken T (2010) A numerical case study on footprint model performance under inhomogeneous flow conditions. Meteorol Z 19:539–547
Mathieu N, Strachan IB, Leclerc MJ, Karipot A, Pattey E (2005) Role of low-level jets and boundary-layer properties on the NBL budget technique. Agric Forest Meteorol 135:35–43
Moriwaki R, Kanda M (2004) Seasonal and diurnal fluxes of radiation, heat, water vapor, and carbon dioxide over a suburban area. J Appl Meteorol 43:1700–1710

Morse AP, Gardiner BA, Marshall BJ (2002) Mechanisms controlling turbulence development across a forest edge. Bound-Layer Meteorol 103:227–251
Prabha T, Leclerc MY, Baldocchi D (2008) Comparison of in-canopy flux footprints between Large-Eddy Simulation and the Langrangian simulation. J Appl Meteorol Climatol 47:2115–2128
Raasch S, Schröter M (2001) PALM—A Large-Eddy simulation model performing on massively parallel computers. Meteorol Z 10:363–372
Rannik Ü, Aubinet M, Kurbanmuradov O, Sabelfeld KK, Markkanen T, Vesala T (2000) Footprint analysis for measurements over heterogeneous forest. Bound-Layer Meteorol 97:137–166
Rannik Ü, Markkanen T, Raittila T, Hari P, Vesala T (2003) Turbulence statistics inside and above forest: influence on footprint prediction. Bound-Layer Meteorol 109:163–189
Raupach MR (1988) Canopy transport processes. In: Steffen WL, Denmead OT (eds) Flow and transport in the natural environment: advances and applications. Springer, Heidelberg, pp 95–127
Raupach MR (1989) A practical Lagrangian method for relating scalar concentrations to source distributions in vegetation canopies. Quart J Roy Meteorol Soc 115:609–632
Rinne J, Taipale R, Markkanen T, Ruuskanen TM, Hellén H, Kajos MK, Vesala T, Kulmala M (2007) Hydrocarbon fluxes above a Scots pine forest canopy: measurements and modeling. Atmos Chem Phys 7:3361–3372
Sarkar U, Hobbs SE (2003) Landfill odour: assessment of emissions by the flux footprint method. Environ Model Softw 18:155–163
Schmid HP, Cleugh HA, Grimmond CSB, Oke TR (1991) Spatial variability of energy fluxes in suburban terrain. Bound-Layer Meteorol 54:249–276
Schmid HP (1994) Source areas for scalars and scalar fluxes. Bound-Layer Meteorol 67:293–318
Schmid HP (1997) Experimental design for flux measurements: matching scales of observations and fluxes. Agric Forest Meteorol 87:179–200
Shaw RH, den Hartog G, Neumann HH (1988) Influence of foliar density and thermal stability on profiles of Reynolds stress and turbulence intensity in a deciduous forest. Bound-Layer Meteorol 45:391–409
Shaw RH, Patton EG (2003) Canopy element influences on resolved—and subgrid-scale energy within a Large-Eddy simulation. Agric Forest Meteorol 115:5–17
Smolander S, Stenberg P (2005) Simple parameterizations of the radiation budget of uniform broadleaved and coniferous canopies. Remote Sens Environ 94:355–363
Sogachev A, Rannik U, Vesala T (2004) Flux footprints over complex terrain covered by heterogeneous forest. Agric Forest Meteorol 127:143–158
Sogachev A, Lloyd J (2004) Using a one-and-a-half order closure model of atmospheric boundary layer for surface flux footprint estimation. Bound-Layer Meteorol 112:467–502
Sogachev A, Leclerc MJ, Karipot A, Zhang G, Vesala T (2005) Effect of clearcuts on footprints and flux measurements above a forest canopy. Agric Forest Meteorol 133:182–196
Sogachev A, Leclerc MY, Zhang G, Rannik U, Vesala T (2008) CO_2 fluxes near a forest edge: a numerical study. Ecol Appl 18:1454–1469
Sogachev A, Leclerc MY (2011) On concentration footprints for a tall tower in the presence of a nocturnal low-level jet. Agric Forest Meteorol 151:755–764
Steinfeld G, Raasch S, Markkanen T (2008) Footprints in homogeneously and heterogeneously driven boundary layers derived from a Langrangian stochastic particle model embedded into Large-Eddy simulation. Bound-Layer Meteorol 129:225–248
Strong C, Fuentes JD, Baldocchi D (2004) Reactive hydrocarbon flux footprints during canopy senescence. Agric Forest Meteorol 127:159–173
Thomson DJ (1987) Criteria for the selection of stochastic models of particle trajectories in turbulent flows. J Fluid Mech 189:529–556
Vermeulen AT, van Loon M, Builtjes PJH, Erisman JW (2001) Inverse transport modelling of non-CO_2 greenhouse gas emissions of Europe. In: Schiermeier G (ed) Air pollution modeling and its applications XIV. Kluwer, New York, pp 631–640

Vermeulen AT (2007) CHIOTTO, final report. ECN—Energy research Centre of the Netherlands, Petten, ECN-E–07-052, p 116

Vesala T, Huotari J, Rannik Ü, Suni T, Smolander S, Sogachev A, Launiainen S, Ojala A (2006) Eddy covariance measurements of carbon exchange and latent and sensible heat fluxes over a boreal lake for a full open-water period. J Geophys Res 111:D11101

Vesala T et al (2008) Surface–atmosphere interactions over complex urban terrain in Helsinki. Finland Tellus B 60:188–199

Wang W, Davis KJ (2008) A numerical study of the influence of a clearcut on eddy-covariance fluxes of CO2 measured above a forest. Agric Forest Meteorol 148:1488–1500

Wang WG, Davis KJ, Cook BD, Butler MP and Ricciuto DM (2006) Decomposing CO2 fluxes measured over a mixed ecosystem at a tall tower and extending to a region: a case study. J Geophys Res 111:G02005.02001–G02005.02014

Weil JC (1989) Stochastic modeling of dispersion in the convective boundary layer. In: Van Dop H (ed) Air pollution modelling and its applications VII. Plenum, New York, pp 437–449

Weil JC and Horst TW (1992) Footprint estimates for atmospheric flux measurements in the convective boundary layer. In: Schwartz SE and Slinn WGN (eds), Precipitation Scavenging and Atmosphere-Surface Exchange, vol 2. Hemisphere Publishing, pp 717–728

Willis GE, Deardorff JW (1976) A laboratory model of diffusion into the convective planetary boundary. Quart J Roy Meteorol Soc 102:427–445

Zhang G, Leclerc M, Karipot A, Duarte H, Mursch-Radlgruber E, Gholz H (2011) The impact of logging on the surrounding flow in a managed forest. Theor Appl Climat 106:511–521

Chapter 5
Model Validation

A rigorous validation of footprint models is an issue of importance that cannot be overstated. This is so to make sure that their practical applications can be successful. Due to the relative limited amount of robust flux footprint tracer experimental data, emerging models are often compared with other models. Already, the Schuepp et al. (1990) model has been the object of multiple cross-comparisons against airborne flux observations or against the Lagrangian simulations either in its original publication or in a companion paper by Leclerc and Thurtell (1990); the early analytical model proposed by Schuepp et al. (1990) made the object of numerous intercomparisons against other footprint models (Leclerc and Thurtell 1990; Horst and Weil 1992; Schmid 1994; Leclerc et al. 1997; Sogachev et al. 2005). 'Benchmark' Lagrangian model by Leclerc and Thurtell (1990) was previously validated against three earlier artificial tracer flux footprint experiments ranging from the simplest outside the roughness layer of smooth homogeneous surfaces to the more difficult rough and within the roughness layer of non-homogeneous rough forest canopies (Finn et al. 1996; Leclerc et al. 2003a; Leclerc et al. 2003b).

There are only a few models which can be selected as reference model. Support for model validations comes to us via the use of LES-based footprint models: such a model reference is invaluable given the wide range of applications of these models to a broad range of atmospheric boundary-layer conditions and over inhomogeneous surfaces. Their application close to the surface is limited due to the validity of the sub-grid scale parameterization. Markkanen et al. (2009) applied such an approach with an LES model deemed to be an "etalon". In addition to the model intercomparisons with the experimental validation have a high priority. Unfortunately, tracer studies are complex and laborious to realize. Sulfur

M. Y. Leclerc and T. Foken, *Footprints in Micrometeorology and Ecology*,
DOI: 10.1007/978-3-642-54545-0_5, © Springer-Verlag Berlin Heidelberg 2014

Table 5.1 Comparison of different footprint models against tracers and other models. For models description see Chap. 3 and Table 1.4

Comparison study	Model to compare	Reference tracer/model
Finn et al. (1996)	Leclerc and Thurtell (1990), Horst and Weil (1992, 1994)	SF_6
Leclerc et al. (1997)	Leclerc and Thurtell (1990), Horst and Weil (1992, 1994)	SF_6
Rannik et al. (2000)	Rannik et al. (2000)	Thomson (1987), Kurbanmuradov and Sabelfeld (2000)
Leclerc et al. (2003b)	Horst and Weil (1992, 1994), Leclerc et al. (1997)	SF_6
Kljun et al. (2003)	Kljun et al. (2002)	Kormann and Meixner (2001)
Kljun et al. (2004)	Kljun et al. (2002)	Wind tunnel
Sogachev et al. (2005)	Sogachev et al. (2002, 2004)	Thomson (1987), Kurbanmuradov and Sabelfeld (2000), Schuepp et al. (1990), Kormann and Meixner (2001)
Steinfeld et al. (2008)	Steinfeld et al. (2008)	Leclerc et al. (1997)
Markkanen et al. (2009)	Rannik et al. (2000; 2003), Kljun et al. (2002)	Steinfeld et al. (2008)
Leclerc et al. (unpublished)	Moeng and Sullivan (1994)	Perfluorocarbons (PFTs)

hexafluoride (SF_6), the traditional tracer, is no longer accepted given its high global warming potential. Other tracers such as perfluorocarbons (PFTs) offer an interesting alternative and only mean concentrations can be measured with the current detection methods; in addition, its technology is limited to a few select laboratories. It is in that perspective that Foken and Leclerc (2004) proposed a validation against natural tracers which using different source areas in different field sites as a proxy. The three methods will be discussed in this chapter. We have limited our discussion here to only such papers which have a high impact in this field (Table 5.1), leaving out studies focused mostly intercomparisons between models according to topic.

For a quantitative comparison, the crosswind averaged 1D footprint functions of the concentration or the flux are compared. The relevant quantitative parameters are the position of the peak, the level of the footprint function of the peak and the extension of the footprint. The latter means that the integral of 50 or 90 % of concentration or flux footprint (effect levels) are also assessed. Due to random uncertainties in the Lagrangian models, the extension of the footprint makes only sense for a limited number of particles released or a defined effect level. More often, the location and the footprint in the maximum (peak) of the footprint function were intercompared. A quantitative comparison of locations and footprint functions for a 2D footprint was recently presented by Markkanen et al. (2009, cf. Sect. 5.4).

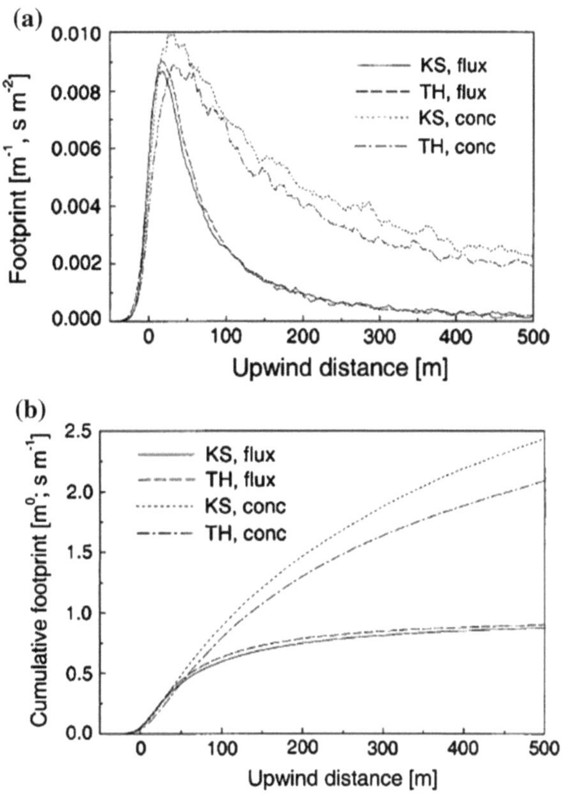

Fig. 5.1 The crosswind integrated **a** footprint function, and **b** its cumulative value, estimated by applying the Lagrangian models of Kurbanmuradov and Sabelfeld (2000, KS) and Thomson (1987, TH), for the observation level 15 m, roughness length 1.5 m, and neutrally stratified flow in the atmospheric surface layer according to Rannik et al. (2000)

5.1 Model Validation Against Other Models

The validation of models against other models is also a very important task. Most of the Lagrangian models based on the well-mixed assumption by Thomson (1987). Therefore, the Kurbanmuradov and Sabelfeld (2000) model, the basis of other models mainly those of Rannik et al. (2000, 2003), was tested against this model. Figure 5.1 shows a good agreement between both models except for the concentration footprint for large upwind distances. This figure also illustrates that the flux footprint is much shorter than the concentration footprint and very well reproduced by both models. The reason is that, close to the measuring point, the trajectories cross the observation level upwards with a positive flux, while for longer distances trajectories moving also downwards leading to a negative contribution to the flux (Rannik et al. 2000).

The comparison against other models was also used to test the model physics. For instance, Rannik et al. (2000) have tested their Lagrangian stochastic simulation with and without along-wind diffusion against the analytical models by Schuepp et al. (1990) and Horst and Weil (1992, 1994). Figure 5.2 shows the

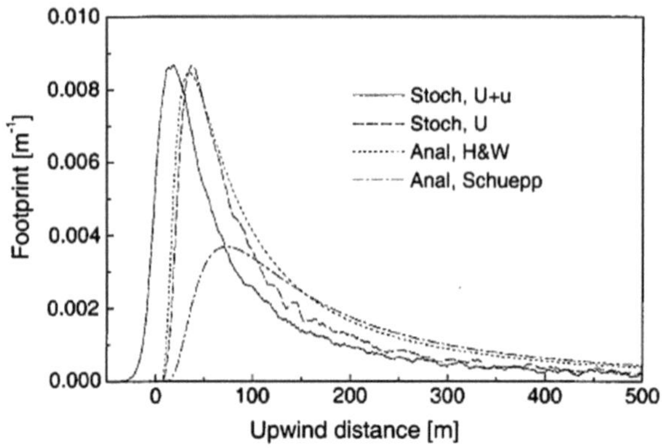

Fig. 5.2 Crosswind integrated flux footprint estimated by the analytical models by Schuepp et al. (1990) and Horst and Weil (1992, 1994, H&W) and stochastic simulation with (U + u) and without (U) along-wind diffusion for the observation level 15 m, roughness length 1.5 m, and neutrally stratified flow in the atmospheric surface layer according to Rannik et al. (2000)

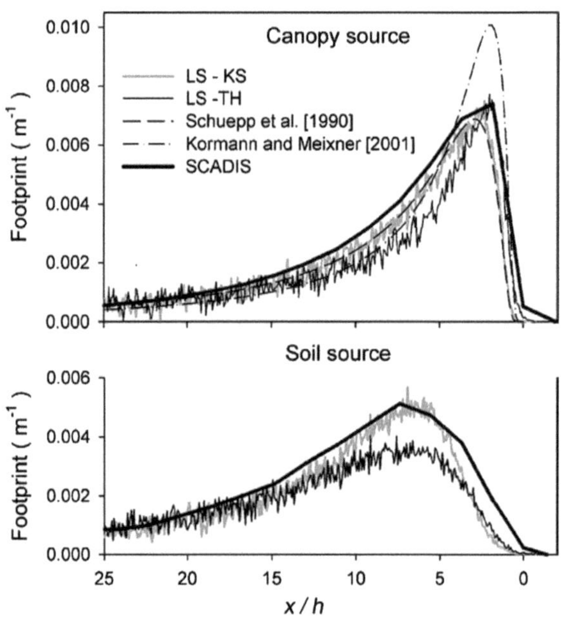

Fig. 5.3 Test of the SCADIS flux footprint model (Sogachev et al. 2002; Sogachev and Lloyd 2004) with footprints derived from both analytical (Schuepp et al. 1990; Kormann and Meixner 2001) and Lagrangian stochastic (Kurbanmuradov and Sabelfeld 2000, LS-KS) and (Thomson 1987, LS-TH) approaches in neutral conditions over a tall homogeneous managed forest ($z = 1.4h_c$) after Sogachev et al. (2005). The distance is normalized by the canopy height x h_c^{-1}, $u_* = 0.46$ m s^{-1}, $d = 9.49$ m, and $z_0 = 1.31$ m, Published with kind permission of © Elsevier, 2005. All Rights Reserved

importance of the along-wind diffusion for a case where the standard deviation of the horizontal wind velocity is of the order of the wind velocity itself, which is the case close to the canopy. Otherwise, the horizontal mean wind velocity is generally much larger than its standard deviation.

The higher-order closure model SCADIS (Sogachev et al. 2002; Sogachev and Lloyd 2004) was compared by Sogachev et al. (2005) with footprints derived from both analytical (Schuepp et al. 1990; Kormann and Meixner 2001) and Lagrangian stochastic (Thomson 1987; Kurbanmuradov and Sabelfeld 2000) approaches in neutral conditions over a tall homogeneous managed pine forest plantation in Florida at a measurement level of 1.4 times of the canopy height. The model footprints exhibit values close to Lagrangian stochastic model results (Fig. 5.3).

Steinfeld et al. (2008) evaluated their Lagrangian simulation (LS) model embedded into an large-eddy simulation code against the work of Leclerc et al. (1997) and found a general agreement when the LS had a subgrid-scale embedded in the turbulence of the LS. They also found that the footprint peak in their model broadly agreed with the results of Leclerc et al. (1997) and the measurements of Finn et al. (1996) with differences in the footprint peak position to be slightly more upstream to the sensor position than the modeled peaks of Leclerc et al. (1997). Steinfeld et al. (2008) attributed this to be possibly because the Leclerc et al. (1997) study did not include the streamwise diffusion on the LS and had a lower resolution. The Steinfeld et al. (2008) model requires that sub-grid scale turbulence be included in the LS for optimum results, as can be seen below in Fig. 5.4. Also Wang and Rotach (2010) compared their LS model with undulating surface against the models by Leclerc et al. (1997) and Steinfeld et al. (2008).

The Steinfeld et al. (2008) study also found that, in neglecting the subgrid-scale parameterisation scheme of turbulent kinetic energy in the embedded Lagrangian simulation model leads, even with the finest resolution, to an underestimation of contributions from near-sensor sources as shown more precise in the cumulative footprint (Fig. 5.5).

Since there is a large demand for results of footprint models in 2D, Markkanen et al. (2009) presented a footprint model validation against the LES which was used as a reference standard (etalon) in a study to evaluate 2D Lagrangian footprints, for observation heights extending throughout the depth of the entire atmospheric boundary layer. As their standard, they used the LES model PALM (Raasch and Schröter 2001) to simulate trajectories of a large number of particles simultaneously with general flow field calculations (Steinfeld et al. 2008). From this data, the footprints are determined in a manner similar to that used in conventional forward Lagrangian models. Markkanen et al. (2009) compared against this Lagrangian footprint model embedded into an LES model the Lagrangian backward simulation footprint model LPDM-B (Kljun et al. 2002) for backward simulations (BW) and the Lagrangian forward simulation model by Rannik et al. (2000, 2003) for forward simulations.

Using the LS forward and backward modes, the models (Rannik et al. 2000; Kljun et al. 2002) examined the sensitivity of the footprint peak with height in the

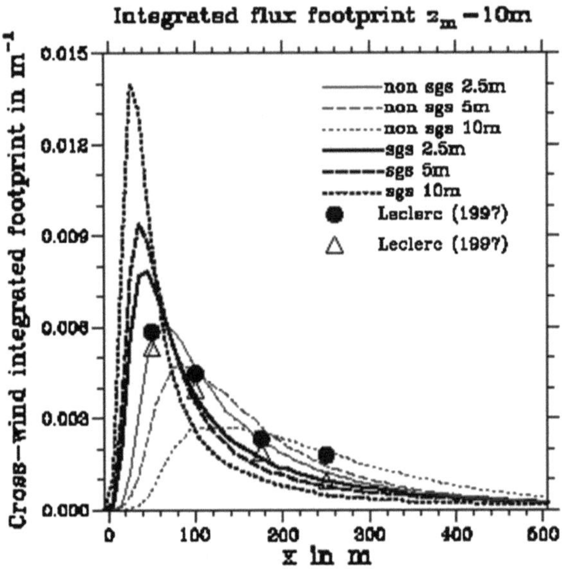

Fig. 5.4 Cross-wind integrated flux footprint for a convective boundary layer similar to that described in Leclerc et al. (1997) for a measurement height of 10 m derived from the six LES runs differing in the application of a subgrid-scale parameterisation scheme (sgs) in the Lagrangian simulation part and in the grid spacing used. For a comparison also the corresponding results derived by Leclerc et al. (1997) from data of the field experiment described in Finn et al. (1996) are shown. (Steinfeld et al. 2008)

Fig. 5.5 Cumulative cross-wind integrated flux footprint for the same case as shown in Fig. 5.4. (Steinfeld et al. 2008)

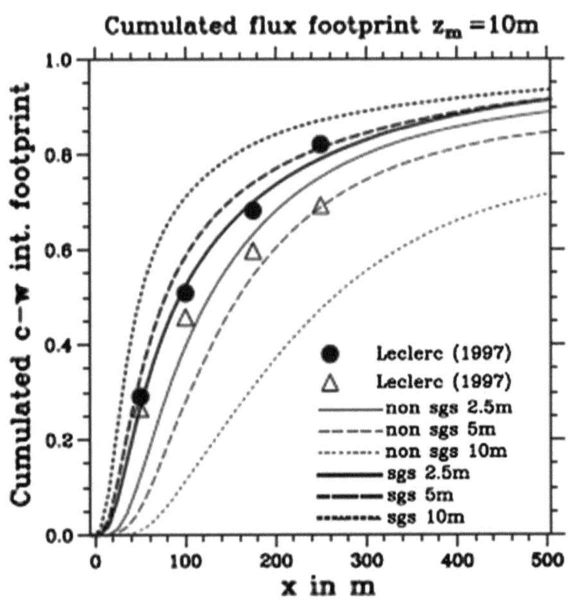

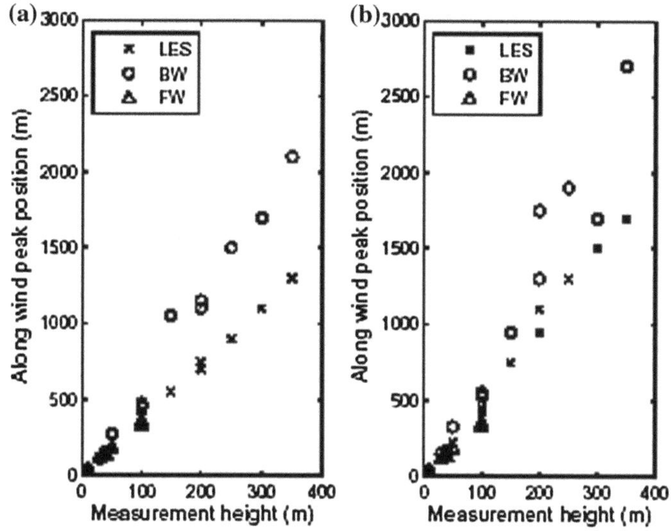

Fig. 5.6 Along-wind peak position of the flux footprint as a function of measurement height for the LES (*crosses*), backward (BW, *circles*) and forward (FW, *triangles*) LS models **a** in the case 1 (*convective*) and **b** in the case 2 (*less unstable*). Results shown only for selected grid resolutions (Markkanen et al. 2009, Published with kind permission of © Copernicus Publications, distributed under the Creative Commons Attribution 3.0 License, 2009. All Rights Reserved)

atmospheric boundary layer, using the LES as their benchmark; they noted that the LS in the backward model's footprint peak departs substantially from their LES counterpart with heights well into the conventional boundary layer. The forward mode, as currently formulated, is a surface layer model so the LS model's peak does not go beyond the top of the surface layer as seen in Fig. 5.6.

5.2 Model Validation and Comparison Against Experimental Data

While much of the efforts related to advance the subject of footprint in a variety of flow over various surfaces i.e. at forest edges, over inhomogeneous surfaces or in complex non-flat terrain have been advancing rapidly, there has been a need to validate the hierarchy of models. The first such study was done by Finn et al. (1996) who conducted a tracer study over a 1–1.5 m tall sagebrush canopy to validate two footprint models. In that experiment, Finn et al. (1996) released sulfur hexafluoride as a passive tracer, and measured the eddy-covariance tracer flux using high-frequency continuous tracer analyzers co-located with sonic anemometers at several distances from a line source and subsequently determined footprint predictions. The tracer flux measurements were made well outside the roughness sub-layer. The measurements were compared against the analytical solution of Horst and Weil (1992, 1994) and the Lagrangian simulation of Leclerc and Thurtell (1990).

Fig. 5.7 Tracer experiment over a peach orchard (*Photograph* by Leclerc)

In a tracer experiment over a peach orchard (Fig. 5.7), Leclerc et al. (2003b) tested several models in the layer outside the roughness sub-layer at the beginning of the summer, and then, due to the rapid orchard growth during summer, for data which were collected within the roughness sub-layer. Both analytical solutions and the Lagrangian simulations of footprints mentioned above tested performed well both within and beyond the roughness sub-layer. In an experiment above a very rough tall managed pine forest plantation, Leclerc et al. (2003a) and Zhang et al. (2010), however, documented that, contributions well outside the footprint envelope contaminate the integrity of flux measurements and that successful footprint modeling applied as long as there are no sharp contrasting temperature differences between the surface of interest (i.e. the pine forest) and the surrounding (the large recently logged swath of land). In these studies, when the wind came from a clearcut located hundreds of meters more than five hundred meters outside the footprint, the tracer fluxes were found to be larger than the modeled footprint fluxes by up to 300 %. Using simultaneously a sodar as a diagnostic tool, the authors noticed that this considerable flux enhancement at the flux tower was accompanied by persistent long lasting vertical motions and both a shift and acceleration of the horizontal flow components during these times. This unexpected flow was part of an organized small-scale circulation—not unlike the land-sea breeze effect—induced by the large fresh hot and bare soil of the clearcut, the latter's temperature (as per the satellite imagery to Landsat records) reaching as much as 20 K higher than the surrounding forest canopy during the day. This result is therefore a tribute to the fortuitous use of fast response continuous eddy-tracer flux measurements, a sodar system, and Landsat maps. Without the combined use of these diagnostic tools, these large errors in the CO_2 flux measurements would normally have gone unnoticed by typical one-point tower CO_2 flux-energy balance measurements alone. This result is of significance since most flux experimentalists work in real, natural, often forested terrain at sites that are less than ideal. The authors therefore recommend that, at many if not most sites, spatial observations of the three-dimensional component of the flow be made in concert with surface-atmosphere exchange point measurements.

5.3 Model Validation with Natural Tracers

The use of a natural tracer experiment over two adjoining surfaces consisting of crops of contrasting fluxes as a means to evaluate footprint models was highlighted earlier by Foken and Leclerc (2004). A schematic layout for a footprint comparison experiment with natural tracers is shown in Fig. 5.8. The fluxes of two contrasting surfaces are measured by single eddy-covariance flux systems. A third system measures the flux from a footprint area, which includes different percentage of the fluxes from the two surfaces depending on the atmospheric stability.

Göckede et al. (2005) used the proposed setup (Fig. 5.8) to validate footprint models using a natural tracer experiment consisting of two dissimilar adjoining surfaces (ploughed field and grassland) with respective fluxes. They tested the FSAM footprint model of Schmid (1997, 2002) and the Lagrangian forward simulation model by Rannik et al. (2000, 2003) and found that the Lagrangian simulation of footprints produce a better performance than the FSAM as shown in the Fig. 5.9 The models vary more dramatically in the near field with the LS performing better. This is due to the natural inclusion of the distance of the sources being included in the trajectories of the particles through the Markov process.

In practice, however, the lack of near-field in the FSAM for the purpose of experiments carried out only a few meters near the ground exerts a minor effect as Göckede et al. (2005) have shown. However, there exists many cases where experiments are carried out using flux systems at greater levels and stable conditions, and in this case the lack of inclusion of the near field in the FSAM could lead to dramatically important errors.

5.4 Classification of the Comparison Results

Most comparisons of footprint models evaluated the crosswind-integrated footprint (Fig. 5.10a). These footprint shapes can be similar despite the fact that the location of the 2D footprint can be very different (Fig. 5.10b). To overcome this deficit, Markkanen et al. (2009) sought to find a more objective way of quantifying similarities and differences. They selected two criteria for this comparison.

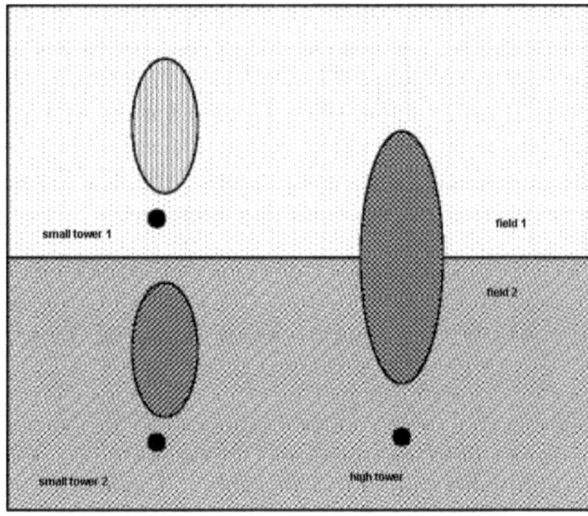

Fig. 5.8 Schematic layout for a footprint comparison experiment with natural tracers according to Foken and Leclerc (2004), Published with kind permission of © Elsevier, 2004. All Rights Reserved

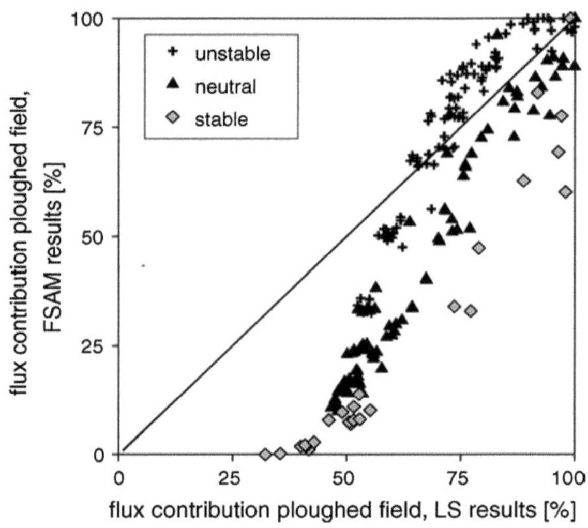

Fig. 5.9 Comparison of the footprint results of the analytic (FSAM, Schmid 1997, 2002) and Lagrangian (LS Rannik et al. 2000, 2003) footprint models for the percentage flux contribution of the ploughed field area at eddy-covariance measurement according to Göckede et al. (2005), Published with kind permission of © Elsevier, 2005. All Rights Reserved

Firstly, for both models, they evaluated the smallest areas contributing 10 (the smallest ellipse in the centre of the footprint), 20, 50 and 80 % (the largest ellipse at the outer border) to the footprints that is $\Omega_p = \Omega_{10}$, Ω_{20}, Ω_{50}, Ω_{80}, respectively. Then they determined the intersection of the two models i.e. both the reference and the validated models ($\Omega_p^{val} \cap \Omega_p^{ref}$) written as $\Omega_p\cap$. In order to compare the equality of predicted footprint functions, the signal predicted by both models originating from $\Omega_p\cap$ can be determined. When both these values are close to or in agreement with the target percentage, the two models agree perfectly. Secondly, of practical relevance, is also the equality of the size of the area of level P by the qualified

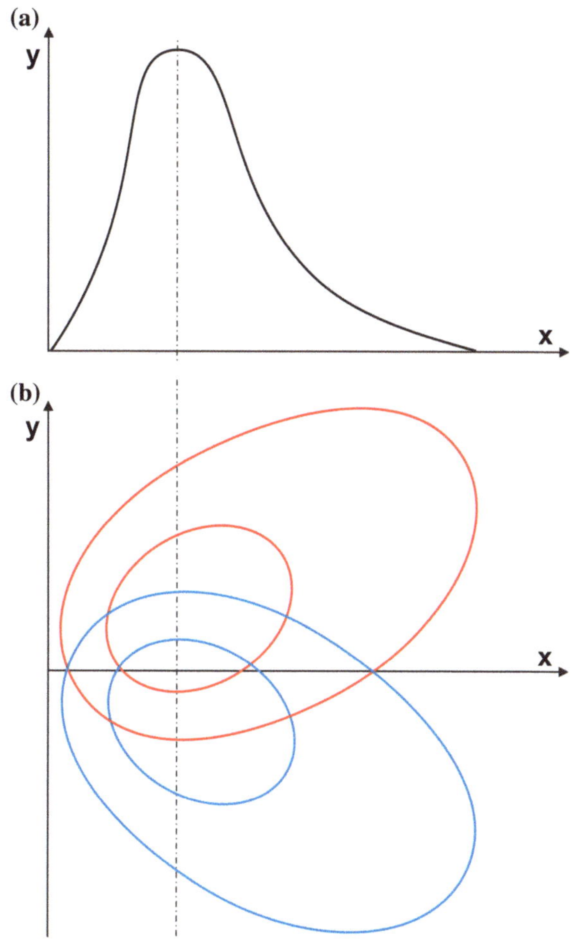

Fig. 5.10 In part **a** the crosswind integrated footprint is shown (1D), which is identical for both 2D footprints shown in part **b**. But the 2D footprints, which are identical in size, cover widely different areas

Table 5.2 Quality categories for footprint comparison according to Markkanen et al. (2009)

Quality comparison	Code	$\frac{\left\vert\Omega_p^{val}-\Omega_p^{ref}\right\vert}{\Omega_p^{ref}}$ size agreement (%)	$\frac{1-\Omega_p^{\cap}}{\Omega_p^{ref}}$ overlapping agreement (%)
High agreement	3	>60	>70
Moderate agreement	2	>40	>50
Low agreement	1	>20	>30
No agreement	0	<20	<30

model Ω_p^{val} and the reference model Ω_p^{ref}. Also, the footprint size has an influence on the first comparison. Therefore, only a combination of both investigations can provide a good measure for comparison. Accordingly, Markkanen et al. (2009) finally presented a classification of the level of model agreement with the

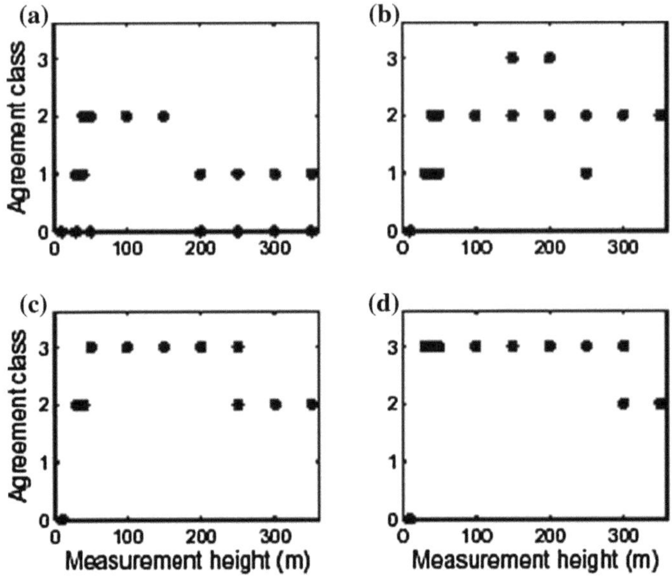

Fig. 5.11 Agreement classes for validation of the flux footprints predicted by the LES parameterization (Steinfeld et al. 2008) without Coriolis force against the parameterization including Coriolis force in the convective case $L = -32$ m. Validations results for (**a**) Ω_{10} (**b**) Ω_{20} (**c**) Ω_{50} and (**d**) Ω_{80}. are shown only for selected grid resolutions ($\Delta x > 0.4\ z_m$) according Markkanen et al. (2009). The small areas around the footprint peak Ω_{10} are often not at the same place and a low quality follows especially for large heights. In contrast the 80 % footprint Ω_{80} overlaps well and a very good quality of agreement is given in nearly all heights. For the classes of agreement see Table 5.2 Published with kind permission of © Copernicus Publications, distributed under the Creative Commons Attribution 3.0 License, 2009. All Rights Reserved

reference. The classification is based both on agreement of sizes of the source areas and on the degree of their overlapping. The size agreement between the examined model and the reference is given as follows:

$$\frac{\left|\Omega_p^{val} - \Omega_p^{ref}\right|}{\Omega_p^{ref}} \tag{5.1}$$

and the degree of overlapping as follows:

$$\frac{1 - \Omega_p^{\cap}}{\Omega_p^{ref}} \tag{5.2}$$

The final agreement class ranging from 0 to 3 (no agreement to good agreement) is consequently determined according to the decision shown in Table 5.2. This method of classification was principally adopted from Rebmann et al. (2005)

and in the updated version by Göckede et al. (2008) for footprint applications for FLUXNET stations. The latter developed a scheme to combine footprints with the land cover data and with the quality check of turbulent fluxes (Foken et al. 2004), for details see Sect. 7.2.4.

An example of the application of this method is given in Fig. 5.11. While for Ω_{80} the agreement of the LES model (Steinfeld et al. 2008) with and without the Coriolis force is very good (high agreement), it was adequate for lower effect levels but low for low and high measuring heights. For the smallest effect level Ω_{10}, the agreement in measuring heights of 100–200 m only class 2. Here, the influence of the first test is more relevant.

References

Finn D, Lamb B, Leclerc MY, Horst TW (1996) Experimental evaluation of analytical and Lagrangian surface-layer flux footprint models. Boundary-Layer Meteorol 80:283–308

Foken T, Göckede M, Mauder M, Mahrt L, Amiro BD, Munger JW (2004) Post-field data quality control. In: Lee X et al (eds) Handbook of micrometeorology: A guide for surface flux measurement and analysis. Kluwer, Dordrecht, pp 181–208

Foken T, Leclerc MY (2004) Methods and limitations in validation of footprint models. Agric Forest Meteorol 127:223–234

Göckede M, Markkaken T, Mauder M, Arnold K, Leps JP, Foken T (2005) Validation of footprint models using natural tracer measurements from a field experiment. Agric Forest Meteorol 135:314–325

Göckede M et al (2008) Quality control of CarboEurope flux data—Part 1: Coupling footprint analyses with flux data quality assessment to evaluate sites in forest ecosystems. Biogeosci 5:433–450

Horst TW, Weil JC (1992) Footprint estimation for scalar flux measurements in the atmospheric surface layer. Boundary-Layer Meteorol 59:279–296

Horst TW, Weil JC (1994) How far is far enough?: the fetch requirements for micrometeorological measurement of surface fluxes. J Atm Oceanic Techn 11:1018–1025

Kljun N, Rotach MW, Schmid HP (2002) A three-dimensional backward Lagrangian footprint model for a wide range of boundary layer stratification. Boundary-Layer Meteorol 103:205–226

Kljun N, Kormann R, Rotach M, Meixner FX (2003) Comparison of the Lagrangian footprint model LPDM-B with an analytical footprint model. Boundary-Layer Meteorol 106:349–355

Kljun N, Kastner-Klein P, Federovich E, Rotach MW (2004) Evaluation of Lagrangian footprint model using data from wind tunnel convective boundary layer. Agric Forest Meteorol 127:189–201

Kormann R, Meixner FX (2001) An analytical footprint model for non-neutral stratification. Boundary-Layer Meteorol 99:207–224

Kurbanmuradov O, Sabelfeld KK (2000) Lagrangian stochastic models for turbulent dispersion in atmospheric boundary layers. Boundary-Layer Meteorol 97:191–218

Leclerc MY, Thurtell GW (1990) Footprint prediction of scalar fluxes using a Markovian analysis. Boundary-Layer Meteorol 52:247–258

Leclerc MY, Shen S, Lamb B (1997) Observations and large-eddy simulation modeling of footprints in the lower convective boundary layer. J Geophys Res 102(D8):9323–9334

Leclerc MY, Karipot A, Prabha T, Allwine G, Lamb B, Gholz HL (2003a) Impact of non-local advection on flux footprints over a tall forest canopy: a tracer flux experiment (Special issue: Advances in micrometeorology: Tribute to G. W. Thurtell). Agric Forest Meteorol 115:19–30

Leclerc MY, Meskhidze N, Finn D (2003b) Comparison between measured tracer fluxes and footprint modeling predictions over a homogeneous canopy of intermediate roughness. Agric Forest Meteorol 117:145–158
Markkanen T, Steinfeld G, Kljun N, Raasch S, Foken T (2009) Comparison of conventional Lagrangian stochastic footprint models against LES driven footprint estimates. Atmos Chem Phys 9:5575–5586
Moeng CH, Sullivan P (1994) A comparison of shear- and buoyancy-driven planetary boundary layer flows. J Atmos Sci 51:999–1022
Raasch S, Schröter M (2001) PALM—A large-eddy simulation model performing on massively parallel computers. Meteorol Z 10:363–372
Rannik Ü, Aubinet M, Kurbanmuradov O, Sabelfeld KK, Markkanen T, Vesala T (2000) Footprint analysis for measurements over heterogeneous forest. Boundary-Layer Meteorol 97:137–166
Rannik Ü, Markkanen T, Raittila T, Hari P, Vesala T (2003) Turbulence statistics inside and above forest: Influence on footprint prediction. Boundary-Layer Meteorol 109:163–189
Rebmann C et al (2005) Quality analysis applied on eddy covariance measurements at complex forest sites using footprint modelling. Theor Appl Climat 80:121–141
Schmid HP (1994) Source areas for scalars and scalar fluxes. Boundary-Layer Meteorol 67:293–318
Schmid HP (1997) Experimental design for flux measurements: matching scales of observations and fluxes. Agric Forest Meteorol 87:179–200
Schmid HP (2002) Footprint modeling for vegetation atmosphere exchange studies: A review and perspective. Agric Forest Meteorol 113:159–184
Schuepp PH, Leclerc MY, MacPherson JI, Desjardins RL (1990) Footprint prediction of scalar fluxes from analytical solutions of the diffusion equation. Boundary-Layer Meteorol 50:355–373
Sogachev A, Menzhulin G, Heimann M, Lloyd J (2002) A simple three dimensional canopy-planetray boundary layer simulation model for scalar concentrations and fluxes. Tellus 54B:784–819
Sogachev A, Lloyd J (2004) Using a one-and-a-half order closure model of atmospheric boundary layer for surface flux footprint estimation. Boundary-Layer Meteorol 112:467–502
Sogachev A, Leclerc MJ, Karipot A, Zhang G, Vesala T (2005) Effect of clearcuts on footprints and flux measurements above a forest canopy. Agric Forest Meteorol 133:182–196
Steinfeld G, Raasch S, Markkanen T (2008) Footprints in homogeneously and heterogeneously driven boundary layers derived from a Langrangian stochastic particle model embedded into large-eddy simulation. Boundary-Layer Meteorol 129:225–248
Thomson DJ (1987) Criteria for the selection of stochastic models of particle trajectories in turbulent flows. J Fluid Mech 189:529–556
Wang W, Rotach M (2010) Flux Footprints Over an Undulating Surface. Boundary-Layer Meteorol 136:325–340
Zhang G, Leclerc MY, Karipot A (2010) Local flux-profile relationships of wind speed and temperature in a canopy layer in atmospheric stable conditions. Biogeosciences 7:3625–3636

Chapter 6
Land Surface: Coupled Footprints

Coupling footprint models to datasets characterizing surface properties is a most useful endeavour when surface-exchange information is sought (fluxes or parameters characterizing a surface of interest). Land-cover maps are particularly helpful in that regard to identify the various land-cover types contributing to the footprint area. Footprint models require information on characteristics of the underlying surface for surface-related properties needed in footprint calculations such as roughness length. Often, area-averaging methods are necessary to determine the input parameters for the model (see Sect. 2.4). In this chapter, principles underpinning the coupling of footprint models are described. The application of the described methods is given in Chap. 8.

6.1 Grid Schema of Surface Characteristics

Coupling footprint model runs with data describing surface characteristics around a measurement site requires dividing the surrounding landscape into discrete matrices (Fig. 6.1). In these matrices, each grid cell contains mean attributes of the area it represents, such as an integer ID that indicates most land cover classes, or values for e.g. averaged roughness length or stand height. The required spatially explicit information can best be prepared using typical Geographical Information System (GIS) software. Alternatively, the matrices can be produced using maps describing e.g. the land-cover structure in the domain.

The dimensions of the model domain must be sufficiently large to enclose large footprint areas in stable night time conditions, while the grid resolution has to be sufficiently high to yield plausible results in convective, unstable conditions with associated small footprints. The local wind climatology should be taken into

M. Y. Leclerc and T. Foken, *Footprints in Micrometeorology and Ecology*,
DOI: 10.1007/978-3-642-54545-0_6, © Springer-Verlag Berlin Heidelberg 2014

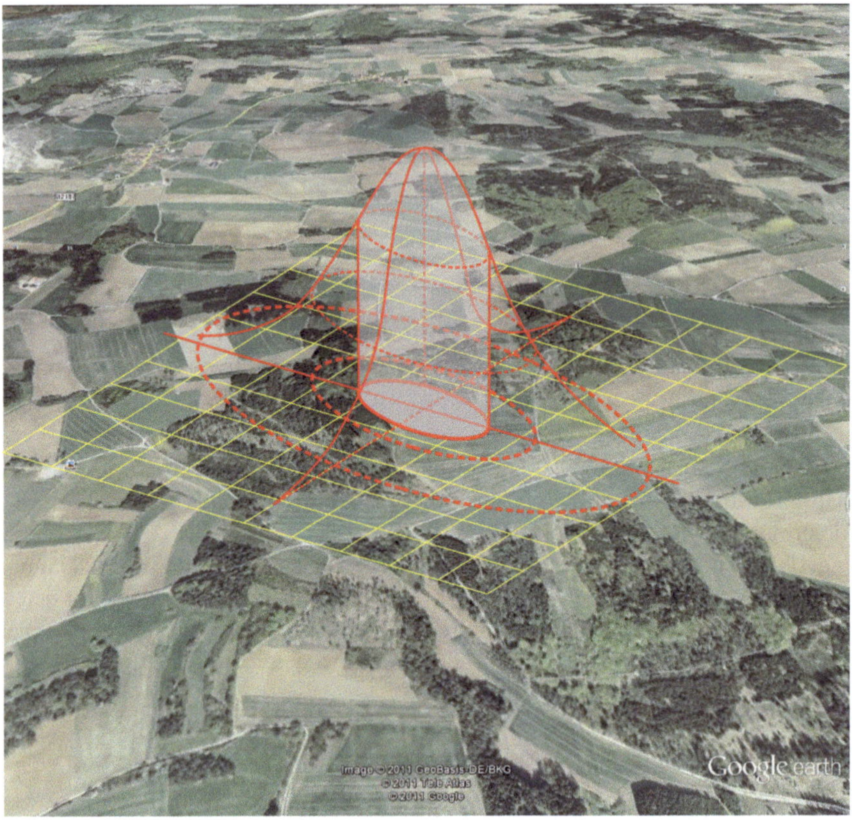

Fig. 6.1 Map with grid elements covering the possible footprint area and the effect levels in up-wind direction from the measuring point at the right side (Published with kind permission of © 2011 GeoBasis DE/BKG © 2011 Tele Atlas © 2011 Google. All Rights Reserved)

consideration to reduce processing time in both preparation steps—the creation of land use maps and footprint simulations. Final settings should be tested with preliminary model runs based on different input parameters. It is generally recommended to reduce the size of the grid elements, since higher resolution enables a realistic representation of smaller scale heterogeneities throughout the model domain, avoiding the application of complex averaging schemes.

Over heterogeneous terrain, land-cover information is important to control and investigate the influence of the underlying surface in the footprint of micrometeorological measurements. Regarding distinguished land cover classes, the scheme to be chosen needs to be customized for the specific objective of each footprint study. Many FLUXNET observation sites target to monitor carbon exchange processes for a specific land-cover type, e.g. mixed forest, so in the simplest case a land-cover map is required that differentiates this target (forest) from other areas (non-forest). This approach can be refined into arbitrary levels of details such as differentiating coniferous from deciduous forests, or dividing one forest type into

Table 6.1 List of land cover classes for the preparation of a land use matrix from topographical map information

Surface	Comment
Forest	If the forest is the target area, the major forest types should be accounted for deciduous, coniferous, mixed and given their own respective class. In addition, it is recommended to separate forest sections with different roughness characteristics (age classes, height)
Settlement	Rural settlements, buildings
Traffic areas	Roads
Water areas	Lakes, rivers
Grassland	Permanent grassland, pasture land
Agricultural areas	Crops of all kind. When possible, several classes should be identified according to their respective roughness and thermal characteristics

age classes. Since different land cover classes are often also associated with different roughness lengths which have an impact on the local flow characteristics and therefore on footprint computation, it is recommended to describe the surrounding landscape as accurately as possible. However, at the bare minimum, land-cover classes listed in Table 6.1 should be distinguished.

Since the size and position of the footprint area changes with wind direction, measurement height and atmospheric stability, these factors must be taken into account when setting up the domain for a specific footprint study. As a general guideline, suitable matrix sizes are $5 \times 5\ km^2$ for tower measurements over tall forests and $0.5 \times 0.5\ km^2$ for experiments (sensor height approximately 2–3 m) over agricultural areas. Concerning the matrix resolution, information should be provided for grid elements of $25 \times 25\ m^2$ or smaller, as typically provided by remote sensing data sources such as Landsat. Besides the benefit of representing small scale heterogeneities adequately, such high resolution matrices are necessary to project the source weight function onto the grid, particularly in the case of smaller footprints in unstable conditions.

When remote sensing data is available, the creation of large model domains with high resolution grids as outlined above should be easily achieved; however, such high-resolution settings are impractical in the case where matrices have to be produced from conventional data such as e.g. topographic maps. In addition, computational demand related to both processing time and memory requirements scale with the total number of grid cells in the model domain, thus an optimized domain setup can help increasing the efficiency of the data processing. Figure 6.2 provides a guideline on customizing the domain size and resolution, based on sensor height.

Concerning the dimensions of the area to be covered by the model domain, the parameter D_{min} defines the minimum fetch in each direction (Fig. 6.2). For example, if D_{min} has a value of 1 km, the resulting minimum-matrix would be a square of $2 \times 2\ km^2$, with the tower located at the center of the area. The second parameter D_{ext} defines an extended fetch distance into the main wind direction. The parameter L_R gives the required resolution for the given matrix size. The minimum settings for D_{min}, D_{ext}, and L_R for specific ranges of the effective

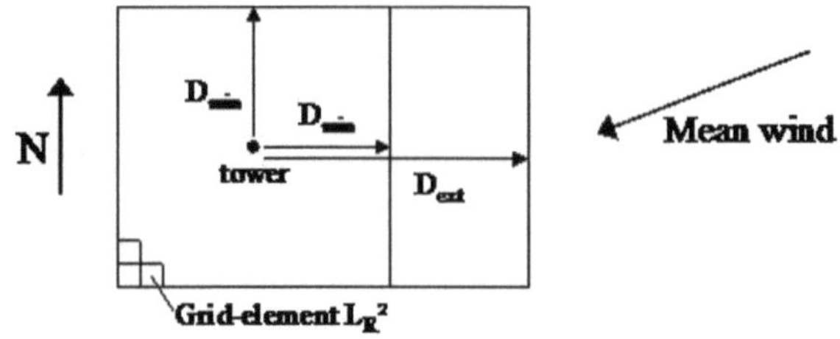

Fig. 6.2 Sketches of the concept of the matrix dimensions defined by parameters D_{min}, D_{ext} and L_R with a stream wisest direction of East-North-East

Table 6.2 Parameter selection for D_{min}, D_{ex}t, and L_R (see Fig. 6.2) for the effective measurement height $z_{m\text{-eff}}$

$z_{m\text{-eff}}$ (m)	D_{min} (m)	D_{ext} (m)	L_R (m)
<4	600	1,000	50
4–7	1,200	1,600	80
7–10	1,400	2,000	100
10–13	1,400	2,200	100
13–16	1,500	2,400	100
16–20	1,800	3,300	150
20–24	2,100	3,300	150
>24	2,400	3,600	150

measurement height $z_{m\text{-eff}}$, typically the height above zero-plane displacement, are given in Table 6.2. A map covering a larger domain, or with a higher resolution, will further enhance the accuracy of the footprint approach in the stable as well as in the convective case.

6.2 Determination of Surface Characteristics

The key surface properties to account for in eddy-covariance footprint computation are the land-cover types. That parameter is required to incorporate the contribution of the designated target vegetation to the footprint. Since land-cover types impact the roughness length, one can see the importance of such an input parameter in footprint models. It is thus essential to characterize different land-cover types with individual roughness length values (Sect. 6.2.1). Furthermore, a methodology to determine the land-cover structure either through topographic maps or remote sensing techniques is needed (Sect. 6.2.2). Finally, the determination of other area averaged input data for footprint models is discussed (Sect. 6.3).

6.2.1 Roughness Length

The quality and success of analyses aiming at identifying the contribution of upwind sources to a point area hinges to a large degree to the judicious choice of

Fig. 6.3 The typical landscape in Southwest England with rows of bushes or stones against wind erosion shows that not the roughness of the single fields but the combined roughness of the structured landscape with of fields and bushes creates the effective roughness of the landscape (*Photograph* by Foken)

an appropriate roughness length classification. This is most easily accomplished using an effective roughness length (Fiedler and Panofsky 1972) for a structured landscape (see also Sect. 2.4.1), including e.g. wake-producing obstacles such as hedges or lines of trees (see Fig. 6.3). However, such parameters are seldom available for surfaces that are being studied, leaving the researcher to determine a mean roughness length through aggregation for individual surface classes. The authors remind the reader to be aware that a simple parameter aggregation (see Sect. 2.4) can only be provided if the roughness lengths of the different surfaces are of similar orders of magnitude.

Roughness lengths of different surface types are available in most of the text-books with reference to a British standard (ESDU 1972). These data, together with other classifications, are given in Table 2.1. Another popular roughness length classification has been published in the European Wind Atlas (Troen and Peterson 1989), which was developed for wind energy applications and as such applies mainly to open terrain (see Table 2.15). The Wind Atlas classification distinguishes only between four general roughness classes and was developed for landscapes with high wind energy potential. Larger forested areas are assigned a roughness length value of 0.4 m—a very low estimation compared to other classification schemes and measured values—reflecting the fact that wind turbines are usually placed far away from large forests. The reader is thus advised to use different roughness length classifications spanning a greater range of z_0 values for landscapes dominated by forests. As a third commonly used classification scheme, the roughness length values proposed by Fiedler, cited in Hasager and Jensen

(1999), are based on micrometeorological field observations made in various land cover types within the region of the Upper Rhine Valley, Germany. The fourth classification scheme cited here (Wieringa 1992) compiles quality-proofed roughness length measurements from several hundred original publications. Finally, the last roughness length classification developed by Davenport et al. (2000) presents effective roughness lengths, assuming a more heterogeneous characteristic of the given land cover types as in the effective values presented by Fiedler and Panofsky (1972). Therefore, their values may be slightly larger than the presented in the other four schemes (Wieringa 1992).

According to a study by Reithmaier et al. (2006), it is critically important to ensure that the dominating land cover type in the area surrounding the tower be correctly classified. Therefore, for studies in predominantly forested areas, we recommend using the classification scheme of Wieringa (1992). In studies over a more heterogeneous landscape, a classification providing effective roughness lengths should be preferred (Davenport et al. 2000). A second criterion to take into account for the choice of a suitable roughness length classification is the footprint model itself. Some footprint models become numerically unstable for very high roughness length values, and there is usually a maximum threshold for the ratio of measuring height z_m and the roughness length z_0. For example, the analytic model by Schmid (1997) cannot be applied for ratios of $z_m/z_0 < 12$. Other models are less sensitive to this ratio, like e.g. the algorithm proposed by Kormann and Meixner (2001) that uses measured wind velocity and friction velocity instead of roughness length as input. Since the roughness length can be calculated from wind speed and friction velocity (e.g. Eq. 2.16), this approach is basically another version of calculating an effective roughness length.

The use of effective roughness lengths can only replace a full-scale flux aggregation scheme when the chosen effective roughness length is valid over the footprint area. Based on readily available classification schemes e.g. Davenport et al. (2000), this condition is seldom fulfilled. Furthermore, in the paper by Fiedler and Panofsky (1972), large scale aggregated values are given only for flat landscapes (0.42 m), landscapes with small hills (0.99 m), and for hilly regions (1.42 m). Therefore, an area averaging roughness lengths is generally necessary in footprint models requiring homogeneous surface properties. Flux aggregation tools such as the one proposed by Hasager and Jensen (1999) overcome this problem by averaging the friction instead of the roughness length itself. Note that the roughness length of a certain natural surface technically also depends on influence of the wind field on the surface; however, such effects should be negligible within this context.

6.2.2 *Remote-Sensing Data*

The remote-sensing approach is the method of choice to analyze and describe land cover structure at a field site. Remote sensing databases have first been used for

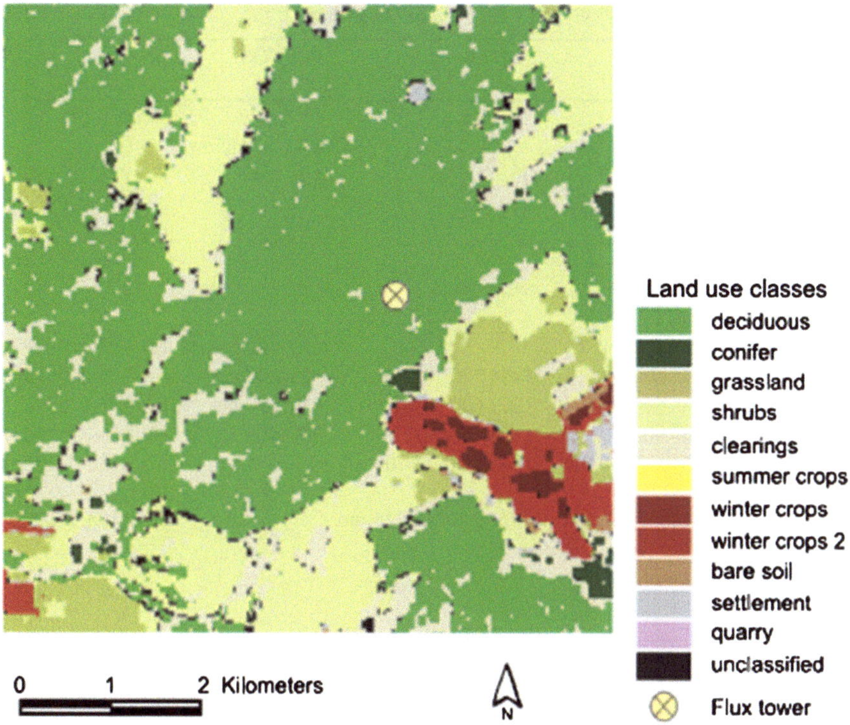

Fig. 6.4 Land use map of the FLUXNET site Hainich DE-Hai based on remote sensing data from Landsat ETM+ with 30 m resolution (Reithmaier et al. 2006)

this purpose during large-scale experiments like FIFE (Sellers et al. 1988) or BOREAS (Sellers et al. 1997) in the 1980s and 1990s. Satellite techniques are a valuable tool to identify surface patterns for footprint analyses at very high resolution of $25 \times 25\ m^2$ or smaller (Fig. 6.4). Resolution and sampling frequency vary by satellite type and employed spectral channels. An overview over available satellite types is given in Table 6.3. Because of the presence of clouds satellites providing high-resolution imagery pass over specific regions only at low temporal resolution. Thus, in many locations, images can only be updated at seasonal intervals, while changes in land use taking place over shorter timescales. Important effects such as e.g. crop harvest, leaf development and leaf fall in deciduous forests and agricultural crops, go undetected.

Satellite remote sensing spectral images have to be corrected to account the atmospheric influence, particularly in the case when more than one image is used to classify the land use. If only one image is available, uncorrected images can be used (Song et al. 2001).

Table 6.3 Remote sensing systems technical specifications

Satellite	Landsat 7 (ETM+)	ASTER	IKONOS	MODIS
Orbital period	16 days	16 days	14 days	Daily
Panchromatic	$15 \times 15\ m^2$	–	$0.85 \times 0.85\ m^2$	
Multispectral	$30 \times 30\ m^2$	$15 \times 15\ m^2$	$4 \times 4\ m^2$	(1,2) $250 \times 250\ m^2$ (3–7) $500 \times 500\ m^2$
Thermal	$60 \times 60\ m^2$	$90 \times 90\ m^2$		
Band (multispectral)				
1	450–520 nm		450–520 nm	(3) 459–479 nm
2	520–600 nm	(B1) 520–600 nm	520–600 nm	(4) 545–565 nm
3 (red)	630–690 nm	(B2) 630–690 nm	630–690 nm	(1) 620–670 nm
4 (NIR)	760–900 nm	(B3) 760–860 nm[a]	760–900 nm	(2) 841–876 nm
5 (SWIR)	1,550–1,730 nm	(B4) 1,600–1,700 nm		(5) 1,230–1,250 nm (6) 1,628–1,652 nm
7 (SWIR)	2,080–2,350 nm	(B5) 2,185–2,225 nm		(7) 2,105–2,155 nm
Band (thermal)				
6	10.4–12.5 μm	(B14) 10.25–10.95 μm (B15) 10.95–11.65 μm		

[a] Nadir, B4: backward scan

The most common spectral index to detect spatial and temporal variability of biomass, and therefore evaluate the distribution of land-cover types, is the Normalized Difference Vegetation Index (NDVI). NDVI is based on the difference of the red reflectance (band 3, Landsat) and the NIR reflectance (band 4):

$$NDVI = \frac{NIR - red}{NIR + red} \tag{6.1}$$

Differences in red and NIR wavelengths are mainly caused by the canopy architecture of green vegetation. Using this contrast of reflectance and absorption, the amount of vegetation present on the surface can be evaluated. It is also necessary to calculate the ratio of the spectral bands 7 und 5 to identify from the reflectance rocks and soils (Richards 1993). However, these bands are not available using IKONOS data. The latter provides high resolution information of canopy cover. This information was successfully used by Kim et al. (2006) to determine the crown cover of a forest.

While the NDVI is chlorophyll sensitive the Enhanced Vegetation Index (EVI) represents better the canopy structure including the leaf area index (Huete et al. 2002):

Table 6.4 Contribution of different land use types in percentage in the area of the FLUXNET site Waldstein-Weidenbrunnen (DE-Bay, 36.18 km^2) for different resolutions of remote sensing images (Reithmaier et al. 2006)

Resolution	15 (m)	30 (m)	50 (m)	75 (m)	100 (m)
Conifer	61.1	61.1	61.0	61.2	61.1
Clearings	12.3	12.2	12.1	12.0	11.9
Grassland	5.6	5.6	5.5	5.7	5.6
Summer crops	6.5	6.5	6.6	6.6	6.7
Winter crops	6.2	6.2	6.4	6.3	6.6
Settlements	4.8	4.9	4.9	5.0	5.0
Quarry	0.3	0.3	0.3	0.4	0.4
Unclassified	3.2	3.2	3.1	3.0	2.7

Note, due to a storm event in 2007 the area of clearings is now much larger as given in this table (Foken et al. 2012)

$$EVI = G\frac{NIR - red}{NIR + C1 \cdot red - C2 \cdot blue + L} \tag{6.2}$$

The factors are C1 = 6, C2 = 7.5, and the gain factor G = 2.5 and L = 1. For the case of the absence of the blue band (MODIS channel 3) the factors can be changed to C1 = 2.4, C2 = 0.

To convert remote sensing spectral bands into land cover types, standard classifiers like e.g. the maximum likelihood classifier (Richards 1993) can be employed (Fig. 6.4). The accuracy of the adopted classifier needs to be determined based on different statistical tests (Smits et al. 1999), as well as in situ comparisons over a test area (ground truthing).

Reithmaier et al. (2006) tested the influence of map resolution on the ability of footprint models to detect the influence of disturbance on eddy-covariance measurements. In a first analysis, they tested how different map resolutions affected the simulated distribution of the land-use types in the larger area. They found that, at their test site, the resolution of the images had no impact on the frequency of the land-use type (see Table 6.4), and the contribution of certain land-cover classes did not shift significantly between map versions. However, in the same study Reithmaier et al. (2006) also showed that the higher level of details maintained in the land cover structure of a remote sensing map can significantly shift footprint results as compared to the use of a very low resolution map (e.g. a 100 × 100 m^2 map read out from topographical maps). As shown in Fig. 6.5, the flux contribution from the target area (here: conifer forest) is much larger using the land use classification of a topographic map than of those by remote sensing data. This shift is caused by removing small-scale heterogeneities such as clearings by applying of a majority filter in coarser maps. Though these areas may appear small and insignificant, their cumulative effect can be important in the average land-use classification. Remote sensing identifies these areas, and thus can provide a more realistic picture of the flux contributions.

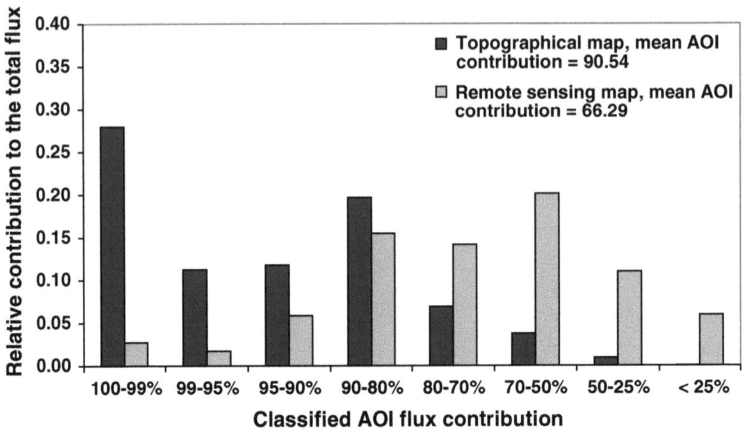

Fig. 6.5 Relative flux contribution of the area of interest (AOI, conifer forest) to the measured flux at the FLUXNET site Waldstein-Weidenbrunnen (DE-Bay) of both, the remote sensing data set (resolution 15 × 15 m^2) and the topographical map (resolution 100 × 100 m^2) according to Reithmaier et al. (2006)

In conclusion, the resolution of land-use classification should be high especially when landscapes are characterized by small-scale heterogeneities. This can be most effectively realized with the help of remote sensing data. Over more homogeneous land covers and footprint calculations for stable stratification with associated larger footprint areas, low resolution maps are acceptable.

Remote sensing data for land use classification were used by several authors (Hasager et al. 2003; Rebmann et al. 2005; Reithmaier et al. 2006). The main principle combined land-use classifications that use roughness lengths and to use averaging procedures for the roughness lengths within the footprint area.

Kim et al. (2006) accumulated the Normalized Difference Vegetation Index (NDVI) over the footprint area by weighting the NDVI of each grid cell with its footprint function f_i:

$$NDVI_f = \sum_{i=1}^{N} (f_i \cdot NDVI_i) \tag{6.4}$$

In the same way, they studied the crown closure or stand density. This is defined as the percentage of the ground covered by vertically projected crown in a stand. They used the IKONOS panchromatic band with 1 m resolution to detect the distribution of trees. This is comparable with the aggregation schema used by Göckede et al. (2004, 2006) and provided in Sect. 6.3.

6.3 Coupling Footprint Results with Surface Information

Procedures to link surface information to footprint areas have been proposed for a matrix of roughness lengths by Grimmond et al. (1998) and later applied by Göckede et al. (2004) for use in footprint investigations. According to Göckede et al. (2004), the source-weight function has to be calculated for each individual step of the time series (e.g. 30 min intervals), and projected onto the land cover matrix according to actual meteorological conditions like wind velocity, stability, or other parameters influencing the output of the type of footprint model chosen. Weighting factors ranging from zero to one reproducing the source weight function were assigned to all matrix cells lying within the concentric 10–90 % isopleths produced by the footprint model, while all matrix cells outside this area were labelled with a weighting factor of zero (Fig. 6.1). Subsequently, for each matrix cell, land-cover information (up to N different land cover types) read out from the matrix was multiplied by the assigned weighting factor (for example, footprint function of the effect level f^P, Eq. 2.94 in combination with Eq. 3.5), and the final roughness length z_0 for the specific measurement was determined as the linear average of these products:

$$z_o = \sum_{i=1}^{N} \left(f_i^P \cdot z_{0i}\right) \tag{6.4}$$

If land cover characteristic such as the roughness length is also an input parameter in the footprint model, the entire process should be repeated iteratively with computed roughness lengths as the new input value, until the difference between input and output roughness length falls below a user-defined threshold. The first model runs for each 30 min-measurements has to be performed with an approximate value for the roughness length. Usually, no more than three iterations are necessary to reach the final roughness length. Because this roughness length was determined as a linear mean, the algorithms performed a parameter aggregation, while roughness length values provided by the matrix are prepared using a non-linear flux aggregation approach (see Sect. 2.4.2). From the physical standpoint, a flux aggregation is better suited method. This approach was used successfully by Göckede et al. (2006) and showed to differ from the simple approach. It can be concluded that the accuracy of the results was significantly improved.

References

Davenport AG, Grimmond CSB, Oke TR, Wieringa J (2000) Estimating the roughness of cities and sheltered country. In: 12th Conference on Applied Climatology, Ashville, NC2000. American Meteorological Society, pp. 96–99

ESDU (1972) Characteristics of wind speed in the lowest layers of the atmosphere near the ground: strong winds. Engineering Sciences Data Unit, London, 35 pp

Fiedler F, Panofsky HA (1972) The geostrophic drag coefficient and the 'effective' roughness length. Quart J Roy Meteorol Soc 98:213–220

Foken T et al (2012) Coupling processes and exchange of energy and reactive and non-reactive trace gases at a forest site—results of the EGER experiment. Atmos Chem Phys 12:1923–1950

Göckede M, Rebmann C, Foken T (2004) A combination of quality assessment tools for eddy covariance measurements with footprint modelling for the characterisation of complex sites. Agric Forest Meteorol 127:175–188

Göckede M, Markkanen T, Hasager CB, Foken T (2006) Update of a footprint-based approach for the characterisation of complex measuring sites. Boundary-Layer Meteorol 118:635–655

Grimmond CSB, King TS, Roth M, Oke TR (1998) Aerodynamic roughness of urban areas derived from wind observations. Boundary-Layer Meteorol 89:1–24

Hasager CB, Jensen NO (1999) Surface-flux aggregation in heterogeneous terrain. Quart J Roy Meteorol Soc 125:2075–2102

Hasager CB, Nielsen NW, Jensen NO, Boegh E, Christensen JH, Dellwik E, Soegaard H (2003) Effective roughness calculated from satellite-derived land cover maps and hedge-information used in a weather forecasting model. Boundary-Layer Meteorol 109:227–254

Huete A, Didan K, Miura T, Rodriguez EP, Gao X, Ferreira LG (2002) Overview of the radiometric and biophysical performance of the MODIS vegetation indices. Remote Sens Environ 83:195–213

Kim J, Guo Q, Baldocchi DD, Leclerc MY, Xu L, Schmid HP (2006) Upscaling fluxes from tower to landscape: overlaying flux footprints on high resolution (IKONOS) images of vegetation cover. Agric Forest Meteorol 136:132–146

Kormann R, Meixner FX (2001) An analytical footprint model for non-neutral stratification. Boundary-Layer Meteorol 99:207–224

Rebmann C et al (2005) Quality analysis applied on eddy covariance measurements at complex forest sites using footprint modelling. Theor Appl Climat 80:121–141

Reithmaier LM, Göckede M, Markkanen T, Knohl A, Churkina G, Rebmann C, Buchmann N, Foken T (2006) Use of remotely sensed land use classification for a better evaluation of micrometeorological flux measurement sites. Theor Appl Climat 84:219–233

Richards JA (1993) Remote sensing digital image analyse—An introduction. Springer, Berlin 340 pp

Schmid HP (1997) Experimental design for flux measurements: matching scales of observations and fluxes. Agric Forest Meteorol 87:179–200

Sellers PJ, Hall FG, Asrar G, Strebel DE, Murphy RE (1988) The first ISLSCP field experiment (FIFE). Bull Amer Meteorol Soc 69:22–27

Sellers PJ et al (1997) BOREAS in 1997: Experiment overview, scientific results, and future directions. J Geophys Res 102:28731–28769

Smits PC, Dellepiane SG, Schowengerdt R (1999) Quality assesment of image classification algorithms for landcover mapping: a review and a proposal for a cost-based approach. Int J Remote Sens 20:1461–1486

Song C, Woodcock CE, Seto KC, Lenney MP, MacOmber SA (2001) Classification and change detection using Landsat TM data: when and how to correct atmospheric effects? Remote Sens Environ 75:230–244

Troen I, Peterson EW (1989) European Wind Atlas. Risø National Laboratory, Roskilde 656 pp

Wieringa J (1992) Updating the Davenport roughness classification. J Wind Eng Ind Aerodyn 41:357–368

Chapter 7
Application of Footprint Models to Different Measurement Techniques

Footprint models were mainly developed to interpret the results of flux measurement techniques. The aim was to replace the *ad hoc* typical 'empirical rule' used in the past to determine optimal measurement conditions. According to this rule, the ratio of the measuring height to that of the undisturbed fetch on the upwind site is of approximately 1:100. Since such simple assumptions were reasonable in pre-footprint time (before 1990), most micrometeorological experiments took place over homogeneous surfaces, typically using short towers over agricultural crops. In those days, the eddy-covariance technique was not used in ecology and in environmental fields. As the eddy-covariance flux method grew in popularity, more sophisticated approaches with realistic assumptions—a footprint analysis—had to be developed and applied along with the need for a theoretical framework explaining its physical underpinnings. The different flux measurement methods are divided into direct ones for which flux footprint models were developed or indirect methods for which mainly footprint models for scalars are relevant. In the following chapter, these measurement techniques are discussed in relation to the use of footprint models.

7.1 Profile Technique

The profile method is based on flux-gradient similarity (see Sect. 2.2.1). It is an indirect method, because the turbulent Prandtl number and universal functions must be determined in comparison with a direct method, such as the eddy-covariance

M. Y. Leclerc and T. Foken, *Footprints in Micrometeorology and Ecology*,
DOI: 10.1007/978-3-642-54545-0_7, © Springer-Verlag Berlin Heidelberg 2014

technique (see Sect. 7.2). Because of the advantages of the eddy-covariance method in the last 15 years, the profile method is seldom often used. Many considerations including internal boundary layers in the footprint area and measurements over heterogeneous surfaces are limitations that have hindered the dissemination and application of this method. More often, simplified methods with only two measuring heights are used mainly in applied meteorology (Agrometeorology etc.). But all approaches are significantly influenced by the footprint because measurements are made in different heights but each sensor must be influenced by the same underlying surface type for all wind velocities and stability conditions.

7.1.1 Profile Technique with Three and More Measuring Levels

The profile method uses approximately 4–6 levels with wind, temperature, humidity or trace gas measurements (Fig. 7.1). The basis for the method lies in the neutral case Eqs. (2.16–2.18). From the measured profile, it is necessary to determine the gradient of the state parameters. In the simplest case, this can be done using a linear approximation, a method also used in approaches with only two measuring heights. Therefore a diagram is necessary with the wind velocity *u*—can by replaced by the temperature or trace gas concentration—on the abscissa and *z* on the ordinate, were the differential $\partial u/\partial z$ can determined by the differences of both, *u* and *z* (Fig. 7.2a).

$$\left(\frac{\partial u}{\partial z}\right)_{z_a} \cong \frac{\Delta u}{\Delta z} = \frac{u_2 - u_1}{z_2 - z_1} \qquad z_a = (z_2 - z_1)/2 \tag{7.1}$$

A much better application of the physical background is a logarithmical approximation with a geometric average of the measurement heights. In this approach, a diagram with *u*, *T*, or *c* on the abscissa and *z* on the vertical ordinate is which has a logarithmical scale (Fig. 7.2b):

$$\left(\frac{\partial u}{\partial \ln z}\right)_{z_m} \cong \frac{\Delta u}{\Delta \ln z} = \frac{u_2 - u_1}{\ln(z_2/z_1)} \qquad z_m = (z_1 \cdot z_2)^{1/2} \tag{7.2}$$

The basis for the profile method in the non-neutral case can be found in Eqs. (2. 24)–(2.26). The simplest way is to use the integrated form of Eq. (2.28) and compares on the ordinate ($\ln z - \psi(z/L)$) and in case of the wind profile on the abscissa u (Fig. 7.2c). The following equation can be used for the momentum and the sensible heat fluxes as an example (Arya 2001):

Fig. 7.1 Measuring tower for profile measurements, *Photograph* Foken

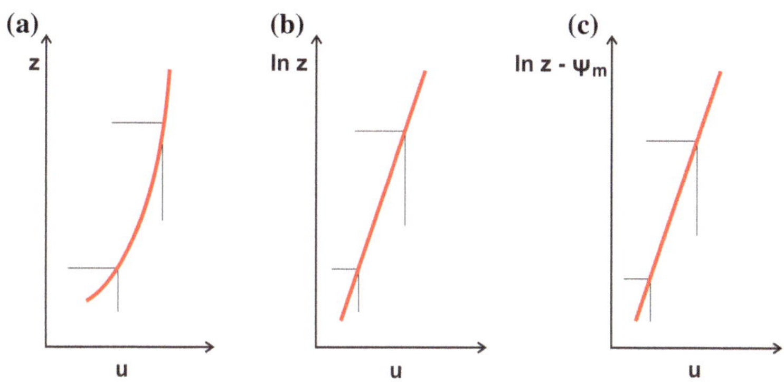

Fig. 7.2 Approximations of the profile function **a** with a linear approximation with Eq. (7.1), **b** with a lin-log approximation for the neutral case with Eq. (7.2), **c** with the lin-log approximation for the non-neutral case with Eq. (7.3). The red line is the measured profile and the thin lines are the range for determination the differences for the gradient

$$\ln z - \psi_m(z/L) = \frac{\kappa}{u_*} \cdot u + \ln z_0 \tag{7.3}$$

$$\ln z - \psi_H(z/L) = \frac{\kappa}{\Pr_t T_*} \cdot T - \frac{\kappa}{\Pr_t T_*} \cdot T_0 + \ln z_0 \tag{7.4}$$

The Obukhov length or the Richardson number (see Sect. 2.2.1) is necessary for this approximation. It can be determined by an iterative solution of Eqs. (7.3) and (7.4). Several approaches can be used for the interpolation of the profile function such as the cubic spline method. Because of possible measurements errors, an overshoot can influence the results. Therefore, the choice of the approximation function should be carefully done, e.g. the spline method by Akima (1970) has often been successfully used.

Often applied is the Nieuwstadt-Marquardt-approach. Therefore, the quadratic cost function as a measure of the differences between measuring values and the profile equation are calculated (Nieuwstadt 1978). The non-linear system of equations of the minimization of the deviations can be solved using the method by Marquardt (1983).

7.1.2 Profile Technique with Two Measuring Levels

The Bowen-ratio method (Bowen 1926) is the most popular approach to determine sensible and latent heat fluxes mainly in agricultural meteorology. The method is based on the Bowen-ratio and the energy balance equation (Fritschen and Fritschen 2005; Foken 2008):

$$Bo = \frac{Q_H}{Q_E} \approx \gamma \cdot \frac{\Delta T}{\Delta e} \tag{7.5}$$

$$-Q_s^* = Q_H + Q_E + Q_G \tag{7.6}$$

The psychrometric constant is $\gamma = 0.667$ K hPa^{-1} for p = 1013 hPa and t = 20 °C. From both Equations follows for the sensible and latent heat flux:

$$Q_H = \left(-Q_s^* - Q_G\right) \frac{Bo}{1 + Bo} \tag{7.7}$$

$$Q_E = \frac{-Q_s^* - Q_G}{1 + Bo} \tag{7.8}$$

The experimental setup consists of measurements at two levels for temperature and humidity and additionally a net radiometer and a soil heat flux plate and soil temperature sensor (Fig. 7.3). The approximation in Eqs. (7.5) to (7.8) depends on several assumptions, which are discussed in more details, e.g. by Ohmura (1982) or Foken (2008). It is essential that the turbulent atmospheric conditions be fulfilled for wind velocities in the upper measurement level of >1 m s^{-1} and/or a difference of the wind velocity between both levels of >0.3 m s^{-1} are necessary. Furthermore, the ratio of both measuring heights should be 4–8 to ensure that the temperature, humidity etc. difference between both levels is significantly larger than the measurement error.

A special version of the Bowen-ratio method is the Modified Bowen-ratio method, which was developed mainly for trace gas fluxes (Businger 1986) and can also be applied for energy fluxes (Liu and Foken 2001). Such a system is shown in

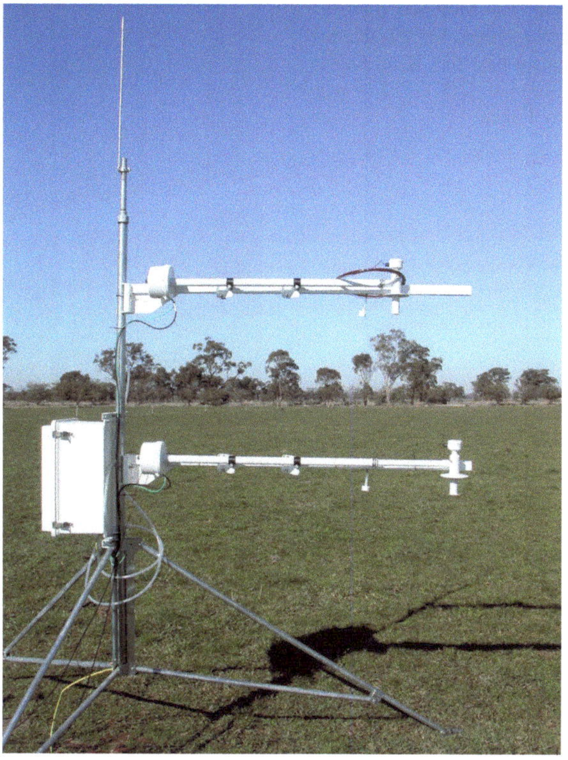

Fig. 7.3 Bowen-ratio system (*Photograph* Campbell Scientific Inc. Logan UT, USA, Published with kind permission of © Campbell Sci. Inc., 2012. All Rights Reserved)

Fig. 7.4. Equation (7.5) and the measurements set up at two levels are analogous to the standard Bowen-ratio method. Only one flux, often the sensible heat flux, is directly measured with the eddy-covariance method (see Sect. 7.2). From the definition of the Bowen-ratio, the latent heat flux can easily be determined. In the case of trace gas fluxes, the measurement of the humidity gradient is replaced by the gradient of the trace gas and the modified Bowen-ratio is defined as the ratio of the sensible to the trace gas flux. Since a sonic anemometer measures the buoyancy flux (see Sect. 7.2), this flux must be transformed into the sensible heat flux (Schotanus et al. 1983; Foken et al. 2012a) or the temperature gradient must be replaced by the gradient of the virtual temperature. With the sonic anemometer, the wind velocity can be controlled and no additional anemometer is necessary.

Additional methods to parameterize the fluxes with measurements at two levels are given by Foken (2008).

Fig. 7.4 Modified Bowen-ratio system for sensible and latent heat flux according to Liu et al. (2001), *Photograph* by Foken

7.1.3 Accuracy and Footprint Issues for Profile Technique

The basis for this method is (1) the assumption that the differences of the measurement signal between two adjacent measuring levels is significant larger than the measurement error of the sensor and (2) the assumption is that the influence of the vertical exchange process on the differences is significant larger than possible effects of different footprint areas on the measured signal on the different measurement levels. Because the last assumption cannot easily be fulfilled, surface characteristics in footprint areas of the different measuring levels should be equal to one another. The consequence may be that, for limited fetch conditions, the range of the measurement height decreases with an increase in stability. Furthermore, no internal boundary layers should influence the profile measurements (see Sect. 8.1). In the case when the differences in the footprint between the levels of the profile method cause differences in the temperature, moisture, and trace gas measurements which are larger as the minimal errors of the system, the error due to the different footprints determine the error of the whole system.

Table 7.1 Minimal measurable flux (20 % error) for energy and trace gases above low, $z_2/z_1 = 8$, and tall, $z_2/z_1 = 1.25$, vegetation for neutral stratification and $u_* = 0.2$ m s^{-1} (dimensions μg m^{-3} and μg s^{-1} m^{-2}), the "*italic*" fluxes are lager as the typical fluxes in the nature (Foken 1998, 2008)

Energy and matter flux	χ_{min}	$\Delta_{\chi,min}$	Flux $z_2/z_1 = 8$	Flux $z_2/z_1 = 1.25$
Sensible heat	0.05 K	0.5 K	0.025 m K s^{-1} 30 W m^{-2}	0.05 m K s^{-1} 60 W m^{-2}
Latent heat	0.05 hPa	0.5 hPa	0.025 hPa K s^{-1} 45 W m^{-2}	0.05 hPa K s^{-1} 90 W m^{-2}
Nitrate particles	0.01	0.1	0.005	0.01
Ammonium particles	0.02	0.2	0.01	*0.02*
CO_2	100	1000	50	100
NO	0.06	0.6	*0.03*	*0.06*
NO_2	0.1	1.0	0.05	0.1
O_3	1.0	10.0	0.5	*1.0*
NH_3	0.014	0.14	0.007	0.014
HNO_3	0.2	2.0	*0.1*	*0.2*
HNO_2	0.25	2.5	*0.125*	*0.25*

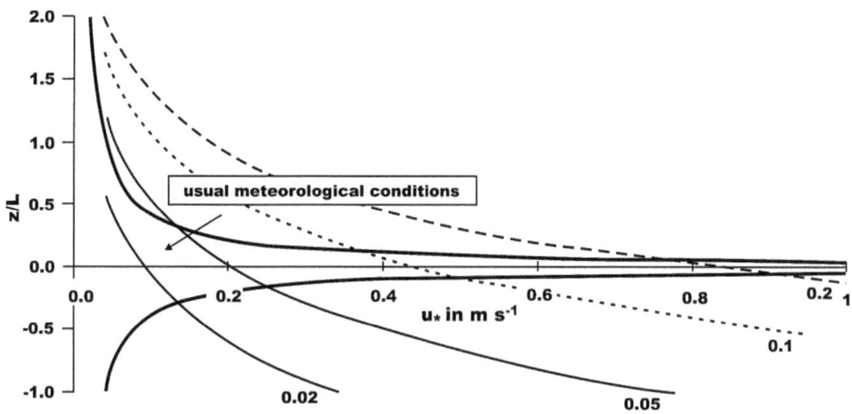

Fig. 7.5 The normalized flux Q_N (numbers are written in the *hyperbolic lines*) depending on stratification and the friction velocity for $z_2/z_1 = 8$ (Foken 2008). The typical range of meteorological measurements is between the *black lines*

7.1.3.1 Accuracy of Profile Measurements

The first assumption can be more easily controlled. According to Foken (1998, 2008), the profile Eqs. (2.16)–(2.18) can be divided into a term depending on the dynamical-thermal turbulence Q_N and another term, namely, the difference of the state parameter between the different measurement levels $\Delta\chi$.

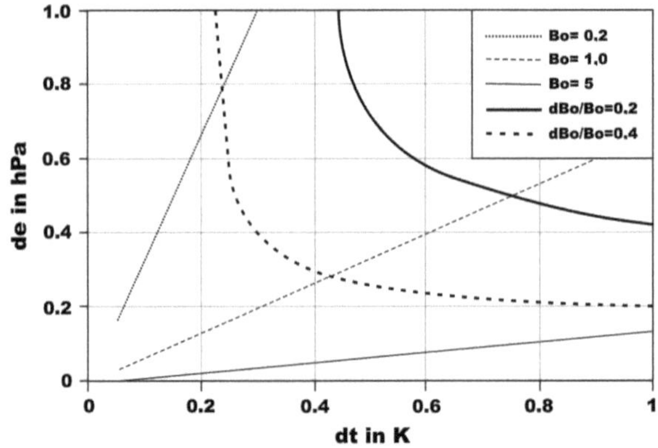

Fig. 7.6 Error of the Bowen-ratio (20 and 40%) determined with the Bowen-ratio method dependent on the temperature and moisture difference in between both levels (Foken et al. 1997). The accuracy of the measuring system is ± 0.05 K and hPa, Published with kind permission of © Zentralanstalt für Meteorologie und Geodynamik, 1997. All Rights Reserved

$$Q_\chi = Q_N\left[u_*, \varphi(z/L), \ln(z-d)\right] \cdot \Delta\chi \tag{7.9}$$

The normalized flux Q_N is shown in Fig. 7.5. The minimal fluxes, which can be measured with an accuracy of 20 %, depend on the tenfold accuracy of the measurement system χ_{min}:

$$Q_{\chi,\min} = Q_N \cdot 10 \cdot \chi_{\min} \tag{7.10}$$

Typical values of measurable fluxes above low and tall vegetation are given in Table 7.1. This table can be used in the following way: Only typical values of the accuracy of the measurement system are given in the Table. With the specification of the system, one can use Eq. (7.10) and find the minimal flux which can be measured with an accuracy of 20 %. Therefore, one has to determine the Q_N value according to meteorological conditions (stratification, friction velocity) from Fig. 7.5. The figure is calculated for a ratio $z_2/z_1 = 8$, which applies only above low vegetation. Going back to Table 7.1, one can see the difference to high vegetation $z_2/z_1 = 1.25$. So this is a simple approach to check for which fluxes which accuracy of the measurements is necessary to make flux measurements with the two levels profile approach. If the number of levels goes up, one can increase also the accuracy and can easily determine whether measurements at one level represent the surface of interest or not. Not included in this system is the influence of the roughness sub layer, which must be taken into account above tall vegetation. Due to a higher mixing above a forest canopy for instance, the gradient is even more reduced, up to a factor of 2 (enhancement factor) and therefore the accuracy of the system must be assumed to be twice as high as what is stated in the Table 7.1 to determine the accuracy of the final flux.

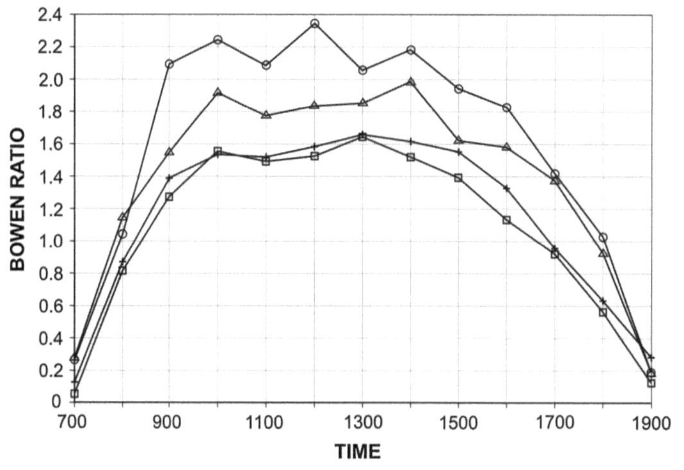

Fig. 7.7 Bowen-ratio measurements by Tanner (1988) with two systems (+, □) at one day in the same level and at a second day in two different levels (△, ◇) (Leclerc and Thurtell 1990, figure was reconstructed)

In the case of a Bowen-ratio system, a simpler approach was given by Foken et al. (1997). Assuming an error in the measurement set up of ± 0.05 K or hPa for temperature and moisture measurements, Fig. 7.6 shows the ranges with a possible error in the Bowen-ratio. Note that the error of the temperature and moisture measurements in the atmosphere is significantly higher than the pure instrumentation error (Dugas et al. 1991). An error in the Bowen ratio of 0.1 is related to a flux error of 10 %. For this case, typical differences between both measurement levels should not only be in the order of 0.5 K and 0.5 hPa but also in a range of the Bowen ratio of about 0.3–1.0. Therefore, the method fails in the case of very dry or humid conditions.

7.1.3.2 Footprint of Profile Measurements

A first example of the influence of the footprint on the Bowen-ratio in the field was discussed in Leclerc and Thurtell (1990) and was the link to develop their Lagrangian simulation (see Sect. 3.2.1). Two systems on the same point and in the same height showed identical Bowen-ratios, but the application of these two systems in different heights showed because of the different footprint of both systems, which included for the upper system a second field with different land use, significant different values (Fig. 7.7).

An illustration of the footprint problem was also given by Schmid (1997). That study shows that in Fig. 7.8 three measuring heights have a different footprint area and covers different types of surfaces. In such a case, the profile approach would not measure only vertical gradients but also horizontal differences of different land cover. Therefore, the profile tower should be placed left of the cross in Fig. 7.8, at

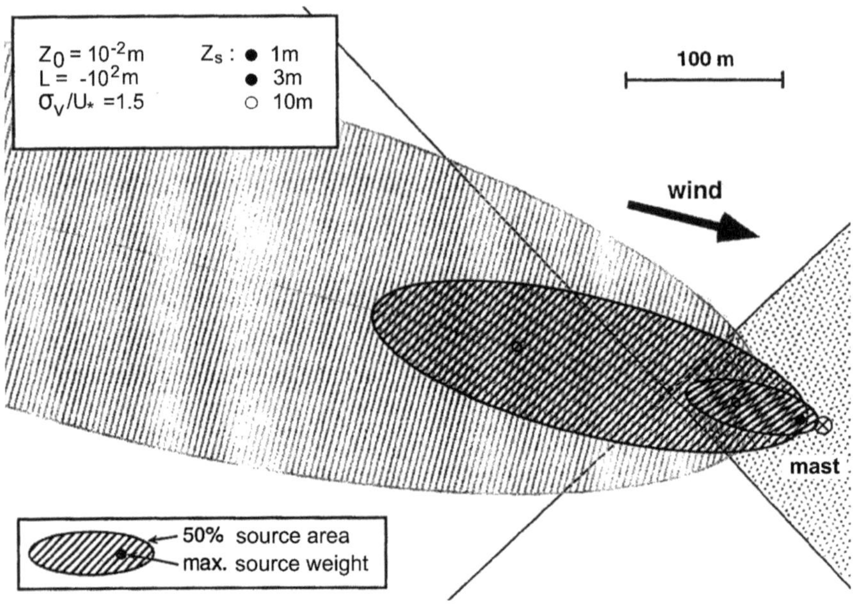

Fig. 7.8 Footprint area for different measuring heights covering different surfaces (Schmid 1997, Published with kind permission of © Elsevier, 1997. All Rights Reserved)

the junction of the four different surface types. Short of using this method, it is likely that the experimentalist will be unable to interpret the data. Sometimes the differences of the underlying surfaces may not be small so that a profile approach appears impossible. The footprint area of all sensors must be identical in size and must complete cover only one surface type. The data can be tested by the representativity test (Nappo et al. 1982).

Horst (1999) pointed out that the simple concentration footprint related approach by Schmid (1997) does not perform the conditions of the profile technique because the gradient approach has a special flux footprint: Based on a previous study by Stannard (1997), Horst (1999) extended his model (Horst and Weil 1992, 1994) to estimate footprint fluxes obtained from micrometeorological profile techniques. He presented a formulation for use with the concentration profile to estimate flux footprints and for fluxes measured using the Bowen-ratio technique.

While a flux footprint can be theoretically derived for concentration measurements made at two or more levels as is the case in Bowen-ratio and profile methods, the reader is therefore reminded that this method, in practice, works only in the special case where tower concentration sensors see a consistency in the emission rate of the surface within their different footprints. The flux footprint for the Bowen-ratio technique is identical to that for a two-level profile measurement only for very limited circumstances. In the more general case, a flux footprint cannot be defined when using the Bowen-ratio technique.

The full derivation is found in Horst (1999), with the resulting flux footprint equation determined from measurements made at two levels as in the Bowen ratio method, can be expressed as

$$\overline{f}^{y} = -\frac{A\,u_{*}\kappa}{\overline{z}U}\frac{\exp^{-(z_2/b\overline{z})^{r}} - \exp^{-(z_1/b\overline{z})^{r}}}{\ln\left(z_2/z_1\right) - \psi\left(z_2/L\right) + \psi\left(z_1/L\right)} \tag{7.11}$$

In the case of the concentration-profile footprint flux estimates, Horst (1999) found the upwind extent of the footprint for concentration-profile flux estimates to be similar to that of the footprint for eddy-covariance flux measurements when the eddy-covariance measurement is made at a height equal to the arithmetic mean of the highest and lowest profile measurement height for stable stratification or the geometric mean for unstable stratification. The resulting expression for a flux footprint determined from a multiple level concentration measurement is

$$\overline{f}^{y} = -\frac{Au_{*}\kappa}{\overline{z}u\varphi_{m}\left(z_m/L\right)}\sum_{j=1}^{n} b_j e^{-(z_j/b\overline{z})^{r}} \tag{7.12}$$

According to the theoretical approach by Horst (1999), the concentration-profile flux footprint depends on the ratio of the highest to the lowest measurement height, but appears to be insensitive to the number of measurement levels. That study also found that the concentration-profile flux footprint extends closer to the measurement location than does the 'equivalent' eddy-covariance flux footprint, with the difference becoming more pronounced as the ratio of the profile measurement heights increases. For the Modified Bowen-ratio system, it can then be concluded that the limiting factor in the footprint is the flux measurements with the sonic anemometer, because the anemometer is installed above the profile measuring levels.

7.2 Eddy–Covariance Technique

7.2.1 Basics of the Eddy-Covariance Method

The eddy covariance method based on the transfer equations for momentum, heat, humidity or trace gases by application of the Reynolds's decomposition (Businger 1982; Stull 1988; Foken 2008; Foken et al. 2012b), which divides a turbulent parameter x into a mean part $\overline{x}$ and into a fluctuating part x' (Fig. 7.9)

$$x = \overline{x} + x'. \tag{7.13}$$

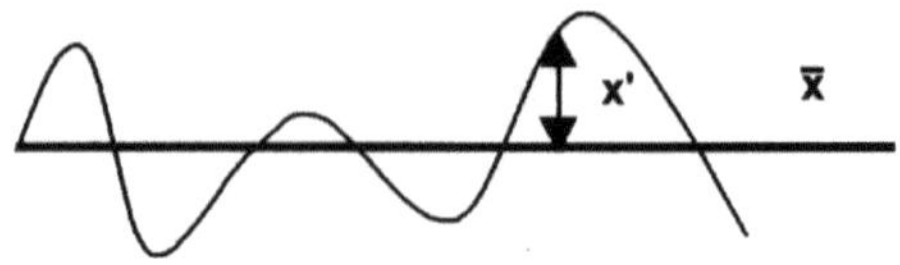

Fig. 7.9 Schematic presentation of Reynolds's decomposition of the value x (Foken 2008)

By neglecting the pressure gradient, molecular/viscous transport, gravity and Coriolis terms which have no significant impact on the eddy-covariance method over flat terrain the equation can be simplified. The coordinate system must be chosen in such a way that the perpendicular, $\overline{v}$, and vertical, $\overline{w}$, wind component are zero and assuming horizontal homogeneity as well as steady-state conditions. For the momentum flux follows finally:

$$\frac{\partial \overline{w'u'}}{\partial z} = 0 \tag{7.14}$$

where $\overline{u'w'}$ is the eddy covariance term for the momentum flux. The eddy-covariance terms are analogous to the former, as for the sensible heat flux $\overline{w'T'}$, for the latent heat flux $\overline{w'q'}$, and $\overline{w'\chi'}$ for the trace gas flux. From Eq. (7.14) follows that, under the preceding assumptions, this flux is constant with height and that it is representative of the vertical flux through a horizontal plane above the surface roughness elements. This approach is called the eddy-covariance method. More details and necessary assumptions are given in the relevant literature (Lee et al. 2004; Foken 2008; Aubinet et al. 2012), but most important are the assumption on steady-state conditions and horizontal homogeneous surfaces. Furthermore, the mean vertical wind velocity must equal zero for the equation of the total flux to hold (Reynolds' postulate; Eq. 2.5), i.e.

$$\overline{wx} = \bar{w}\bar{x} + \overline{w'x'} \tag{7.15}$$

Therefore, the flux can only be determined with the covariance term provided this assumption is fulfilled. The assumption will be fulfilled by an coordinate transformation where recently the planar-fit method is recommended (Wilczak et al. 2001). This method is applied for longer periods like weeks or months to avoid strong influences of single burst and gusts.

According to Eq. (7.15) the turbulent fluctuations of the components of the wind vector and of scalar parameters must be measured at a high sampling frequency so that the turbulence spectra can be extended to 10–20 Hz. The measuring devices used for such purposes are sonic anemometers for the wind components and sensors that can measure scalars with the required high resolution in time. The latter are often optical measurement methods. The sampling time depends on atmospheric stability, wind velocity, and measuring height. Such a measurement complex is shown in Fig. 7.10. According to the theory, the method is a direct one without any empirical function. Nevertheless, the simplifications given above and instrumental problems need a set of corrections. Literature is available regarding

Fig. 7.10 Measuring complex for eddy-covariance measurements consistent on a sonic anemometer CSAT3 and an IR gas analyser LiCor 7500 (*Photograph* by Foken)

these issues (Haugen 1973; Kaimal and Finnigan 1994; Lee et al. 2004; Foken 2008; Aubinet et al. 2012).

Because the eddy-covariance technique is often not applied in homogeneous terrain, the influence of different underlying surface conditions must be taken into account in the data interpretation. This was the main reason behind the development of footprint techniques.

7.2.2 1D Eddy-Covariance Method

Eddy-covariance measurements can be used to estimate fluxes of energy, heat, water vapor, and gases between the ecosystem and the atmosphere. The method was described above in such a way that the measurement above the canopy represents the flux between the atmosphere and the ecosystem (Fig. 7.11). This 1D net ecosystem flux is the sum of the eddy-covariance measurements (term II) and the change of the storage (term I). Term V is a sink or source term.

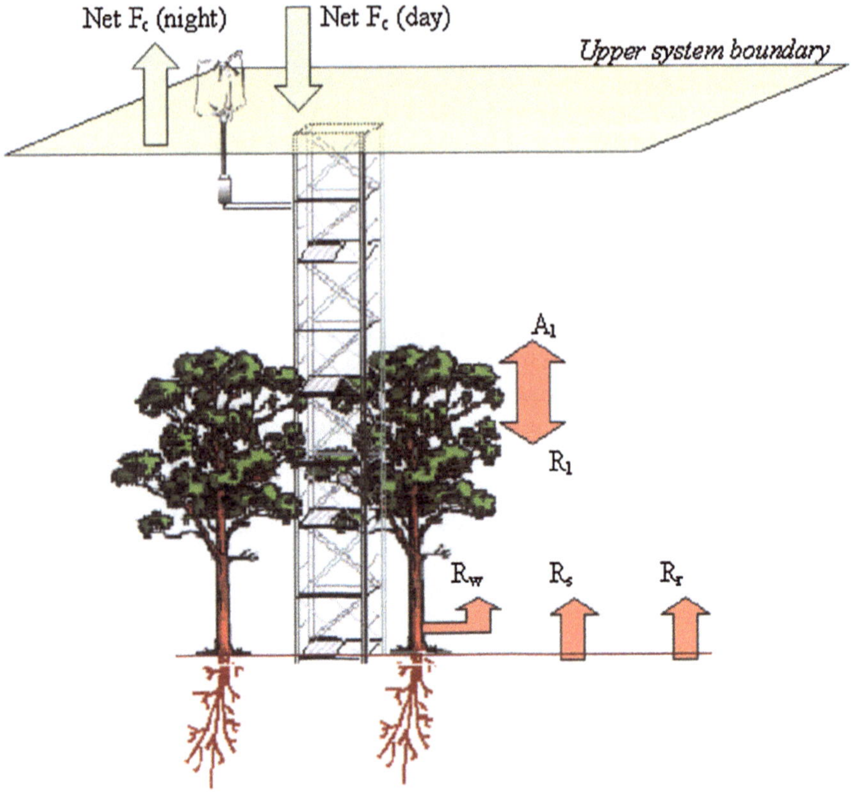

Fig. 7.11 Determination of the net ecosystem exchange with the eddy-covariance method with the assumption of a point measurement (Moncrieff 2004) with A: assimilation and R: different respiration pathways

$$\underbrace{\int_0^{z_m} \overline{\rho_d} \frac{\overline{\partial \chi}}{\partial t}}_{I} + \underbrace{\overline{\rho_d} \overline{w'\chi'}|_h}_{II} = \underbrace{\overline{\sum}}_{V} \tag{7.16}$$

For this assumption all flux footprint models including all analytical models can be applied.

7.2.3 Generalized Eddy-Covariance Method (3D)

In reality, the ecosystem is more complex and an equation for a volume element must be formulated (Fig. 7.12). The expression for the ecosystem exchange is

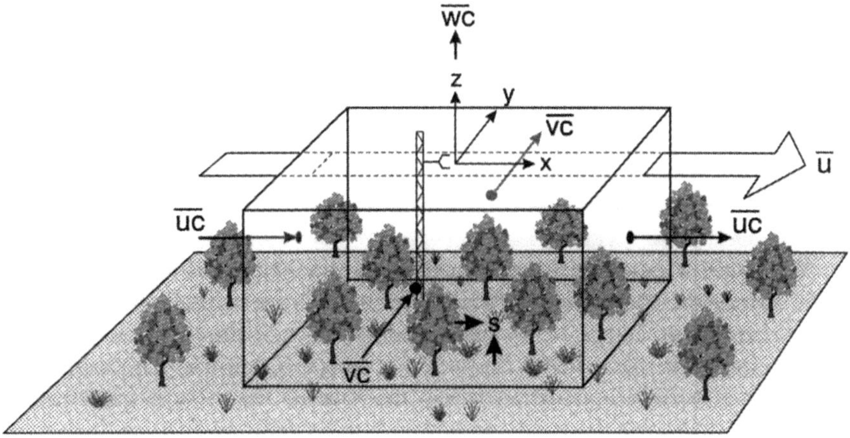

Fig. 7.12 Schematic image of integration of Eq. (7.17) on a control volume in homogeneous terrain (Finnigan et al. 2003)

$$\underbrace{\int_0^{z_m} \overline{\rho_d} \frac{\overline{\partial \chi}}{\partial t} dz}_{I} + \underbrace{\overline{\rho_d} \overline{w'\chi'}\Big|_{z_m}}_{II} + \underbrace{\int_0^{z_m} \left[\overline{\rho_d u} \frac{\Delta \overline{\chi_x}}{\Delta x} + \overline{\rho_d v} \frac{\Delta \overline{\chi_y}}{\Delta y} \right]}_{III} + \underbrace{\int_0^{z_m} \left[\overline{\rho_d w} \frac{\Delta \overline{\chi}}{\Delta z} \right]}_{IV} = \underbrace{\overline{\sum}}_{V} \tag{7.17}$$

This includes besides the terms I and II also the horizontal advection (term III) and the vertical advection (term IV). The often neglected convergence or divergence of the horizontal flux is not included in Eq. (7.17). This flux may be averaged over time and integrated both horizontally over the area and vertically, from the ground to instrument height z_m (Fig. 7.12). This approach is now called the "generalized eddy-covariance" method (Foken et al. 2012b). Simple analytic footprint models cannot handle this volume average but for Lagrangian models with in-canopy turbulence parameterization, this is partly possible (Baldocchi 1997; Rannik et al. 2000, 2003; Lee 2004).

It is extremely difficult and cost intensive to measure the terms III and IV: recent and carefully planned special advection experiments fell short of expectations (Aubinet 2008). Often the terms III and IV are of the same order with the different sign and the influence on the net ecosystem exchange is negligible. Nevertheless, possible effects of the advection should be tested for a specific site at least with a special designed short-term experiment. Because of these problems, the generalized eddy-covariance technique is often reduced to the 1D version.

Table 7.2 Possible combination of single quality flags into a flag of the general data quality

Flag of the general data quality		Steady state test according to Eq. (7.20)	Integral turbulence characteristics according to Eq. (7.21)
High quality	1	1	1–2
	2	2	1–2
	3	1–2	3–4
Reasonable	4	3–4	1–2
Quality	5	1–4	3–5
	6	5	≤5
Bad quality	7	≤6	≤6
	8	≤8	≤8
		≤8	6–8
Not to use	9		One flag equal to 9

7.2.4 *Quality Control of Eddy-Covariance Data*

The analysis of the data quality of the eddy-covariance method is an important issue and is one that can also be combined with the footprint technique as discussed in Sect. 8.2.3. In contrast to standard meteorological measurements, there are only a few papers available addressing quality control of eddy-covariance measurements (Foken and Wichura 1996; Vickers and Mahrt 1997). Quality control of eddy-covariance should include not only tests for instrument errors and problems with the sensors, but also evaluate how closely the conditions fulfil the theoretical assumptions underlying the method. Because the latter depends on meteorological conditions, eddy-covariance quality control tools must be a combination of a typical test for high resolution time series and an examination of the turbulent conditions. The most relevant tests are on steady-state conditions and on the fulfilment of turbulent conditions, which are given here only briefly. For details see Foken et al. (2004, 2012a).

The steady-state test used by Foken and Wichura (1996) is based on developments attributed to Russian scientists (Gurjanov et al. 1984). It compares the statistical parameters determined for the averaging period and for short intervals within this period. For instance, the time series for the determination of the covariance of the measured signals w (vertical wind) and x (horizontal wind component or scalar) of about 30 min duration will be divided into M = 6 intervals of about 5 min. N is the number of data points comprised in the short interval (N = 6,000 for 20 Hz scanning frequency and a 5 min interval):

$$\left(\overline{x'w'}\right)_i = \frac{1}{N-1}\left[\sum_j x_j \cdot w_j - \frac{1}{N}\left(\sum_j x_j \cdot \sum_j w_j\right)\right]$$
$$\overline{x'w'} = \frac{1}{M}\sum_i \left(\overline{x'w'}\right)_i \qquad (7.18)$$

This value will be compared with the covariance determined for the whole interval:

$$\overline{x'w'} = \frac{1}{M \cdot N - 1}\left[\sum_{k=1}^{M \cdot N} x_k \cdot w_k - \frac{1}{M \cdot N}\left(\sum_{k=1}^{M \cdot N} x_k \cdot \sum_{k=1}^{M \cdot N} w_k\right)\right] \tag{7.19}$$

The authors proposed that the time series is steady state if the normalized difference between both covariances (parameter of relative non-stationarity)

$$RN_{Cov} = \left|\frac{\overline{(x'w')}_{Eq.(7.18)} - \overline{(x'w')}_{Eq.(7.19)}}{\overline{(x'w')}_{Eq.(7.19)}}\right| \tag{7.20}$$

is less than 30 %. This value has been found by long experience but is in good agreement with other test parameters including those of other authors (Foken and Wichura 1996). Otherwise, the data quality is likely to be lower.

The test on developed turbulent conditions based on the flux-variance similarity (Panofsky and Dutton 1984). This similarity means that the ratio of the standard deviation of a turbulent parameter and its turbulent flux is nearly constant or a function of stability. These so-called integral turbulence characteristics are basic similarity characteristics of atmospheric turbulence and are discussed in Sect. 2.2. 6. These functions depend on stability and are given in Eqs. (2.62) and (2.63). The test can be done for the integral turbulence characteristics of both parameters used to determine the covariance. Similar to Eq. (7.20) both measured and the modelled parameters can be compared according to

$$ITC_\sigma = \left|\frac{\left(\sigma_x/X_*\right)_{model} - \left(\sigma_x/X_*\right)_{measurement}}{\left(\sigma_x/X_*\right)_{model}}\right| \tag{7.21}$$

If the test parameter ITC_σ is less than 30 %, a well developed turbulence can be assumed.

The quality tests given above open the possibility to also flag the quality of a single measurement (Foken and Wichura 1996; Foken et al. 2004). For these tests, the definition of flags is possible and can be combined to an overall flag (Table 7.2). The user of such a scheme must know the appropriate use of the flagged data. The presented scheme was classified by micrometeorological experience so classes 1–3 can be used for fundamental research, such as the development of parameterisations. Classes 4–6 are available for general use such as for continuously running systems of the FLUXNET programme. Classes 7 and 8 are only for orientation. It is often preferable to use such data rather than a gap filling procedure, but then these data should not differ significantly from the data located before and after these data in the time series. Data of class 9 should be excluded

Table 7.3 Eddy-covariance software tools with embedded footprint models (Foken et al. 2012a)

Software	TK3	EddySoft	EdiRE	ECO2S
	University of Bayreuth	Max-Planck-Institute Jena	University of Edinburgh	IMECC-EU University of Tuscia
Footprint tools	Kormann and Meixner (2001)	Schuepp et al. (1990)	Schuepp et al. (1990)	Kljun et al. (2004), Schuepp et al. (1990)

under all circumstances. The combination of the flagging system with the footprint analysis is given in Sect. 8.2.3.

Some of the eddy-covariance software tools include simple footprint tools for data quality control. Table 7.3 gives an overview.

7.3 Scintillometer Technique

The scintillometer (Hill et al. 1980; Hill 1997) is an optical instrument consisting of a infrared laser which measures the scintillation of the light in the atmosphere due to the movements of turbulent eddies. Essentially, scintillometers are separated into two classes (DeBruin 2002) the large aperture scintillometer (LAS) and the small aperture scintillometer (DBSAS, Fig. 7.13). The LAS has a measuring path length of several kilometres. In contrast, the DBSAS works with two laser beams over a distance of about 100 m (Andreas 1989). Temperature or humidity inhomogeneities (IR scintillometer for sensible heat flux or microwave scintillometer for latent heat flux) cause a scintillation of the measuring beam which can be evaluated. These systems can determine also the path-length-averaged turbulence scale and are also able to determine the friction velocity when a stability dependence is taken into account (Thiermann and Grassl 1992). Note that scintillometers are not able to determine the sign of the sensible heat flux. Additional measurements (temperature gradient) are necessary. In the footprint analysis, it should be stated that—in contrast with more standard measurements seeking to obtain the footprint at one location—the scintillometer-based footprint method requires that it be determined over one measuring path.

The instrument measures the refraction structure function parameter, $C_n^{\;2}$,

$$C_n^2 = \left(79.2 \cdot 10^{-6} \frac{p}{T^2}\right)^2 C_T^2 \tag{7.22}$$

and offers a method for the determination of the sensible heat flux $\overline{w'T'} = T_* \cdot u_*$ which is a function of the temperature structure function parameter, $C_T^{\;2}$ (Wyngaard et al. 1971).

Fig. 7.13 Small aperture scintillometer DBSAS, receiver unit, at Svalbard, Norway. The laser source is about 20 m away in the background (*Photograph* by Lüers, Published with kind permission of © Dr. habil. Lüers, 2012. All Rights Reserved)

$$\frac{C_T^2\, z^{2/3}}{T_*^2} = \left\{ \begin{array}{ll} 5\left(1+6.4\frac{-z}{L}\right)^{-2/3} & for\ z/L \le 0 \\ 5\left(1+3\frac{z}{L}\right) & for\ z/L > 0 \end{array} \right\} \tag{7.23}$$

Scintillometers have the highest sensitivity in the middle of the measurement path rather than near the transmitter and receiver. This must be taken into account for footprint analyses of the measurement sector (Meijninger et al. 2002; Göckede et al. 2005). The influence of the source area at different positions within the scintillometer path must be normalized with a weighting factor obtained by a bell-shaped weighting function (Thiermann, personal communication),

$$W(x) = A \cdot x^{11/6}(P-x)^{11/6} \tag{7.24}$$

where $W(x)$ is the weighting factor for position x in m along the measurement path with a total length P in m. A is a scaling factor that is of no importance in footprint studies.

To modify the application of footprint models for line measurements such as scintillometers, a superposition of multiple models must be implemented in software comparing land cover maps with footprints (see Chap. 6) along the measurement path. All the models must we multiplied with a path dependent factor weighting $W(x)$. The number of model runs depends on the path length and the spacing of the land cover map. In Sect. 7.4, this aggregation schema is shown for

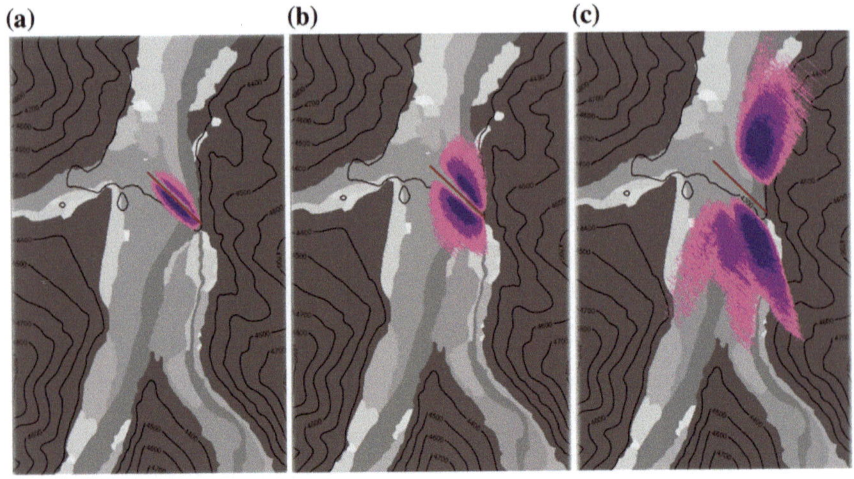

Fig. 7.14 Analysis of footprint climatology for installation of a Large Aperture Scintillometer at 20 m height on the basis of a wind climatology for unstable **a** neutral **b** and stable **c** stratification (Babel and Foken 2009, unpublished study) in a mountain valley, E-W-distance is 4 km. The contour lines are for 80, 50 and 20 % of the footprint. The different *grey areas* are different land cover classes

aircraft measurements. This schema must be modified using the weighting function according to Eq. (7.24).

In Fig. 7.14, an example of footprint climatology analysis is shown for LAS even before the installation of the instrument. The basis for this lies in the local wind climatology and the assumption of three different stability classes. It can easily be seen that only in the unstable and neutral case, the beam is over uniform terrain. Such an analysis can help to identify the optimum scintillometer location.

7.4 Airborne Measurement Technique

The first airborne study based on the data sets by Desjardins et al. (1989) was connected with the first papers pertaining to the theme of footprints (Schuepp et al. 1990). The application of airborne measurements for area-averaged turbulent fluxes becomes important in comprehensive experiments over heterogeneous landscapes like FIFE (Sellers et al. 1988), HAPEX-MOBILHY (André et al. 1990) or BOREAS (Sellers et al. 1997). Airborne fluxes showed a very heterogeneous picture and it was the first task to combine this picture with the underlying surface.

During BOREAS, the idea of application of the footprint tool became progressively evident, but first fluxes along a fly lag were compared with surface characteristics like the NDVI or tower measurements (Desjardins et al. 1997). The method to apply the footprint approach was shown and applied by Chen et al. (1999): The flux measured using airborne measurements is the sum of all fluxes of

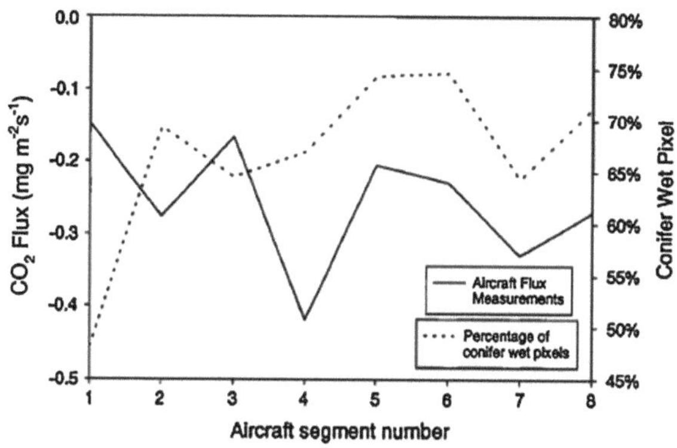

Fig. 7.15 Carbon dioxide flux measurements by an aircraft on a single line separated into eight 2 km long segments. Besides the flux the percentage of the pixels for the dominant conifer wet cover type in the footprint area is shown (Chen et al. 1999, Published with kind permission of © American Geophysical Union (Wiley), 1999. All Rights Reserved)

different flight legs—typically 2 km in most studies—and depending on a weighting function C_{kj} for each flight leg k and land use type j with M fly lags. The flux of each land use type and each flight leg is with F_j the averaged flux of each land use type:

$$F_{kj} = \sum_{k=1}^{M} C_{kj} F_j \tag{7.25}$$

The weighting function C_{kj} can be determined as

$$C_{kj} = \sum_{i=1}^{N} n_{ij} f_i \tag{7.26}$$

with i as the number of pixels in the upwind side of the flight leg and n_{ij} as the land-use type of this pixel. f_i is the footprint function which gives the weight of the pixel in the distance i. The footprint function in the case of the paper by Chen et al. (1999) was described by Kaharabata et al. (1997). An example of this first paper is given in Fig. 7.15.

This schema was too difficult to use because it was based on fluxes for each land-use type F_j measured on towers which were not representative of the whole area. Ogunjemiyo et al. (2003) proposed for the flux F_{ik} for each flight leg segment and each pixel i in this segment the following relation

$$F_{ik} = \sum_{j=1}^{K} \psi_{ijk} d_{ijk} \tag{7.27}$$

Table 7.4 Overview of key aircraft studies using footprint-related flux calculations

Airborne study	Footprint model	Remark
Desjardins et al. (1989) and Schuepp et al. (1990)	Schuepp et al. (1990)	National Aeronautical Establishment and Agriculture Canada, 1986
Schuepp et al. (1992)	Schuepp et al. (1990)	BOREAS-experiment in Canada
Chen et al. (1999)	Modification of Horst and Weil (1992, 1994) and Kaharabata et al. (1997)	BOREAS-experiment in Canada
Samuelsson and Tjernström (1999)	Schuepp et al. (1990) with modification convective conditions by Mahrt et al. (1994)	NOPEX-experiment in Sweden
Ogunjemiyo et al. (2003)	Kaharabata et al. (1997)	BOREAS-experiment in Canada, multiple regression model for aggregation
Gioli et al. (2004)	Hsieh et al. (2000)	Comparison with European FLUXNET sites
Kirby et al. (2008)	Kljun et al. (2004)	
Mauder et al. (2008)	Kljun et al. (2004) in a 2D version similar to Kormann and Meixner (2001)	GEWEX study MAGS 1999
Hutes et al. (2010)	Hsieh et al. (2000) in the 2D version by Detto et al. (2006)	Linear flux aggregation
Metzger et al. (2013)	Kljun et al. (2004) in a 2D version by Metzger et al. (2012)	China, Inner Mongolia

where d_{ijk} is the spatially averaged flux density and ψ_{ijk} a weighting function for the land cover type j. Because the flux density is difficult to measure, Ogunjemiyo et al. (2003) proposed a nonlinear multiple regression which is rarely used in most airborne data analyses. The number of pixels in the upwind direction was determined that up to 98 % of total estimated flux contribution could be included. Similar limitations comparable with the effect level approach were used by most authors. The weighting function can be determined according to

$$\psi_{ijk} = \frac{1}{R} \sum_{j}^{K} \sum_{i}^{M} I_{ijk} a_i, \tag{7.28}$$

where R is the number of pixels i in the flight segment j. $k_{ijk} = 1$ if the cover type k represents the pixel i, otherwise $I_{ijk} = 0$. a_i is a normalized footprint function for the pixel i

$$a_i = \frac{\int_{i=i_a}^{i=i_a+1} f di}{\int_{i=1}^{N} f di} \tag{7.29}$$

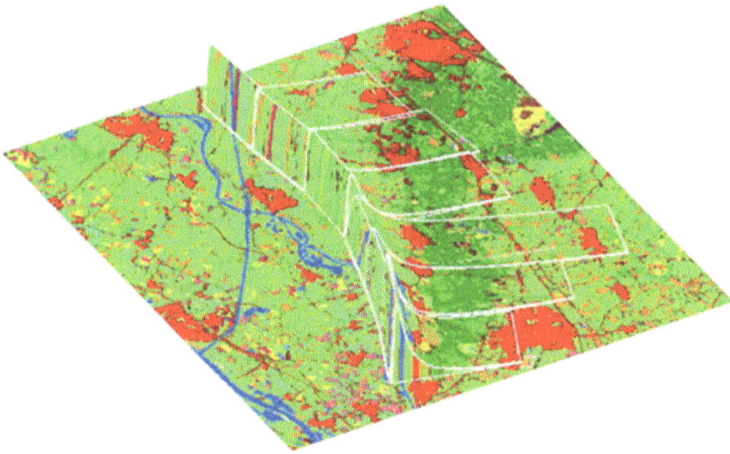

Fig. 7.16 Footprints of consecutive flux estimates along a flight track projected onto a land use map of central Netherlands. The vertical dimension gives the footprint weight, the horizontal bounds the footprint area where 90 % of the flux emanates from (Hutjes et al. 2010, Published with kind permission of © Elsevier, 2010. All Rights Reserved)

For the footprint function, Ogunjemiyo et al. (2003) used the function given by Kaharabata et al. (1997). In most aircraft studies, the above schema is applied with slight modifications (see Table 7.4). A nice example for the footprint analysis is shown by Hutjes et al. (2010) in Fig. 7.16. In contrast to earlier studies (e. g. Chen et al. 1999), the authors determine for each flight leg a footprint dependent on actual wind velocity and stability.

A deficit in aircraft footprint studies is the application of analytical approaches in homogeneous surfaces. In future studies, Lagrangian backward models or LES models should be applied. Up until now the model by Kljun et al. (2004), based on the Lagrangian backward model by Kljun et al. (2002), is used as 2D model has recognized this and included a crosswind component to their footprint models (Mauder et al. 2008; Metzger et al. 2013).

Based on aircraft investigations, Desjardins et al. (1994) pointed out that spatial variability in the flux observations can be viewed as a response to systematic changes in the flux footprint. The influence of these changes on the observed flux along a flight line can be determined from footprint investigations in combination with a linear mixing matrix. This concept has been expressed in a numerical (Chen et al. 1999) and in a regression form (Ogunjemiyo et al. 2003; Hutjes et al. 2010). Metzger et al. (2013) used these basic ideas to define an environmental response function (ERF), which relates flux observations (responses) to surface and basic meteorological properties (drivers). In their study, the land surface temperature (LST) and the enhanced vegetation index (EVI) are used as proxies for the spatial distribution of sources and sinks for sensible and latent heat, respectively. Figure 7.17 shows a low-level ($< 0.05\ z_i$) flight line of a weight-shift microlight aircraft (Metzger et al. 2011, 2012), superimposed over a land cover classification, LST and EVI from MODIS,

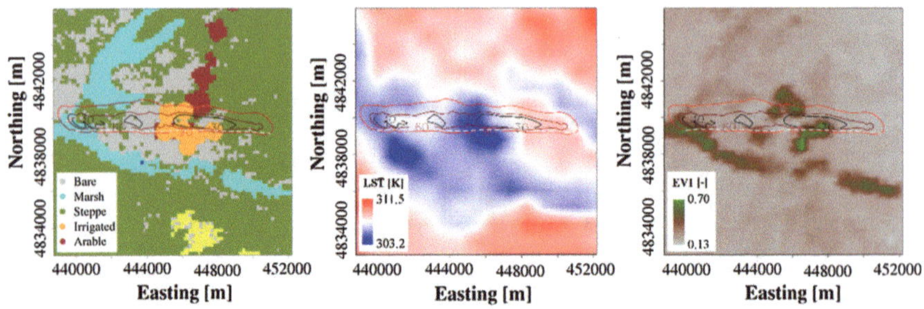

Fig. 7.17 Flight along a pattern over Northeast China on 8 July 2009, 12:16–12:24 BST (*white dashed line*) from (Metzger et al. 2013). The composite flux footprint along the flight line (30, 60, 90 % *contour lines*) is superimposed over maps of land cover (*left panel*), land surface temperature (LST, *center panel*), and enhanced vegetation index (EVI, *right panel*) Published with kind permission of © Copernicus Publications, distributed under the Creative Commons Attribution 3.0 License, 2007. All Rights Reserved

respectively. Generally speaking, the medium spatial resolution (250–1000 m) of the MODIS data is less-than-ideal, but more importantly, the MODIS satellite data enables considering temporal changes in the surface properties. In contrast to prior attempts, Metzger et al. (2013) used (i) time-frequency analysis to increase the sample size along a flight line (ii) continuous and contemporary representations of the land cover, and (iii) a non-parametric machine learning technique (Elith et al. 2008) to determine the ERFs. The resulting parameter of determination between the drivers and the fluxes of sensible and latent heat was surprisingly high with $R^2 \approx 0.99$. Provided the ERFs are well calibrated using direct flux measurements, this method can be used to determine spatially resolved turbulent fluxes remote sensing data to within 20 % accuracy.

References

Akima H (1970) A new method of interpolation and smooth curve fitting based on local procedures. J Assc Comp Mach 17:589–602

André J-C, Bougeault P, Goutorbe J-P (1990) Regional estimates of heat and evaporation fluxes over non-homogeneous terrain, examples from the HAPEX-MOBILHY programme. Bound-Layer Meteorol 50:77–108

Andreas EL (1989) Two-wavelength method of measuring path-averaged turbulent surface heat fluxes. J Atm Oceanic Tech 6:280–292

Arya SP (2001) Introduction to micrometeorology. Academic Press, San Diego 415 pp

Aubinet M (2008) Eddy covariance CO_2 flux measurements in nocturnal conditions: an analysis of the problem. Ecol Appl 18:1368–1378

Aubinet M, Vesala T, Papale D (2012) Eddy covariance: a practical guide to measurement and data analysis. Springer, Dordrecht 438 pp

Baldocchi D (1997) Flux footprints within and over forest canopies. Bound-Layer Meteorol 85:273–292

Bowen IS (1926) The ratio of heat losses by conduction and by evaporation from any water surface. Phys Rev 27:779–787

Businger JA (1982) Equations and concepts. In: Nieuwstadt FTM, Van Dop H (eds) Atmospheric turbulence and air pollution modelling: a course held in the hague, 21-25 September 1981. D. Reidel Publishing Company, Dordrecht, pp 1–36

Businger JA (1986) Evaluation of the accuracy with which dry deposition can be measured with current micrometeorological techniques. J Appl Meteorol 25:1100–1124

Chen JM, Leblanc SG, Cihlar J, Desjardins RL, MacPherson IJ (1999) Extending aircraft- and tower-based CO_2 flux measurements to a boreal region using a Landsat thematic mapper land cover map. J Geophys Res 104(D14):16,859–816,877

DeBruin HAR (2002) Introduction: renaissance of scintillometry. Bound-Layer Meteorol 105:1–4

Desjardins RL, MacPherson JI, Schuepp PH, Karanja F (1989) An evaluation of aircraft flux measurements of CO_2, water vapor and sensible heat. Bound-Layer Meteorol 47:55–69

Desjardins RL, MacPherson JI, Schuepp PH, Hayhoe HN (1994) Airborne flux measurements of CO_2, sensible, and latent heat over the hudson bay lowland. J Geophys Res Atmos 99:1551–1561

Desjardins RL et al (1997) Scaling up flux measurements for the boreal forest using aircraft-tower combination. J Geophys Res 102(D24):29125–29133

Detto M, Montaldo N, Albertson JD, Mancini M, Katul G (2006) Soil moisture and vegetation controls on evapotranspiration in a heterogeneous mediterranean ecosystem on sardinia. Italy Water Resour Res 42:W08419

Dugas WA, Fritschen LJ, Gay LW, Held AA, Matthias AD, Reicosky DC, Steduto P, Steiner JL (1991) Bowen ratio, eddy correlation, and portable chamber measurements of sensible and latent heat flux over irrigated spring wheat. Agric For Meteorol 56:12–20

Elith J, Leathwick JR, Hastie T (2008) A working guide to boosted regression trees. J Anim Ecol 77:802–813

Finnigan JJ, Clement R, Malhi Y, Leuning R, Cleugh HA (2003) A re-evaluation of long-term flux measurement techniques, part I: averaging and coordinate rotation. Bound-Layer Meteorol 107:1–48

Foken T, Wichura B (1996) Tools for quality assessment of surface-based flux measurements. Agric For Meteorol 78:83–105

Foken T, Richter SH, Müller H (1997) Zur Genauigkeit der Bowen-Ratio-Methode. Wetter Leben 49:57–77

Foken T (1998) Genauigkeit meteorologischer Messungen zur Bestimmung des Energie- und Stoffaustausches über hohen Pflanzenbeständen. Ann Meteorol 37:513–514

Foken T, Göckede M, Mauder M, Mahrt L, Amiro BD, Munger JW (2004) Post-field data quality control. In: Lee X et al (eds) Handbook of micrometeorology: a guide for surface flux measurement and analysis. Kluwer, Dordrecht, pp 181–208

Foken T (2008) Micrometeorology. Springer, Berlin, 308 pp

Foken T, Leuning R, Oncley SP, Mauder M, Aubinet M (2012a) Corrections and data quality. In: Aubinet M et al (eds) Eddy covariance: a practical guide to measurement and data analysis. Springer, Dordrecht, pp 85–131

Foken T, Aubinet M, Leuning R (2012b) The eddy-covarianced method. In: Aubinet M et al (eds) Eddy covariance: a practical guide to measurement and data analysis. Springer, Dordrecht, pp 1–19

Fritschen LJ, Fritschen CL (2005) Bowen ratio energy balance method. In: Hatfield JL, Baker JM (eds) Micrometeorology in agricultural systems. American Society of Agronomy, Madison, pp 397–405

Gioli B et al (2004) Comparison between tower and aircraft-based eddy covariance fluxes in five European regions. Agric For Meteorol 127:1–16

Göckede M, Markkaken T, Mauder M, Arnold K, Leps JP, Foken T (2005) Validation of footprint models using natural tracer measurements from a field experiment. Agric For Meteorol 135:314–325

Gurjanov AE, Zubkovskij SL, Fedorov MM (1984) Mnogokanalnaja avtomatizirovannaja sistema obrabotki signalov na baze EVM (automatic multi-channel system for signal analysis with electronic data processing). Geod Geophys Veröff R II 26:17–20

Haugen DA (ed) (1973) Workshop on micrometeorology. American Meteorological Society, Boston, 392 pp

Hill R (1997) Algorithms for obtaining atmospheric surface-layer from scintillation measurements. J Atm Oceanic Tech 14:456–467

Hill RJ, Clifford SF, Lawrence RS (1980) Refractive index and absorption fluctuations in the infrared caused by temperature, humidity and pressure fluctuations. J Opt Soc Am 70:1192–1205

Horst TW, Weil JC (1992) Footprint estimation for scalar flux measurements in the atmospheric surface layer. Bound-Layer Meteorol 59:279–296

Horst TW, Weil JC (1994) How far is far enough?: the fetch requirements for micrometeorological measurement of surface fluxes. J Atm Oceanic Tech 11:1018–1025

Horst TW (1999) The footprint for estimation of atmosphere-surface exchange fluxes by profile techniques. Bound-Layer Meteorol 90:171–188

Hsieh C-I, Katul G, Chi T-W (2000) An approximate analytical model for footprint estimation of scalar fluxes in thermally stratified atmospheric flows. Adv Water Res 23:765–772

Hutjes RWA, Vellinga OS, Gioli B, Miglietta F (2010) Dis-aggregation of airborne flux measurements using footprint analysis. Agric For Meteorol 150:966–983

Kaharabata SK, Schuepp PH, Ogunjemiyo S, Shen S, Leclerc MY, Desjardins RL, MacPherson JI (1997) Footprint considerations in BOREAS. J Geophys Res 102(D24):29113–29124

Kaimal JC, Finnigan JJ (1994) Atmospheric boundary layer flows: their structure and measurement. Oxford University Press, New York 289 pp

Kirby S, Dobosy R, Williamson D, Dumas E (2008) An aircraft-based data analysis method for discerning individual fluxes in a heterogeneous agricultural landscape. Agric For Meteorol 148:481–489

Kljun N, Rotach MW, Schmid HP (2002) A three-dimensional backward Lagrangian footprint model for a wide range of boundary layer stratification. Bound-Layer Meteorol 103:205–226

Kljun N, Calanca P, Rotach M, Schmid HP (2004) A simple parameterization for flux footprint predictions. Bound-Layer Meteorol 112:503–523

Kormann R, Meixner FX (2001) An analytical footprint model for non-neutral stratification. Bound-Layer Meteorol 99:207–224

Leclerc MY, Thurtell GW (1990) Footprint prediction of scalar fluxes using a Markovian analysis. Bound-Layer Meteorol 52:247–258

Lee X (2004) A model for scalar advection inside canopies and application to footprint investigation. Agric For Meteorol 127:131–141

Lee X, Massman WJ, Law B (eds) (2004) Handbook of micrometeorology: a guide for surface flux measurement and analysis. Kluwer, Dordrecht, 250 pp

Liu H, Foken T (2001) A modified Bowen ratio method to determine sensible and latent heat fluxes. Meteorol Z 10:71–80

Mahrt L, Sun J, Vickers D, MacPherson JI, Pederson JR, Desjardins RL (1994) Observations of fluxes and inland breezes over a heterogeneous surface. J Atmos Sci 51:2484–2499

Marquardt D (1983) An algorithm for least-squares estimation of nonlinear parameters. J Soc Ind Appl Math 11:431–441

Mauder M, Desjardins R, MacPherson I (2008) Creating surface flux maps from airborne measurements: application to the Mackenzie area GEWEX study MAGS 1999. Bound-Layer Meteorol 129:431–450

Meijninger WML, Green AE, Hartogensis OK, Kohsiek W, Hoedjes JCB, Zuurbier RM, DeBruin HAR (2002) Determination of area-averaged water vapour fluxes with large aperture and radio wave scintillometers over a heterogeneous surface: Flevoland field experiment. Bound-Layer Meteorol 105:63–83

Metzger S, Junkermann W, Butterbach-Bahl K, Schmid HP, Foken T (2011) Corrigendum to "Measuring the 3-D wind vector with a weight-shiftmicrolight aircraft" published in atmospheric measuring technique, 4:1421–1444, 1515–1539

Metzger S, Junkermann W, Mauder M, Beyrich F, Butterbach-Bahl K, Schmid HP, Foken T (2012) Eddy-covariance flux measurements with a weight-shift microlight aircraft. Atmos Meas Tech 5:1699–1717

Metzger S et al (2013) Spatial resolution and regionalization of airborne flux measurements using environmental response functions. Biogeochemistry 10:2193–2217

Moncrieff J (2004) Surface turbulent fluxes. In: Kabat P et al (eds) Vegetation, water, humans and the climate: a new perspective on an interactive system. Springer, Berlin, pp 173–182

Nappo CJ et al (1982) The workshop on the representativeness of meteorological observations, June 1981, Boulder CO. Bull Am Meteorol Soc 63:761–764

Nieuwstadt FTM (1978) The computation of the friction velocity u_* and the temperature scale T_* from temperature and wind velocity profiles by least-square method. Bound-Layer Meteorol 14:235–246

Ogunjemiyo SO, Kaharabata SK, Schuepp PH, MacPherson IJ, Desjardins RL, Roberts DA (2003) Methods of estimating CO_2, latent heat and sensible heat fluxes from estimates of land cover fractions in the flux footprint. Agric For Meteorol 117:125–144

Ohmura A (1982) Objective criteria for rejecting data for Bowen ratio flux calculations. J Climate Appl Meteorol 21:595–598

Panofsky HA, Dutton JA (1984) Atmospheric turbulence: models and methods for engineering applications. Wiley, New York 397 pp

Rannik Ü, Aubinet M, Kurbanmuradov O, Sabelfeld KK, Markkanen T, Vesala T (2000) Footprint analysis for measurements over heterogeneous forest. Bound-Layer Meteorol 97:137–166

Rannik Ü, Markkanen T, Raittila T, Hari P, Vesala T (2003) Turbulence statistics inside and above forest: Influence on footprint prediction. Bound-Layer Meteorol 109:163–189

Samuelsson P, Tjernström M (1999) Airborne flux measurements in NOPEX: comparison with footprint estimated surface heat fluxes. Agric For Meteorol 98–99:205–225

Schmid HP (1997) Experimental design for flux measurements: matching scales of observations and fluxes. Agric For Meteorol 87:179–200

Schotanus P, Nieuwstadt FTM, DeBruin HAR (1983) Temperature measurement with a sonic anemometer and its application to heat and moisture fluctuations. Bound-Layer Meteorol 26:81–93

Schuepp PH, Leclerc MY, MacPherson JI, Desjardins RL (1990) Footprint prediction of scalar fluxes from analytical solutions of the diffusion equation. Bound-Layer Meteorol 50:355–373

Schuepp PH, MacPherson JI, Desjardins RL (1992) Adjustment of footprint correction for airborne flux mapping over the FIFE site. J Geophys Res 97(D17):18455–18466

Sellers PJ, Hall FG, Asrar G, Strebel DE, Murphy RE (1988) The first ISLSCP field experiment (FIFE). Bull Am Meteorol Soc 69:22–27

Sellers PJ et al (1997) BOREAS in 1997: experiment overview, scientific results, and future directions. J Geophys Res 102:28(769)731–728

Stannard DI (1997) A theoretically based determination of Bowen-ratio fetch requirements. Bound-Layer Meteorol 83:375–406

Stull RB (1988) An introduction to boundary layer meteorology. Kluwer Academic Publisher, Dordrecht, 666 pp

Tanner BD (1988) Use requirements for Bowen ratio and eddy correlation determination of evapotranspiration. In: Proceedings of the 1988 speciality conference of the irrigation and drainage devisions, ASCE Lincoln, Nebraska, 19–21 July 1988

Thiermann V, Grassl H (1992) The measurement of turbulent surface layer fluxes by use of bichromatic scintillation. Bound-Layer Meteorol 58:367–391

Vickers D, Mahrt L (1997) Quality control and flux sampling problems for tower and aircraft data. J Atm Oceanic Tech 14:512–526

Wilczak JM, Oncley SP, Stage SA (2001) Sonic anemometer tilt correction algorithms. Bound-Layer Meteorol 99:127–150

Wyngaard JC, Izumi Y, Collins SA (1971) Behavior of the refractive-index-structure parameter near the ground. J Opt Soc Am 61:1646–1650

Chapter 8
Practical Applications of Footprint Techniques

This chapter is intended for readers interested in applying footprint techniques without becoming a specialist of the method. This format was chosen given the importance of footprint techniques to users from the fields of ecology, agricultural and forest meteorology, engineers, hydrologists and air quality specialists resorting to meteorological measurements as part of their work. This chapter thus focuses on topics where practical solutions can be easily applied.

8.1 Selection of Flux Measurement Sites

To ensure the quality of micrometeorological and especially of flux measuring data, a complete quality assurance procedure for meteorological measurements as described in Sect. 7.2.4 must be performed (Shearman 1992; DeFelice 1998). This quality evaluation should include the careful choice of the instrument location. The latter depends on criteria such as the measurement technique, the instrument types, and the characteristics of the surrounding terrain.

The starting point for the selection of a suitable measurement location should be a review of the dataset requirements (Munger et al. 2012). For example, the first and fundamental question is whether a user expects a continuous high-quality dataset, or if it will be sufficient to have good measurements only for selected wind directions or times of the day. In the first case, finding a suitable site will be much more difficult because large homogeneous areas with good measurement conditions in all wind directions are rare, particularly in landscapes affected by management and with natural disturbances. In the second case, provided it is acceptable to exclude only just a single sector (ideally the one with the smallest frequency of wind directions), the size of the area with homogeneous surface

M. Y. Leclerc and T. Foken, *Footprints in Micrometeorology and Ecology*,
DOI: 10.1007/978-3-642-54545-0_8, © Springer-Verlag Berlin Heidelberg 2014

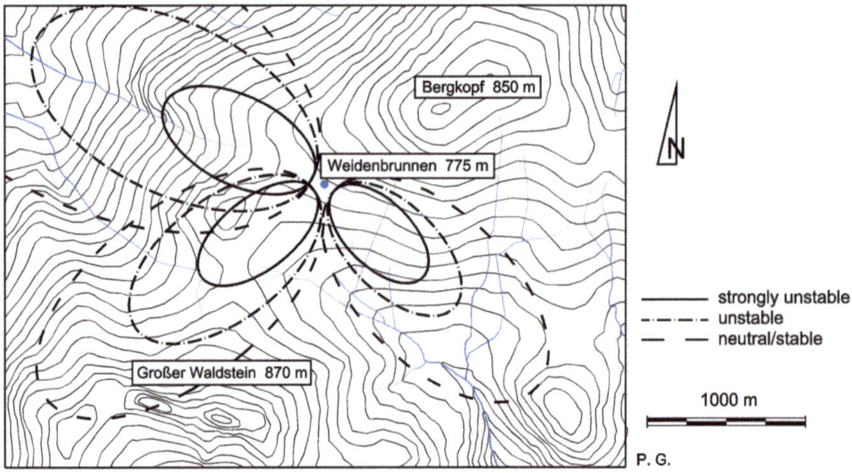

Fig. 8.1 Footprints for different wind directions, related to different roughnesses, and stabilities for the FLUXNET site DE-Bay (Waldstein-Weidenbrunnen) for a data set in summer 1998 (Foken and Leclerc 2004, Published with kind permission of © Elsevier, 2004. All Rights Reserved)

characteristics required to ensure high quality data is significantly reduced, thus increasing the number of suitable candidate sites.

To identify the importance of certain wind direction sectors to the measurements made at a candidate site, it will generally suffice to use climatological information records from a nearby weather station to create a wind rose. More customized (and more reliable) results can be achieved through test measurements of wind velocity and direction at the planned location itself, ideally covering different periods of the year. Though local measurements require more resource investments, this approach is highly recommended particularly if the planned measurement station is located within complex terrain, i.e. topography with steep slopes and heterogeneous land cover structure. If only daytime measurements (those with the highest fluxes) are required, the footprint area that should be homogeneous and flat to ensure high quality observations will be relatively small (some 100 m, depending on the measurement height); however, during nighttime or wintertime conditions with stable stratification of the atmospheric boundary layer, the footprint area increases by a vast amount, generally by several kilometres (Fig. 8.1).

When selecting a measurement location according to 'good' and 'poor' wind sections, the impact of the instruments and/or the tower structure itself on data quality should also be considered. If the way the instrumentation is set up below the top of the tower implies that data from a certain direction is likely to be impacted by flow distortion effects, this 'poor' wind sector should be the one with the lowest wind direction frequencies.

Once the basic requirements for the planned measurement program have been formulated, the first step to narrow down potential site locations is to check the distances to the nearest discontinuity in surface conditions in different wind directions. Such discontinuities can be either in the surface roughness (e.g. a transition from grassland to forest), or in thermal conditions of the surface (e.g. from a dark surface to a light-colored one), both of which can generate internal boundary layers (see Sect. 2.3). If the tower position is too close to a discontinuity, internal boundary layers can seriously degrade the quality of micrometeorological measurements; a second relevant consideration is that footprint simulations may considerably degrade with respect to the quality of their results. It is therefore critical to check for possible internal boundary layers using the following fetch/height relationship to roughly estimate the height of the new equilibrium layer downwind of a roughness change depending on fetch x (Raabe 1983; Jegede and Foken 1999) neglecting weak stability effects (Savelyev and Taylor 2005).

$$z_m - d < \delta = 0.3\sqrt{x} \tag{8.1}$$

with the measurement height z_m, the zero-plane displacement d, the height of the new equilibrium layer δ, and the fetch distance to the nearest roughness transition, x. The relationship is valid for neutral and unstable well-mixed conditions. For calm and stably stratified conditions, this relation would be inaccurate. Yet, at present no suitable alternative is available.

An example describing the analysis is shown over an agricultural field in Fig. 8.2. If the height of new equilibrium layer falls below the measuring height, the measurements must be excluded. This is because in that case the sensor is decoupled from the fluxes of the agricultural field itself. To minimize data losses, the choice of the position of a measurement system on a field should take into consideration the wind climatology at the site, matching short fetches with directions that have the lowest frequencies of occurrence.

In the case given in Fig. 8.2, the instruments were placed in the northern part of the field because wind directions from the sector N-NE were rare. The distance to the next roughness discontinuity, a wind break consisting of small bushes, was chosen to be about ten times the height of these obstacles to still allow for an acceptable data quality during daytime conditions. A simple rule of thumb for observations of good quality to determine the distance to the next obstacle is two times of the obstacle height (WMO 1981) for conventional meteorological measurements, and five to ten times for flux measurements (Foken 2008). However, these are just rough guidelines, and a full footprint analysis should always be performed for a final fetch evaluation, e.g. in this case of the setup in Table 8.1 to determine whether or not the NW sector must be excluded from further analysis.

A flow chart summarizing the necessary steps to prepare and perform a footprint analysis for a given site is shown in Fig. 8.3. The work flow includes the preparation steps discussed above, the selection of a land cover map (see also Chap. 6) and finally the application of a footprint model. As a rule of thumb, for low measurement heights over short vegetation, an analytical footprint model can provide source-weight functions of sufficient accuracy, while for high vegetation and/or complex topography, more sophisticated approaches such as Lagrangian Stochastic models should be used. These models are described in detail in Sect 3.2.

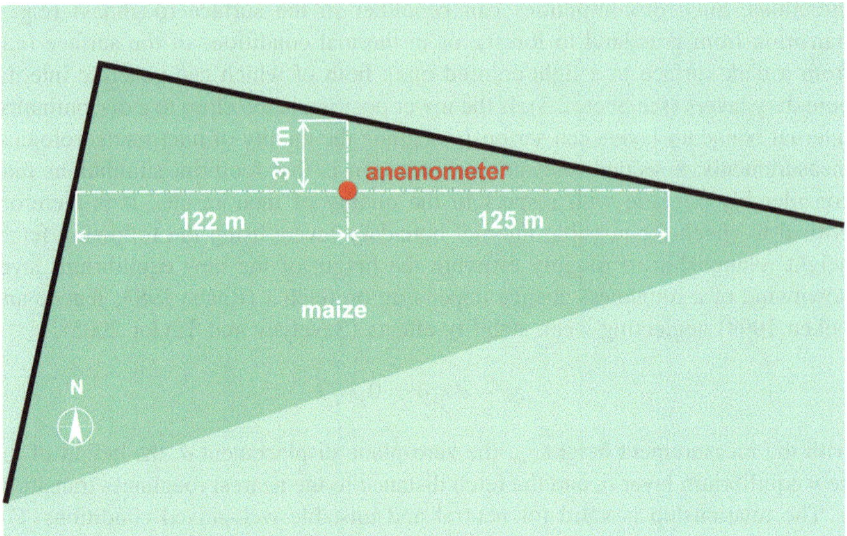

Fig. 8.2 Field (maize) site of the LITFASS-2003 experiment related to Table 8.1

Table 8.1 Fetch x and height of the new equilibrium layer δ for a maize field during the LITFASS-2003 experiment (Mauder et al. 2006)

	30°	60°	90°	120°	150°	180°	210°	240°	270°	300°	330°	360°
x in m	29	41	125	360	265	203	211	159	122	81	36	28
δ *in m*	1.6	1.9	3.4	5.7	4.9	4.3	4.4	3.8	3.3	2.7	1.8	1.6

For the given measurement height of 2.7 m follows that the grey wind directions must be excluded from the data analysis (see Fig. 8.2)

Analytical footprint models (Sect. 3.1) have been widely applied to characterize the 'field of view' of eddy-covariance measurements (Schmid 1997; Rebmann et al. 2005). Their popularity is mainly based on their relative mathematical simplicity (e.g. Schmid 2002) that allows integrating them into eddy-covariance processing software packages without high additional computational expense, or even estimate fetch through spreadsheet applications. This simplicity is what makes them attractive as a component in site evaluation tools, since in particular network studies require the processing of tens of thousands of footprint estimates. Analytic footprint models are often restricted to rather narrow ranges of input parameters. Since analytical footprint estimates tend to be larger than stochastic ones due to the neglect of along wind diffusion, and at the same time the land cover structure is usually more heterogeneous with increasing distance from the tower (tower location are commonly selected to be homogeneous at least in the near field), the site evaluations based on analytical models provide a conservative estimate of quality results.

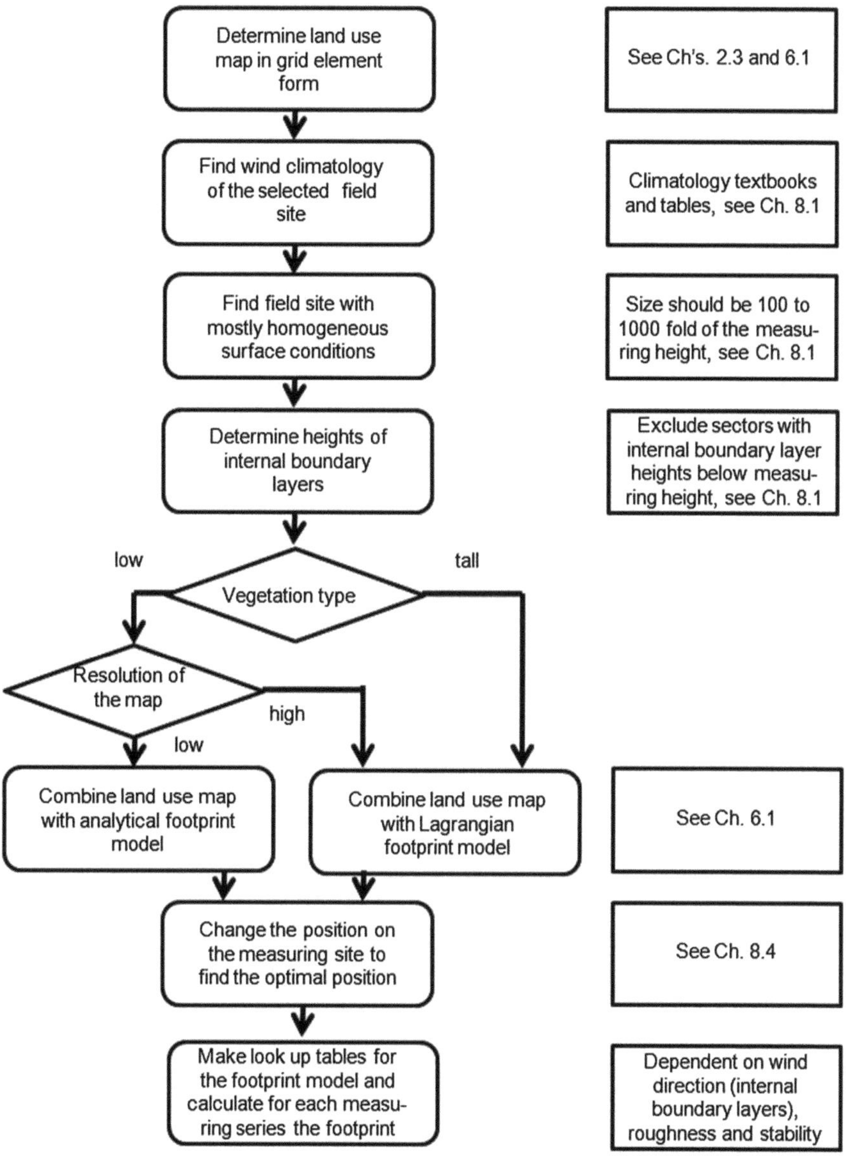

Fig. 8.3 Schema to find the best position of the measuring point for measurements in the surface layer and to apply footprint modeling

Lagrangian Stochastic (LS) footprint models offer more ways of adaptation to local measurement conditions, which is particularly valuable for studies over tall vegetation or for non-homogeneous surfaces. However, gains in accuracy achieved by e.g. the consideration of within-canopy transport (Baldocchi 1997; Rannik et al.

2003; Sogachev et al. 2005b; Klaassen and Sogatchev 2006; Prabha et al. 2008), sources at multiple levels (Markkanen et al. 2003), or along wind diffusion (Rannik et al. 2000), come along with significantly increased computational expense, which plays a major role for site evaluation concepts that cover multiple sites over a timeframe of several months. Also, the quality of the simulations is dependent on a reliable description of vertical turbulence profiles under various atmospheric conditions (Göckede et al. 2007), while high quality datasets to describe these profiles are rarely available. Application of Lagrangian simulation models in extensive site evaluation studies therefore usually calls for simplifications in the setup, such as the use of profiles of the wind velocity and its standard deviation that are not customized for each specific forest stand, and the pre-calculation of source weight functions for specific combinations of atmospheric stability, measurement height and terrain roughness (Göckede et al. 2006).

If local measurements of typical fluxes and turbulence conditions are not yet available at the time of the selection of the tower position, synthetic input datasets have to be created that approximate the expected terrain and atmospheric conditions as closely as possible. For this purpose, it is usually sufficient to run a footprint analysis for three major classes of stability of stratification, using e.g. $z/L \sim -0.3$ to -0.5 for unstable stratification, $z/L \sim -0.1$ to 0.1 for neutral stratification, and $z/L \sim 0.3$ to 0.5 for stable stratification. Additional input parameters can either be derived for these stability classes (e.g. the friction velocity), or estimated from literature (e.g. roughness length values for given types of vegetation). This way, a set or 'typical' meteorological conditions can be constructed that allows analyzing the fetch conditions for major wind sectors. The aim is the creation of a table like Table 8.2 which shows the contribution of the target area for each wind sector (i.e. the land cover type to be sampled) on the flux footprint for different stability ranges. Such a table will give sufficient insight into the expected measurement conditions at candidate sites, and can later be refined when local measurements will be available.

The interpretation of the footprint results comprised in this Table should be based on the contribution of the surface of interest (target area) to the total fluxes measured, as e.g. suggested by Göckede et al. (2008). They recommended a classification scheme based on the degree of homogeneity of flux sources within the source area by defining thresholds of target area flux contributions and defined four different classes:

- Class 1: Homogeneous measurements, with 95 % or more of the flux emitted by the target land cover type
- Class 2: Representative measurements (80–95 %)
- Class 3: Acceptable measurements (50–80 %)
- Class 4: Disturbed measurements (<50 %)

The instruments should be installed at a location that ensures a composition of flux sources in the fetch that falls within classes 1 and 2 most of the time, i.e. the target land cover should ideally contribute >80 % of the fluxes all the time. To highlight this rule, classes 3 and 4 are shaded in grey in Table 8.2. If short fetches

Table 8.2 Flux contribution in percent of the target land cover type dependent on wind direction and stability for a maize field during the LITFASS-2003 experiment (Mauder et al. 2006, see Fig. 8.2)

	30°	60°	90°	120°	150°	180°	210°	240°	270°	300°	330°	360°
stable	26	37	76	97	93	84	86	81	76	61	37	26
neutral	56	67	100	100	100	100	100	100	100	88	67	56
unstable	76	87	100	100	100	100	100	100	100	98	87	76

Class 1 and 2 are given in white and class 3 is grey shaded and the class 4 which should not be used in dark grey (classes according to Göckede et al. 2008)

and heterogeneous source areas cannot be avoided at a candidate site, results as shown in Tables 8.1 and 8.2 should be combined with information on the local wind climatology to minimize the contribution from 'poor' sectors.

The output from simple fetch analyses and more complex footprint studies are well correlated (Tables 8.1 and 8.2). This is not surprising because Horst (1999) found that the typical horizontal and vertical scales used in footprint analysis and internal boundary layers concepts are nearly identical. Differences are mainly caused by the simplification included in the fetch estimation approach based on Eq. (8.1). However, results also highlight that a footprint analysis can yield a superior level of detail and quantitative results, which may greatly facilitate the evaluation of potential observation sites.

8.2 Interpretation of Flux Data

20–40 years ago, most flux experiments took place over homogeneous surfaces. Then footprint issues did not bear as high a relevance in data interpretation. However, starting with experiments like FIFE (Sellers et al. 1988), HAPEX-MOBILHY (André et al. 1990), and KUREX-88 (Tsvang et al. 1991) a growing number of eddy-covariance sites was placed in areas with a heterogeneous surface structure. The expansion of the eddy-covariance network coincided with a shift from ideal, homogeneous sites to complex, heterogeneous terrain (Schmid 2002). Most of the recently developed flux sites are organized in networks such as the international network called FLUXNET (Baldocchi et al. 2001) or in continental counterparts such as AmeriFlux and AsiaFlux or in national networks such as MEDFLU, ChinaFlux. A large portion of these about (over six hundred sites) are located in less than ideal terrain, including many forested sites characterized by a mixture of conifer and deciduous forest types and clearings.

Therefore, footprint models are necessary for a reliable interpretation of data collected within heterogeneous landscapes. However, a fundamental problem for this application is that most footprint models—particularly the simplest of use—were originally developed for homogeneous surfaces (see Chaps. 1, 3 and 4). Accordingly, their application in heterogeneous terrain does not fulfill the theory behind the model, and additional uncertainties have to be taken into account. This shortcoming should be considered in most practical footprint applications, but nevertheless the application is possible in most cases. This was shown by Foken and Leclerc (2004) for the validation of footprint models in heterogeneous landscape and by Markkanen et al. (2010) who compared different footprint models at an abrupt change of surface characteristics. In the following sub-sections, the concepts of footprint climatology (Amiro 1998) and the quality of eddy-covariance data depending on the footprint are described.

8.2.1 Footprint Climatology

An integral part of the footprint-based quality assessment approach is the average source weight function created over a longer measurement period, i.e. the footprint climatology (e.g. Amiro 1998). In Amiro's analysis, the source weight functions of individual eddy-covariance measurements were summed up for $10 \times 10\ \mathrm{m}^2$ pixel sizes within 2° radial lines around a center point based on the one-dimensional footprint model by Horst and Weil (1992). This was done for different time periods and reflects the local wind climatology. The result is shown in Fig. 8.4 for a one-month period and the summer season. The comparison of the footprint area with the size of the catchment area gives an impression how the measurements reflect the fluxes of the catchment in different periods.

With the two-dimensional footprint models (Schmid 1997; Rannik et al. 2003), it is possible to calculate footprint climatologies for individual grid cells rather than integrated over a wind direction sector (Göckede et al. 2004, 2006). For such a more detailed analysis, each 30-min source-weight function is projected onto the gridded land-cover map, assigning a weighting factor to each grid cell that represents its relative contribution to actual measurements. The distribution of flux contributions from the different land cover types can then be obtained by accumulating these weights arranged in different land-cover types. The application of footprint analysis to a larger dataset reveals patterns in the composition of the footprint that depend on wind sector and stability regime. This information is particularly valuable in case a dataset represents a certain 'target land-cover type', e.g. in site intercomparisons. It can also be an invaluable contribution to a rigorous interpretation dataset used in ecophysiological models. For such applications, the footprint results can be used to provide the percentage contribution of the specified target land-cover type to the total flux, and measurements that fail to reach a user-specified minimum threshold can be discarded from the database (e.g. Nagy et al. 2006).

Figure 8.5 demonstrates the variability of footprint climatologies under different environmental conditions, highlighting also the change in the composition of land-cover types within the footprint with varying conditions. In these two-dimensional visualizations, the white contours indicate the 3-dimensional topography of the footprint climatologies, with the most influential terrain areas located in the center of the concentric rings.

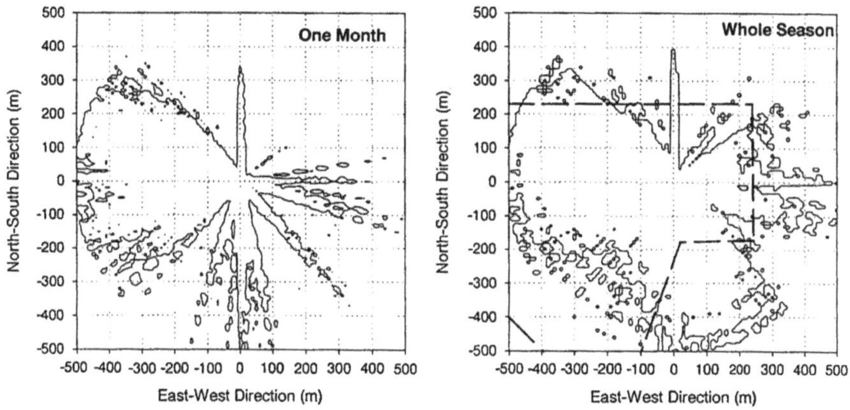

Fig. 8.4 Footprint climatology for 90 % of the total footprint of a boreal forest catchment in southern Manitoba, Canada, for July (*left*) and April 28–Oct. 6, 1985 (*right*). The catchment boundary is shown as a *dashed line* on the *right panel* (Amiro 1998)

Footprint climatologies are of a high interest for many practical applications. As Stoughton et al. (2000) have shown for the Duke Forest FACE (Free Air—Carbon dioxide Enrichment) experiment, it was possible to quantify the influence of the adjacent areas on the experimental plots dependent on time of day and tree height even with a simple analytical model.

8.2.2 Covering the Area of Interest

In the evaluation of the spatial representativeness of an eddy-covariance flux dataset, footprint climatologies integrated to land cover maps (Chap. 6) as shown in Fig. 8.5 already provide a first impression on the potential impact of terrain heterogeneity on the observations. The most prominent land cover classes within the area encircled by the white isolines will also dominate the flux measurements. "Upwind disturbances in the wind path", such as clearings in a forest, or different forest age classes (chronosequences) such as in a forest plantation, will have a significant impact on the quality of the dataset if they are situated within a central position of the footprint climatology effect level rings.

This information can be used to characterize the variability in the flux time series caused by a changing field-of-view of the sensors, and ideally the total flux can be decomposed into flux contributions from different biomes (Soegaard et al. 2003; Wang et al. 2006; Barcza et al. 2009) or land-cover types.

If data from a homogeneous flux source is required, for example to train a model to be used in a specific biome such as a coniferous forest, the footprint can act as a filter to indicate which measurements provide the 'true' forest signal, and

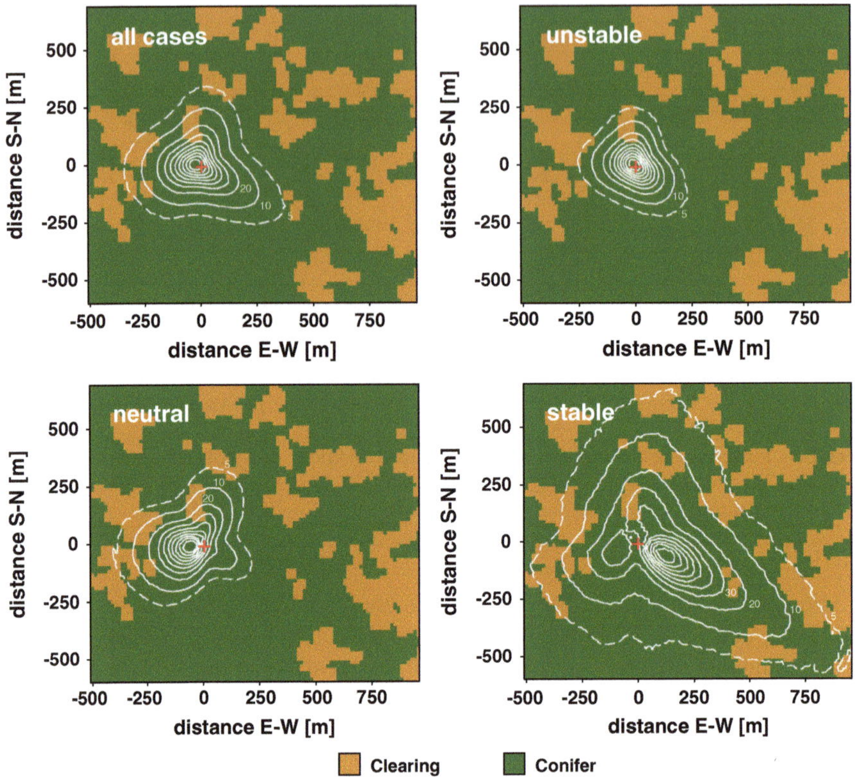

Fig. 8.5 *Top-down view* on footprint climatologies (*white lines*), accumulated for different regimes of atmospheric stability and obtained for the FLUXNET DE-Bay Waldstein-Weidenbrunnen site. Panels give footprint climatologies for all cases (*top left*), unstable (*top right*), neutral (*bottom left*) and stable (*bottom right*) stratification. Values are in percentages to the peak of the function, with *solid lines* ranging from 90 to 10 %, and the *dashed line* as 5 % of the maximum. High values indicate a high relative contribution of the specific area to the fluxes measured in the given observation period. *Colors* in the background indicate land cover classes. Distances to the tower position (*red cross*) are given in m. The analysis based on a nearly 3 month data set in summer 2003. Note, due to a storm event in 2007 the area of clearings is now much larger as given in this figure (Foken et al. 2012).

which are 'contaminated' by e.g. clearings or water bodies (Rebmann et al. 2005; Göckede et al. 2008). In network intercomparison studies, it is recommended to classify the homogeneity of flux sources within the source area by defining thresholds of target area flux contributions as already given in Sect. 8.1. This can be applied for site intercomparisons to show what percentage of the total dataset at each site could e.g. be classified as homogeneous or representative measurements, which can serve as an indicator how well the sites could be compared, or how suitable they are for model training focusing on a specific biome. An example of European FLUXNET sites is shown in Table 8.3.

Table 8.3 Representativeness results for the specified target land cover type

FLUXNET site code	Class 1, >95 % of flux	Class 2, >80 % of flux	Class 3, >50 % of flux
BE-Vielsam	41.5	92.3	99.9
BE-Brasschaat	20.6	42.5	98.9
CZ-Bily Kriz 1	99.7	100.0	100.0
DE-Hainich	83.5	99.9	100.0
DE-Tharandt	18.4	90.5	99.9
DE-Bayreuth	19.0	64.0	100.0
DE-Wetzstein	56.9	100.0	100.0
DK-Soroe	0.0	9.5	99.3
ES- Las Majadas del Tietar	96.3	99.9	100.0
FI-FI-Hyy	9.1	59.0	99.9
FI-Sodankyla	7.6	97.4	99.7
FR-Hesse Forest	11.7	84.5	99.0
FR-Le Bray	33.2	68.5	99.3
FR-Puechabon	99.9	100.0	100.0
IL-Yatir	43.0	92.7	100.0
IT-Collelongo	89.9	100.0	100.0
IT-Renon/Ritten	86.6	99.9	100.0
IT-Roccarespampani 1	0.3	86.2	99.3
IT-Zerbolo-Parco Ticino	0.1	53.7	96.7
IT-San Rossore	0.0	55.9	98.0
NL-Loobos	72.8	99.9	100.0
PT-Espirra	3.6	97.6	99.7
PT-Mitra	99.9	100.0	100.0
UK-Griffin	99.2	100.0	100.0

Values indicate the percentage of 30-min measurements for each site that fall into the classes 1 (homogeneous measurements, with 95 % or more of the flux emitted by the target land cover type), 2 (representative measurements, 80–95 %) and 3 (Acceptable measurements, 50–80 %), see also Sect. 8.1 (Göckede et al. 2008)

8.2.3 Footprint-Dependent Data Quality Control

The eddy-covariance technique used to measure turbulent fluxes between surface and atmosphere is restricted to basic theoretical assumptions (see Sect. 7.2). Deviations from these assumptions will increase measurement uncertainty, and thus have a negative impact on overall data quality. Since horizontal homogeneity in the wind field and of the surface is recommended, clearings in a forest, fields with different crop types in an agricultural area, or obstacles like buildings or trees in an otherwise open grassland, all potentially disturb the atmospheric flow, and trigger the above mentioned deviations from ideal conditions that cause data quality to decrease (e.g. Panin et al. 1998; Schmid and Lloyd 1999; Baldocchi et al. 2005). Evaluating the influence of such terrain heterogeneity on eddy-covariance measurements through footprint modeling can therefore serve as an important component in the overall eddy-covariance data quality assessment strategy (Foken et al. 2004).

A comprehensive quality assessment framework to include footprint analyses into eddy-covariance data quality assessment schemes was first introduced by Göckede et al. (2004). Their approach built on a two-dimensional analytic flux footprint model (FSAM, Schmid 1997) and was successfully applied by Rebmann et al. (2005) to 18 sites of the European FLUXNET network. An upgraded version of this framework (Göckede et al. 2006), which aimed at a more reliable performance and broader applicability, replaced the analytic footprint model by a forward Lagrangian Stochastic trajectory model (Rannik et al. 2003). This software tool provided the results for an extensive quality control study for 25 European forested sites (Göckede et al. 2008). An overview was recently given by Rannik et al. (2012).

Maps of data quality assessment results linked to footprint analyses hold the potential of identifying general instrumentation problems, wind sectors with flow distortion or with upwind surface property changes under different conditions of atmospheric stability, or even the influence of single obstacles in the near field of a sensor. Potential effects will show up as structures in the spatial maps produced by the quality assessment framework, e.g. a single wind sector with reduced data quality for a specific atmospheric stability regime. Such structures are often caused by subtle trends which might easily be missed in a standard database filter. Such "bad" situations can be flagged to strengthen the database.

For data quality assessment (see Sect. 7.2.4) maps, any measured parameter (scalars and fluxes) can be linked to footprint analyses and data quality flags. To ensure representative findings, footprint analyses for data quality assessment should use a database of several months (at least 2–3) of meteorological measurements, so that several thousand half-hourly averaged observations are available. The correct interpretation of the findings relies on a good sample of the local wind climatology, and sufficient coverage of different atmospheric stability conditions for all wind sectors. The analysis will be strengthened by choosing a database that covers a period of the year with high absolute values of exchange fluxes between surface and atmosphere.

A classic example for this application would be the visualization of spatial structures in the mean vertical wind component. Other examples include visualizing the flux fields for sensible or latent heat, which may indicate spatially variable sources for these parameters.

An additional, powerful way to apply footprints to eddy-covariance quality assessment is to link data quality to terrain features. In this scenario, the footprint results are then coupled to other approaches used to evaluate flux data quality. The choice of the method to assign flux data quality, as well as the definition and resolution of quality classes, can be chosen by the user and customized for each study, as long as the quality ratings are numeric to allow aggregation. Göckede et al. (2006, 2008) applied a scheme proposed by Foken and Wichura (1996) in the revised version as presented by Foken et al. (2004), which assigns quality flags between 1 (best) and 9 (worst) for the fluxes of momentum, sensible and latent

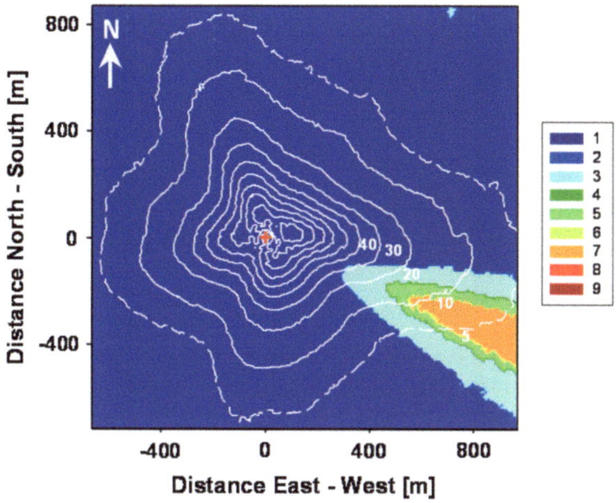

Fig. 8.6 Example of an isolated wind sector with reduced data quality, taken from Göckede et al. (2008). Background colors give the median quality rating (1 = best) of the momentum flux during stable stratification ($z/L > 0.0625$; z: measurement height in m; L: Obukhov length in m) at the FLUXNET site Wetzstein site (DE-Wet). See caption of Fig. 8.5 for further details. Published with kind permission of © Copernicus Publications, distributed under the Creative Commons Attribution 3.0 License, 2008. All Rights Reserved

heat, and CO_2. To create spatial maps of the data quality, the quality flags for each individual 30-min measurement are projected onto a discrete grid, storing relative influence and quality flag results for each grid cell in a database. After processing the entire dataset, this information can be converted into a frequency distribution of data quality for each cell, which in turn yields the overall quality rating as the median of the distribution. Visualization of the results helps reveal spatial patterns in data quality, such as isolated wind sectors with significantly reduced quality ratings compared to neighboring regions (Fig. 8.6). Such patterns may hint at terrain structures in that specific wind sector which have a negative impact on atmospheric measurement conditions, or might be caused by flow distortion induced by the instrumental setup. Observations of multidirectional reduction in data quality for specific subsets of the measurements (Fig. 8.7) can indicate instrumental problems, such as water in the tubing of a closed-path infrared gas analyzer that only condenses during lower temperatures at night. Whatever the cause of the reduced data quality, relevant wind sectors or stability regimes can be flagged and removed from the database to improve overall data quality.

For the visualization of spatial structures in ancillary parameters such as the mean vertical wind speed or the friction velocity, the procedure resembles the one described above for quality flag analysis, only that observational data replace the data quality ratings. This application allows exploring spatial effects for a large number of parameters which hold the potential to help interpret cases of low data quality, or identify instrumental problems. An example of this type of analysis included into the framework by Göckede et al. (2006) is the visualization of spatial

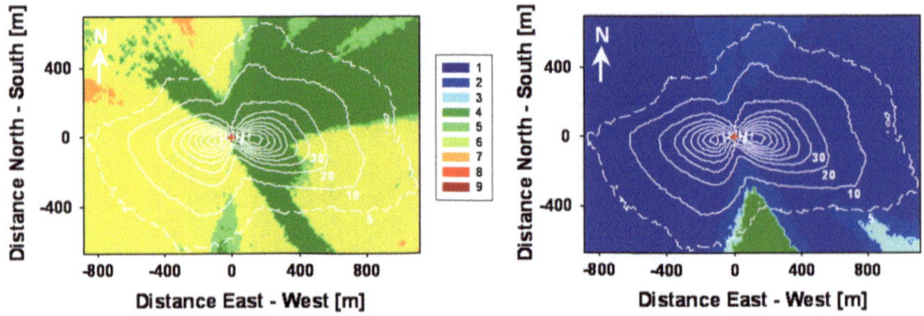

Fig. 8.7 Comparison of spatial data quality of latent heat flux (*left panel*) and the CO_2 flux (*right panel*) during stable conditions, taken from Göckede et al. (2008). Background colors give the median quality rating (1 = best) obtained for the FLUXNET site Sorø (DK-Sor). See captions of Fig. 8.5 for further details. Published with kind permission of © Copernicus Publications, distributed under the Creative Commons Attribution 3.0 License, 2008. All Rights Reserved

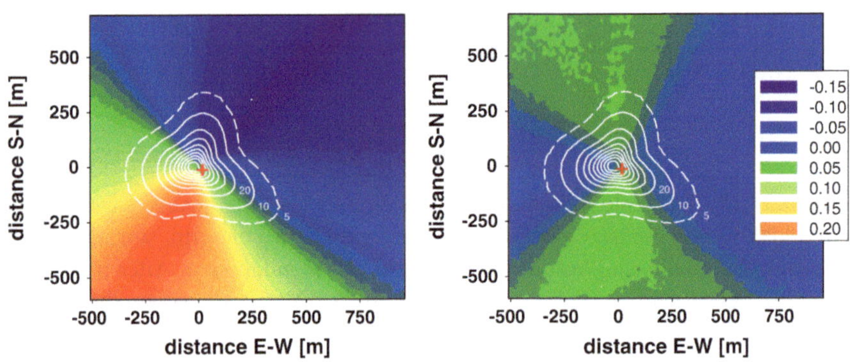

Fig. 8.8 Spatial map of the mean vertical wind component before (*left panel*) and after (*right panel*) application of the Planar-Fit coordinate rotation. Results taken from site analysis of the FLUXNET site Waldstein-Weidenbrunnen (DE-Bay). See captions of Fig. 8.5 for further details

structures in the vertical wind component before and after application of the planar-fit coordinate rotation (Wilczak et al. 2001). These results indicate tilt and distortion of the initial wind field, and the effectiveness of the coordinate rotation to correct the flow conditions to a mean vertical wind of zero, as required for eddy-covariance measurements. Figure 8.8 gives an example of structures in the vertical wind field before and after rotation. In this case, the absolute deviations from the ideal value of zero could be significantly reduced in using the planar-fit method, but spatial patterns still remain in the corrected dataset, because the complex terrain at this site produces a slightly curved wind field that cannot be completely corrected for with a single set of rotation angles. In a similar fashion, sectors with particularly low friction velocities during nighttime could be identified to highlight advection-prone conditions. The visualization of heterogeneities in the sources for momentum, heat, or CO_2, surrounding the site is also possible; however, in this application, the impact of external drivers such as temperature or radiation on the flux variability needs to be taken into account through additional filters.

8.3 Upscaling Point Measurements Using Footprint Models

In many cases, it is necessary to link eddy-covariance measurements to flux observations or simulations of considerably different spatial resolution, such as upscaling to remote-sensing information grids (Kim et al. 2006; Reithmaier et al. 2006; Chen et al. 2008) or aircraft data (Ogunjemiyo et al. 2003; Kustas et al. 2006), or downscaling for comparison to soil chamber measurements (Davidson et al. 2002; Reth et al. 2005; Myklebust et al. 2008).

One of the most relevant applications is the upscaling of eddy-covariance data to the grid size of a mesoscale model or to the pixel size of remote sensing images. Several papers have been published in peer-reviewed literature on this research topic, but only few of them consider footprint technologies. Chen et al. (2009b) determined monthly and annual uncertainties in eddy-covariance fluxes through variability in the footprint climatology to estimate the bias between spatially-explicit ecological models and tower-based remote sensing at finer scales. Furthermore, Chen et al. (2009a) compared remotely sensed carbon fluxes with eddy-covariance fluxes and their footprints.

The general problem is illustrated in Fig. 8.9. As long the footprint covers a homogeneous area (left panel of Fig. 8.9), the observed flux F_{obs} is equal to the flux of the target area F_{tar}. To upscale the fluxes on a grid element the contribution of the fluxes of the other land cover types must be determined, and the total flux of the entire grid cell will be a superposition of those determined through flux averaging (Sect. 2.4). In the case when the footprint covers also different land cover types F_{sur} (right panel of Fig. 8.9) the observed flux is given as

$$F_{obs} = a \cdot F_{tar} + (1 - a)F_{sur} \tag{8.2}$$

where a is the footprint fraction related to the target area. Such influences can be quantified for the grid level as point-to-area representativeness after Nappo et al. (1982) or as target representativeness in relation to the footprint of the measurements after Schmid (1997).

This allows an upscaling schema based on Fig. 8.9). For the case when the footprint is composed entirely of the target area, a SVAT model will be calibrated for this surface. In case other land cover types also contribute to the fluxes measured, the SVAT model will also be applied, but these events will be assigned a lower accuracy because of missing calibration and validation. When footprint-related flux errors (derived by the source weight function) exceed model uncertainty, no flux needs to be modeled. Instead, a quality flag for representativeness

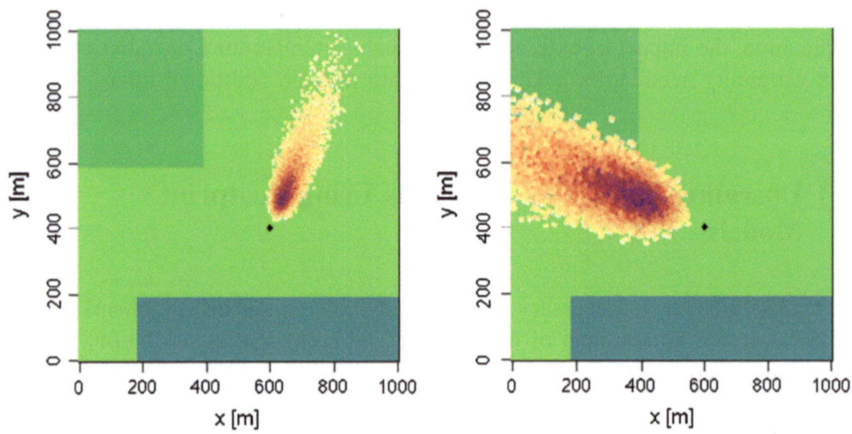

Fig. 8.9 Example of changing footprint scenarios over heterogeneous landscapes for different wind directions and stabilities. The *greenshaded* areas reflect different land uses, the *black dot* indicates the tower position and the source weight function derived by a Lagrangian model is illustrated by *small dots* ranging from high contribution (*purple*) to low contribution (*yellow*) after Babel (personal communication, Published with kind permission of © Dr. Babel, 2012. All Rights Reserved)

will be returned from the comparison of the weighing factors derived by footprint analysis with the fixed land use distribution within the grid cell. Otherwise, fluxes for the target area and the adjacent areas will be modeled by changing the land-use characteristic parameter and calculating a grid representative flux from the target flux and the modeled adjacent flux according to the given land cover distribution of the grid cell. The upscaling issue using footprint calculation is still an active area of research (Biermann et al. 2014).

8.4 Additional Practical Application

8.4.1 Air Pollution Application and Trace Gas Fluxes

Footprint simulations are used mainly to interpret micrometeorological measurements, especially those tied to ecological research. Accordingly, most models for scalars and fluxes are focused on this field of research. However, all of these models are based on more general physical assumptions and algorithms, and therefore closely connected to air pollution modelling. For example, the Eulerian approach used in analytical footprint models is largely identical to the stochastic approach (Blackadar 1997; Arya 1999). Micrometeorological footprint models usually have just a more specific formulation for the atmospheric surface and boundary layer. In the following sub-sections, some concepts are developed for which 'classic' footprint models may be applicable.

The simulation of air pollution and trace gas fluxes is a natural field of application because footprint models do not differ from air pollution models. For example, the mechanisms to transport the inert gases is the same as that for carbon dioxide concentrations (Kaharabata et al. 1999; Sarkar and Hobbs 2003). For reactive trace gases, possible sources or sinks in the atmosphere due to chemical reactions must be included into a footprint model (Strong et al. 2004; Rinne et al. 2007). For more details see Sect. 4.6.

8.4.2 Wind-Energy Application

The World Meteorological Organization addressed the problem of the underlying surface on the windward site of a wind power plant already 30 years ago, but mainly in relation to possible internal boundary layers (WMO 1981). In later years, the European Wind Atlas (Troen and Peterson 1989) and the adjunct software package became the standard instrument for site selection. It includes a hydrodynamic atmospheric flow model as well as additional algorithms to consider the influence of obstacles. Newer and more sophisticated software packages to evaluate site quality for wind energy turbines mostly build upon results from numerical mesoscale models or Large-Eddy Simulation. However, the authors of this book are aware of only a few studies (Hierteis et al. 2000; Hasager et al. 2006; Foken 2013, Fig. 8.10) that employ footprint models used in selection optimal wind turbine locations.

Footprint models could provide an easy-to-use tool to analyze the surface characteristics in a target region such as that of a wind farm to identify the best positions for wind turbines based on the local structure of aerodynamic obstacles. The use of footprint climatologies can thus provide helpful complementary information for this purpose. Since analytical models are limited to model domains with a homogeneous surface, Lagrangian backward models (Kljun et al. 2002) should be preferred. These models are well applicable in the lower part of the boundary layer up to heights of about 200–300 m (Markkanen et al. 2009).

The main purpose to include footprint analyses into the portfolio of software tools to evaluate sites for wind energy turbines is to better assess the impact of the surface conditions on the potential energy yield. This is an issue of considerable importance in this field because the wind velocity scales with the roughness of the surrounding terrain, and wind power is proportional to the cube of the wind velocity. For example, Fig. 8.11 demonstrates that the wind velocity in a clearing surrounded by dense, thick woods is significantly reduced compared to wind fields

Fig. 8.10 Wind power research station near Sassendorf region Bamberg/ Germany (Photograph by Foken)

in open terrain. In addition, the turbulent kinetic energy is increased, further reducing the energy yield.

An analysis of the characteristics of the surface roughness elements in a target area for wind turbines under consideration of wind direction and stability could identify the position exposed to the lowest surface roughness. Such an investigation should be performed in addition to the application of the usual wind field models. A proposal for such a schema is shown in Fig. 8.12. The most difficult part is the consideration of atmospheric stability, since this parameter is rarely available at candidate sites. If such information are not available, Fig. 8.12 proposes Obukhov lengths for the relevant stability classes. To investigate the frequency of the different stability classes, the classification by Pasquill can easily be applied depending on wind velocity and radiation class (Blackadar 1997; Foken 2008).

The application of footprint technologies can play a major role in optimizing the site selection for wind power turbines, particularly if the target region is situated off the coastal areas where heterogeneous land cover structures and complex topography may lead to significant differences in the local wind climatology across very short distances.

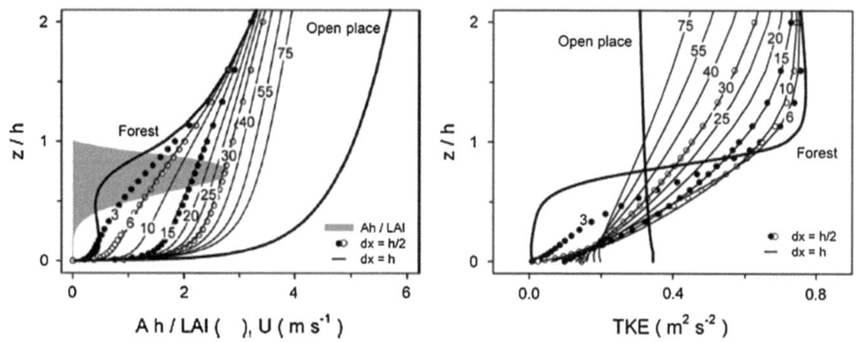

Fig. 8.11 Vertical profiles of the mean wind speed (*left side*) and turbulent kinetic energy (*Right side*), in the middle of a clear cut of different diameters depending on the canopy height 3 h ≤ D ≤ 75 h (D-increasing is shown by *arrow*). Two reference vertical profiles of U (*heavy solid lines*) are also shown: *1* "Forest"—for undisturbed homogeneous forest with the same structural characteristics as the forest surrounding the modeled clear cut and *2* "Open place"—for an "ideal" open place without any vegetation (Panferov and Sogachev 2008)

8.5 Easily Applicable Footprint Models

To potential users of footprint models, the access to freely available tools is an important issue. This, however, is usually only offered for analytical models. In Sect. 3.1.3, the links to these models are given. One of the most popular models due to its wide range of application is the model by Kormann and Meixner (2001) in the form presented by Neftel et al. (2008). Users with a solid background in meteorology will appreciate the versatility of the SCADIS model (Sogachev et al. 2002; Sogachev and Lloyd 2004; Sogachev and Sedletski 2006), which is commercially available through the main author.

For the generation of footprint climatologies or footprint related quality controls of the eddy-covariance method also the tool by Göckede et al. (2004, 2006) is available (Babel 2014). Part of the input required to run their software package is based to the output files of the eddy-covariance software of the University of Bayreuth (TK2(3), Mauder and Foken 2004, 2011); however, these quality ratings for eddy-covariance measurements can be replaced by other quality assessment schemes, if desired.

Most of the more sophisticated footprint models, such as Lagrangian stochastic algorithms or higher-order closure models, will not be readily available to be downloaded on the Internet. However, for users who feel that one of these models is particularly suited for their envisioned application, we highly recommend getting in contact with authors of publications based on these models, since many of them will be willing to share their tools using a fair use policy or a similar agreement.

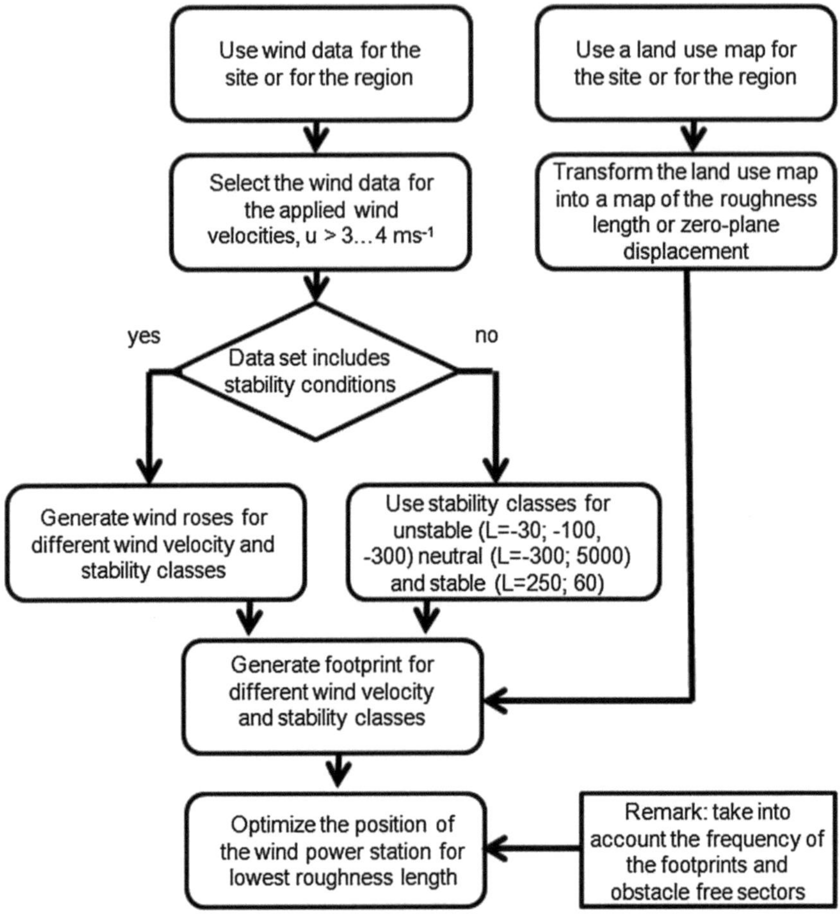

Fig. 8.12 Schema to find the best position for wind power stations based on wind roses, stability selections and footprint analysis. Finally the roughness in the most frequent footprint sector must be minimized (Foken 2012, Published with kind permission of © Deutsches Windenergieinstitut, 2012. All Rights Reserved)

8.6 Limits of Footprint Application

The application of most footprint models is theoretically restricted to horizontally homogeneous flow conditions, which can only be obtained if the tower is surrounded by perfectly uniform terrain with respect to topography, aerodynamic roughness, and sources for sensible and latent heat. This is particularly the case for analytical and forward Lagrangian stochastic (LS) models which are easy-to-use and flexible enough to be applied to multiple sites over longer timeframes. The paramount objective of footprint-based site evaluation, however, is to characterize the influence of terrain heterogeneities on flux measurements. Therefore, such tools will always violate the area of applicability that has been defined for the employed footprint model (Vesala et al. 2008). An exception is made of course to the unusual case that the terrain is perfectly flat, and heterogeneities are only of the sources of "passive" scalars like CO_2 but not the flow conditions. These problems can only be avoided through the use of backward Lagrangian simulation footprint models (Kljun et al. 2002) or closure approaches (Sogachev et al. 2005a, b) that can explicitly handle inhomogeneous flow conditions, but their requirements for setting up the model domain might make extensive network studies impossible.

All footprint results obtained outside the area of applicability of the underlying models are subject to increased uncertainty: Complex topography and step changes in roughness or heat flux source strength alter the atmospheric flow conditions (Schmid and Oke 1990; Klaassen et al. 2002; Leclerc et al. 2003; Foken and Leclerc 2004), so source area predictions based on the assumption of homogeneous transport will be biased (Finnigan 2004). This uncertainty will only slightly impact qualitative site evaluation results like the identification of a wind sector with reduced data quality, but quantitative findings like the percentage flux contribution of a certain land-cover type have to be evaluated carefully. A general error estimate cannot be provided, since the deviations from ideal flow conditions depend on the relative location of flow obstacles (e.g. trees, buildings, barn silos, ravines) with respect to the sensor position, and the local wind climatology, so that they need to be reassessed in every case study.

Footprint studies assessing long-term averaged properties, like a representative footprint climatology or the mean data quality for a specific sector, are likely to be biased by problems related to simulating source weight function under stable stratification at night. Analytical models, like e.g. the FSAM model (Schmid 1997) used in the framework by Göckede et al. (2004), are often restricted to input parameter ranges that exclude parts of the stable stratification range, effectively discriminating against nighttime measurements as these have large source areas and tend towards lower flux quality ratings. Lagrangian stochastic models are less numerically unstable than analytical ones in situations as such, but the representativeness of these footprints is also questionable in case of weak and intermittent turbulence, or even flow conditions dominated by wave motions. Excluding data with these conditions in the dataset leads to a systematic shift towards higher data quality and smaller footprint climatologies, compared to a treatment of the complete dataset. However, at least in case of the Lagrangian stochastic models, the majority of the excluded data fails to fulfill the theoretical assumptions used in eddy-covariance data processing either, so these data also would discarded in the assessment of the net carbon budget.

In an attempt to better characterize the problems linked to the application of footprint models in heterogeneous flow conditions, Markkanen et al. (2009) classified the agreement of different footprint models with a LES study (Steinfeld et al. 2008). Their results categorized the correlation between models in relation to the contribution of different flux sources to the total flux, and also considered the location of "disturbing" grid elements relative to the location of the peak of the source weight function. Factors influencing the model output, such as the horizontal grid resolution, or the sensor level, were taken into account in a sensitivity study. The analysis of data quality described in the previous chapters can generally be improved by comparing the overall accuracy of the chosen footprint model with a reference model that is better suited for application in heterogeneous conditions. Such an approach would strengthen data quality analysis and allow reaching robust conclusions on site characteristics (e.g. Göckede et al. 2008).

Still, certain conditions of atmospheric surface layer flows exist that exclude the conduction of suitable footprint analyses. A classical example of this type of setup is a heterogeneous landscape with tall vegetation and steep topography. None of the current footprint models currently consider advection and non-turbulent flow, while advection can increase the size of the footprint area up to 300 % (Leclerc et al. 2003). This effect can be neglected if only the effect levels up to 80 or 90 % of the footprint are of importance, but for practical issues, one must be aware that there may be significant flux contributions from outside the 'typical' footprint areas. This is especially important when, for air pollution studies, even small concentration can have a significant influence on the measuring point.

Furthermore, there are some phenomena above tall vegetation which have a significant influence on energy and trace gas exchange like coherent structures (Bergström and Högström 1989; Collineau and Brunet 1993a, b) or the mixing layer (Raupach et al. 1996; Finnigan 2000) in the roughness sub-layer (Garratt 1978). In certain conditions, coherent structures can contribute 20 % of the total measured flux (Thomas and Foken 2007). Flow conditions as such can only be modeled in sophisticated Lagrangian models or with Large-Eddy Simulation, and even these sophisticated approaches are still unable to find a solution that is generally applicable. For this reason, any footprint simulation above tall vegetation must take into account a considerable level of uncertainty. Moreover, flow conditions in the atmospheric boundary layer play a role in dictating turbulent fluxes and therefore its footprint. One prominent example is the presence of low-level jets at night, which can cause an increase of turbulent fluxes (Mathieu et al. 2005; Karipot et al. 2008) that cannot be accounted for by simple atmospheric flow models and footprint approaches. Breaking gravity waves cause similar effects.

References

Amiro BD (1998) Footprint climatologies for evapotranspiration in a boreal catchment. Agric Forest Meteorol 90:195–201

André J-C, Bougeault P, Goutorbe J-P (1990) Regional estimates of heat and evaporation fluxes over non-homogeneous terrain, examples from the HAPEX-MOBILHY programme. Bound-Layer Meteorol 50:77–108

Arya SP (1999) Air pollution meteorology and dispersion. Oxford University Press, New York 310 pp

Babel W (2014) An R routine for the simplified application of a footprint-based characterisation of a complex measuring site for flux measurements. Arbeitsergebn, Univ Bayreuth, Abt Mikrometeorol 57:30 pp. ISSN 1614-8916

Baldocchi D (1997) Flux footprints within and over forest canopies. Bound-Layer Meteorol 85:273–292
Baldocchi D et al (2001) FLUXNET: a new tool to study the temporal and spatial variability of ecosystem-scale carbon dioxide, water vapor, and energy flux densities. Bull Amer Meteorol Soc 82:2415–2434
Baldocchi DD, Krebs T, Leclerc MY (2005) "Wet/dry daisyworld": a conceptual tool for quantifying the spatial scaling of heterogeneous landscapes and its impact on the subgrid variability of energy fluxes. Tellus B 57:175–188
Barcza Z, Kern A, Haszpra L, Kljun N (2009) Spatial representativeness of tall tower eddy covariance measurements using remote sensing and footprint analysis. Agric Forest Meteorol 149:795–807
Bergström H, Högström U (1989) Turbulent exchange above a pine forest. II. Organized structures. Bound-Layer Meteorol 49:231–263
Blackadar AK (1997) Turbulence and diffusion in the atmosphere. Springer, Berlin 185 pp
Biermann T, Babel W, Ma W, Chen X, Thiem E, Ma Y and Foken T (2014) Turbulent flux observations and modelling over a shallow lake and a wet grassland in the Narm Co basin, Tibetan Plateau. Theor Appl Climat 116:301–316
Chen B, Chen JM, Mo G, Black A and Worthy DEJ (2008) Comparison of regional carbon flux estimates from CO2 concentration measurements and remote sensing based footprint integration. Global Geochem Cycles 22:GB2012
Chen B, Black TA, Coops NC, Hilker T, Trofymov JA, Morgenstern K (2009a) Assesssing tower flux footprint climatology and scaling between remotely sensed and eddy covariance measurements. Bound-Layer Meteorol 130:137–167
Chen B, Ge Q, Fu D, Liu G, Yu G, Sun X, Wang S, Wang H (2009b) Upscaling of gross ecosystem production to the landscape scale using multi-temporal Landsat images, eddy covariance measurements and a footprint model. Biogeosci Discuss 6:11317–11345
Collineau S, Brunet Y (1993a) Detection of turbulent coherent motions in a forest canopy. Part I: Wavelet analysis. Bound-Layer Meteorol 65:357–379
Collineau S, Brunet Y (1993b) Detection of turbulent coherent motions in a forest canopy. Part II: Time-scales and conditional averages. Boundary-Layer Meteorol 66:49–73
Davidson EA, Savage K, Verchot LV, Navarro R (2002) Minimizing artifacts and biases in chamber-based measurements of soil respiration. Agric Forest Meteorol 113:21–37
DeFelice TP (1998) An introduction to meteorological instrumentation and measurement. Prentice Hall, Upper Saddle River 229 pp
Finnigan J (2000) Turbulence in plant canopies. Ann Rev Fluid Mech 32:519–571
Finnigan J (2004) The footprint concept in complex terrain. Agric Forest Meteorol 127:117–129
Foken T, Wichura B (1996) Tools for quality assessment of surface-based flux measurements. Agric Forest Meteorol 78:83–105
Foken T, Göckede M, Mauder M, Mahrt L, Amiro BD, Munger JW (2004) Post-field data quality control. In: Lee X et al (eds) Handbook of micrometeorology: a guide for surface flux measurement and analysis. Kluwer, Dordrecht, pp 181–208
Foken T, Leclerc MY (2004) Methods and limitations in validation of footprint models. Agric Forest Meteorol 127:223–234
Foken T (2008) Micrometeorology. Springer, Berlin 308 pp
Foken T (2012) Application of footprint models for the fine-tuning of wind power locations on inland areas. DEWI Mag 40:51–54
Foken T, et al. (2012) Coupling processes and exchange of energy and reactive and non-reactive trace gases at a forest site—results of the EGER experiment. Atmos Chem Phys 12:1923–1950
Foken T (2013) Application of footprint models for wind turbine locations. Meteorol Z 22:111–115
Garratt JR (1978) Flux profile relations above tall vegetation. Quart J Roy Meteorol Soc 104:199–211

Göckede M, Rebmann C, Foken T (2004) A combination of quality assessment tools for eddy covariance measurements with footprint modelling for the characterisation of complex sites. Agric Forest Meteorol 127:175–188
Göckede M, Markkanen T, Hasager CB, Foken T (2006) Update of a footprint-based approach for the characterisation of complex measuring sites. Bound-Layer Meteorol 118:635–655
Göckede M, Thomas C, Markkanen T, Mauder M, Ruppert J, Foken T (2007) Sensitivity of Lagrangian stochastic footprints to turbulence statistics. Tellus 59B:577–586
Göckede M et al (2008) Quality control of CarboEurope flux data – Part 1: Coupling footprint analyses with flux data quality assessment to evaluate sites in forest ecosystems. Biogeoscience 5:433–450
Hasager CB, Bartelmie RJ, Christiansen MB, Nielson M, Pryor SC (2006) Quantifying offshore wind resources from satellite wind maps: study area the North Sea. Wind Energ 9:63–74
Hierteis M, Svoboda J and Foken T (2000) Einfluss der Topographie auf das Windfeld und die Leistung von Windkraftanlagen. DEWEK 2000, Wilhelmshaven, 07.-08.06.2000 2000. Deutsches Windenergie-Institut, pp 272–276
Horst TW, Weil JC (1992) Footprint estimation for scalar flux measurements in the atmospheric surface layer. Bound-Layer Meteorol 59:279–296
Horst TW (1999) The footprint for estimation of atmosphere-surface exchange fluxes by profile techniques. Bound-Layer Meteorol 90:171–188
Jegede OO, Foken T (1999) A study of the internal boundary layer due to a roughness change in neutral conditions observed during the LINEX field campaigns. Theor Appl Climat 62:31–41
Kaharabata SK, Schuepp PH, Fuentes JD (1999) Source footprint considerations in the determination of volatile organic compound fluxes from forest canopies. J Appl Meteorol 38:878–884
Karipot A, Leclerc MY, Zhang G, Lewin KF, Nagy J, Hendrey GR, Starr D (2008) Influence of nocturnal low-level jet on turbulence structure and CO_2 flux measurements over a forest canopy. J Geophys Res 113:D10102
Kim J, Guo Q, Baldocchi DD, Leclerc MY, Xu L, Schmid HP (2006) Upscaling fluxes from tower to landscape: overlaying flux footprints on high resolution (IKONOS) images of vegetation cover. Agric Forest Meteorol 136:132–146
Klaassen W, van Breugel PB, Moors EJ, Nieveen JP (2002) Increased heat fluxes near a forest edge. Theor Appl Climat 72:231–243
Klaassen W, Sogatchev A (2006) Flux footprint simulation downwind of a forest edge. Bound-Layer Meteorol 121:459–473
Kljun N, Rotach MW, Schmid HP (2002) A three-dimensional backward Lagrangian footprint model for a wide range of boundary layer stratification. Bound-Layer Meteorol 103:205–226
Kormann R, Meixner FX (2001) An analytical footprint model for non-neutral stratification. Bound-Layer Meteorol 99:207–224
Kustas WP, Anderson MC, French AN, Vickers D (2006) Using a remote sensing field experiment to investigate flux-footprint relations and flux sampling distributions for tower and aircraft-based observations. Adv Water Res 29:355–368
Leclerc MY, Karipot A, Prabha T, Allwine G, Lamb B, Gholz HL (2003) Impact of non-local advection on flux footprints over a tall forest canopy: a tracer flux experiment (Special issue: Advances in micrometeorology: tribute to GW Thurtell). Agric Forest Meteorol 115:19–30
Markkanen T, Rannik Ü, Marcolla B, Cescatti A, Vesala T (2003) Footprints and fetches for fluxes over forest canopies with varying structure and density. Bound-Layer Meteorol 106:437–459
Markkanen T, Steinfeld G, Kljun N, Raasch S, Foken T (2009) Comparison of conventional Lagrangian stochastic footprint models against LES driven footprint estimates. Atmos Chem Phys 9:5575–5586
Markkanen T, Steinfeld G, Kljun N, Raasch S, Foken T (2010) A numerical case study on footprint model performance under inhomogeneous flow conditions. Meteorol Z 19:539–547
Mathieu N, Strachan IB, Leclerc MJ, Karipot A, Pattey E (2005) Role of low-level jets and boundary-layer properties on the NBL budget technique. Agric Forest Meteorol 135:35–43

Mauder M, Foken T (2004) Documentation and instruction manual of the eddy covariance software package TK2. Arbeitsergebn, Univ Bayreuth, Abt Mikrometeorol, ISSN 1614-8916. 26:42 pp

Mauder M, Liebethal C, Göckede M, Leps J-P, Beyrich F, Foken T (2006) Processing and quality control of flux data during LITFASS-2003. Bound-Layer Meteorol 121:67–88

Mauder M, Foken T (2011) Documentation and instruction manual of the eddy covariance software package TK3. Arbeitsergebn, Univ Bayreuth, Abt Mikrometeorol, ISSN 1614-8916. 46:58 pp

Munger JW, Loescher HW, Luo H (2012) Measurement, tower, and site design considerations. In: Aubinet M et al (eds) Eddy covariance: a practical guide to measurement and data analysis. Springer, Dordrecht, pp 21–58

Myklebust MC, Hipps LE, Ryel RJ (2008) Comparison of eddy covariance, chamber, and gradient methods of measuring soil CO_2 efflux in an annual semi-arid grass, Bromus tectorum. Agric Forest Meteorol 148:1894–1907

Nagy MT, Janssens IA, Curiel Yuste J, Carrara A and Ceulemans R (2006) Footprint-adjusted net ecosystem CO_2 exchange and carbon balance components of a temperate forest. Agric Forest Meteorol 139:244–360

Nappo CJ et al (1982) The workshop on the representativeness of meteorological observations, June 1981, Boulder. Colo. Bull Amer Meteorol Soc 63:761–764

Neftel A, Spirig C, Ammann C (2008) Application and test of a simple tool for operational footprint evaluations. Environ Pollut 152:644–652

Ogunjemiyo SO, Kaharabata SK, Schuepp PH, MacPherson IJ, Desjardins RL, Roberts DA (2003) Methods of estimating CO_2, latent heat and sensible heat fluxes from estimates of land cover fractions in the flux footprint. Agric Forest Meteorol 117:125–144

Panferov O, Sogachev A (2008) Influence of gap size on wind damage variables in a forest. Agric Forest Meteorol 148:1869–1881

Panin GN, Tetzlaff G, Raabe A (1998) Inhomogeneity of the land surface and problems in the parameterization of surface fluxes in natural conditions. Theor Appl Climat 60:163–178

Prabha T, Leclerc MY, Baldocchi D (2008) Comparison of in-canopy flux footprints between Large-Eddy Simulation and the Langrangian simulation. J Appl Meteorol Climatol 47:2115–2128

Raabe A (1983) On the relation between the drag coefficient and fetch above the sea in the case of off-shore wind in the near shore zone. Z Meteorol 33:363–367

Rannik Ü, Aubinet M, Kurbanmuradov O, Sabelfeld KK, Markkanen T, Vesala T (2000) Footprint analysis for measurements over heterogeneous forest. Bound-Layer Meteorol 97:137–166

Rannik Ü, Markkanen T, Raittila T, Hari P, Vesala T (2003) Turbulence statistics inside and above forest: Influence on footprint prediction. Bound-Layer Meteorol 109:163–189

Rannik Ü, Sogachev A, Foken T, Göckede M, Kljun N, Leclerc MY, Vesala T (2012) Footprint analysis. In: Aubinet M et al (eds) Eddy covariance: a practical guide to measurement and data analysis. Springer, Berlin, pp 211–261

Raupach MR, Finnigan JJ, Brunet Y (1996) Coherent eddies and turbulence in vegetation canopies: the mixing-layer analogy. Bound-Layer Meteorol 78:351–382

Rebmann C et al (2005) Quality analysis applied on eddy covariance measurements at complex forest sites using footprint modelling. Theor Appl Climat 80:121–141

Reithmaier LM, Göckede M, Markkanen T, Knohl A, Churkina G, Rebmann C, Buchmann N, Foken T (2006) Use of remotely sensed land use classification for a better evaluation of micrometeorological flux measurement sites. Theor Appl Climat 84:219–233

Reth S, Göckede M, Falge E (2005) CO_2 efflux from agricultural soils in Eastern Germany—comparison of a closed chamber system with eddy covariance measurements. Theor Appl Climat 80:105–120

Rinne J, Taipale R, Markkanen T, Ruuskanen TM, Hellén H, Kajos MK, Vesala T, Kulmala M (2007) Hydrocarbon fluxes above a Scots pine forest canopy: measurements and modeling. Atmos Chem Phys 7:3361–3372

Sarkar U, Hobbs SE (2003) Landfill odour: assessment of emissions by the flux footprint method. Environ Model Softw 18:155–163

Savelyev SA, Taylor PA (2005) Internal boundary layers: I. Height formulae for neutral and diabatic flow. Bound-Layer Meteorol 115:1–25
Schmid HP, Oke TR (1990) A model to estimate the source area contributing to turbulent exchange in the surface layer over patchy terrain. Quart J Roy Meteorol Soc 116:965–988
Schmid HP (1997) Experimental design for flux measurements: matching scales of observations and fluxes. Agric Forest Meteorol 87:179–200
Schmid HP, Lloyd CR (1999) Spatial representativeness and the location bias of flux footprints over inhomogeneous areas. Agric Forest Meteorol 93:195–209
Schmid HP (2002) Footprint modeling for vegetation atmosphere exchange studies: a review and perspective. Agric Forest Meteorol 113:159–184
Sellers PJ, Hall FG, Asrar G, Strebel DE, Murphy RE (1988) The first ISLSCP field experiment (FIFE). Bull Amer Meteorol Soc 69:22–27
Shearman RJ (1992) Quality assurance in the observation area of the meteorological office. Meteorol Mag 121:212–216
Soegaard H, Jensen NO, Boegh E, Hasager CB, Schelde K, Thomsen A (2003) Carbon dioxide exchange over agricultural landscape using eddy correlation and footprint modelling. Agric Forest Meteorol 114:153–173
Sogachev A, Menzhulin G, Heimann M, Lloyd J (2002) A simple three dimensional canopy-planetray boundary layer simulation model for scalar concentrations and fluxes. Tellus 54B:784–819
Sogachev A, Lloyd J (2004) Using a one-and-a-half order closure model of atmospheric boundary layer for surface flux footprint estimation. Bound-Layer Meteorol 112:467–502
Sogachev A, Panferov O, Gravenhorst G, Vesala T (2005a) Numerical analysis of flux footprints for different landscapes. Theor Appl Climat 80:169–185
Sogachev A, Leclerc MJ, Karipot A, Zhang G, Vesala T (2005b) Effect of clearcuts on footprints and flux measurements above a forest canopy. Agric Forest Meteorol 133:182–196
Sogachev A, Sedletski A (2006) Footprint calculator, operational manual. In: Kulmala M et al (eds.) Proceedings of BACCI, NECC and FCoE activities 2005, Report series in aerosol science 81B, vol 81, Helsinki
Steinfeld G, Raasch S, Markkanen T (2008) Footprints in homogeneously and heterogeneously driven boundary layers derived from a Langrangian stochastic particle model embedded into large-eddy simulation. Bound-Layer Meteorol 129:225–248
Stoughton TE, Miller DR, Yang X, Hendrey GM (2000) Footprint climatology estimation of potential control ring contamination at the Duke Forest FACTS-1 experiment site. Agric Forest Meteorol 100:73–82
Strong C, Fuentes JD, Baldocchi D (2004) Reactive hydrocrbon flux footprints during canopy senescence. Agric Forest Meteorol 127:159–173
Thomas C, Foken T (2007) Organised motion in a tall spruce canopy: temporal scales, structure spacing and terrain effects. Bound-Layer Meteorol 122:123–147
Troen I, Peterson EW (1989) European wind Atlas. Risø National Laboratory, Roskilde 656 pp
Tsvang LR, Fedorov MM, Kader BA, Zubkovskii SL, Foken T, Richter SH, Zelený J (1991) Turbulent exchange over a surface with chessboard-type inhomogeneities. Bound-Layer Meteorol 55:141–160
Vesala T, Kljun N, Rannik U, Rinne J, Sogachev A, Markkanen T, Sabelfeld K, Foken T, Leclerc MY (2008) Flux and concentration footprint modelling: state of the art. Environ Pollut 152:653–666
Wang WG, Davis KJ, Cook BD, Butler MP, Ricciuto DM (2006) Decomposing CO2 fluxes measured over a mixed ecosystem at a tall tower and extending to a region: a case study. J Geophys Res 111:G02005.02001–G02005.02014
Wilczak JM, Oncley SP, Stage SA (2001) Sonic anemometer tilt correction algorithms. Bound-Layer Meteorol 99:127–150
WMO (1981) Meteorological aspects of the utilization of wind as an energy source. WMO, Techn Note, 175, 180 pp

Chapter 9
Looking Forward to the Next Generation of Footprint Models

When we surveyed the exhaustive list of developments in footprint modeling (Foken and Leclerc 2004) ten years ago, we were most enthusiastic to see the rapid and vigorous progress in the footprint specialty of micrometeorology. Owing to the technical challenges of tracer methods, further progress related to footprint validation has slowed. This may bode well for the present volume given that a certain level of knowledge can be recognized and the book was not written in a period of fast progress. We are not suggesting that all aspects of footprint research have been addressed but instead the major developments have already been achieved. New developments in this arena are expected to lie in the field of applications. Let us underline this with key statements:

We had hoped that differences between analytical and Lagrangian type of footprint models (Kljun et al. 2003; Vesala et al. 2008a), can be solved with advances in the field of artificial tracer studies. The latter should include multiple tracer techniques in three-dimensional footprint studies, with tracers placed at several positions on the soil surface, in the understory and in the crown space of a forest canopy. Tracer studies are undoubtedly a powerful but also an expensive tool to study many influences on flux or concentration footprints. Such studies are always very specific and costly but practical relevance failed in the last years for such studies. We had also hoped that natural tracers (Foken and Leclerc 2004), this means the identification of heterogeneities in the footprint area by the measured fluxes, can be a less expensive tool because they can be included into on-going flux field campaigns. But this was only applied in a study by Göckede et al. (2005) and was on the brink of some other experiments. The really problem was the missing reference for model comparison like etalon instruments in sensor calibration

M. Y. Leclerc and T. Foken, *Footprints in Micrometeorology and Ecology*,
DOI: 10.1007/978-3-642-54545-0_9, © Springer-Verlag Berlin Heidelberg 2014

technique. The Lagrangian footprint model embedded into a LES model (Steinfeld et al. 2008) was probably one step in this direction. It was used by Markkanen et al. (2009) as a reference for the validation of forward and backward Lagrangian models. The result was the same as for the comparison of analytical and Lagrangian models (Kljun et al. 2003), not only as a visual result but as a quantitative result with quality flagging. Therefore, one must conclude that both the footprint peak and the different factors influencing the footprint behavior are similar amongst the various models though the level of sophistication may impact the results though only slightly. This is more relevant in stable then in unstable stratification. All further applications and developments will be based on this statement.

The main progress in the coming years is expected to revolve around the practical application of footprint models. Already simple footprint models are implemented into software tools like for eddy covariance measurements. These are the analytical footprint model applied to the neutral atmospheric surface layer by Schuepp et al. (1990) or for the diabatic surface layer and lower atmospheric boundary layer by Korman and Meixner (2001). The latter is also available as a easily to handle tool (Neftel et al. 2008, see also Sect. 3.1.3.5). The significant better Lagrangian models are too expensive in handling for general users. Therefore it was a great progress by Kljun et al. (2004) to make her backward Lagrangian model (Kljun et al. 2002) online available (see Sect. 3.1.4), unfortunately only as a 1-dimensional version. Mauder et al. (2008) made a 2D version of this model and also H.P. Schmid made a proposal how to make a two dimensional extension (Metzger et al. 2012). Hopefully this version will soon also be online available. The higher order closure model by Sogachev and Lloyd (2004) was also free available (Vesala et al. 2010) but can now only be commercially applied. From the availability of easy to use Lagrangian or higher order closure models the progress in practical footprint application depends. The routine application of footprint models as part of signal processing package in the field for in situ determination of the reliability of the dataset would gain increasing importance. Nowadays the footprint tool is mainly familiar to the flux community, but as shown in Chaps. 7 and 8 and a myriad of possibilities are emerging. This may be a wide area in atmospheric measurements, air pollution, use of renewable energies etc. The consulting and the ideas of the footprint community are necessary to make this important widely known and applicable.

Nevertheless some developments in footprint modeling are still necessary, primarily in the application of the interpretation of fluxes over heterogeneous areas. Small heterogeneities in thermal (Markkanen et al. 2010) and roughness conditions including possible internal boundary layers have not really a significant effect on the calculated footprint of the backward Lagrangian model by Kljun et al. (2002). More important are probably forest areas (Foken and Leclerc 2004; Vesala et al. 2008a). Here the turbulence structure of the roughness sublayer, i.e. the layer of air influenced by the presence of a neighboring rough surface, as in a canopy layer, violates similarity principles used in surface-layer scaling. Yet, most eddy-covariance measurements over forests are contained within the roughness sublayer

(Raupach and Thom 1981), a layer which exhibits some departure from surface-layer similarity. Hence the mean wind speed does not follow a log-linear profile and the observed eddy-diffusivity can be two to four times greater than the similarity value (Raupach et al. 1986; Su et al. 1998). Inside tall forest canopies, the treatment required to derive footprint functions derived from realistic analytical solutions become even more a distant possibility because of the added complexity of having to contend with the presence of multiple vertically distributed sources and sinks. Yet, even approximate solutions would be helpful as increasingly experimentalists seek to determine in-canopy footprints (Göckede et al. 2007). Not only these basic studies in and above idealized tall vegetation are necessary. Gaps in canopies, flow obstacles such as isolated trees or distant tall man-made or pest and wind throw structures, and their effect on footprint functions (and measured fluxes) should be investigated, with special precautions taken under nighttime conditions. Here are only a few studies available at forest edges (Klaassen and Sogatchev 2006; Sogachev and Leclerc 2011).

A similar problem are footprints in urban canopies (Vesala et al. 2010). Here the footprint problem is up to now only addressed (Vesala et al. 2008b) and even large recent urban experiments did not applied footprint technology (Rotach et al. 2005). The reason lies in the difficult definitions of the source area and the height or zero-plane displacement. We are not sure, if footprint technology or better LES will solve the problems of source areas and its footprint in urban meteorology in the future.

The application of footprint technology in air chemistry, mainly for the transport of particles or of reactive trace gases is only presented by few studies (Strong et al. 2004; Rinne et al. 2007). Often backward trajectories are applied in this field, like for the interpretation of the footprint of tall towers (Gloor et al. 2001).

There are still some meteorological situations where the application of footprint models is difficult. Greater consideration ought to be given to the presence of discontinuities upwind beyond the footprint region, particularly in calm conditions or very stable conditions. Recent efforts along the lines of quantifying atmospheric stability effects inside a canopy are being made (Zhang et al. 2010). These new developments should provide a step forward toward the future development and use of analytical footprint models describing in-canopy flux footprints.

As it was already shown in recent studies (Steinfeld et al. 2008) that the Large-Eddy Simulation is a formidable tool which should be used to investigate complex flux footprints originating from patchy terrain and other three-dimensional sources and sinks. The formulation of flow statistics in the LES makes it an ideal tool as an alternative to expensive and time-consuming experiments, and other footprint model formulations could be tested against this method like in Markkanen et al. (2009). Numerical closure models (Belcher et al. 2012) and LES (Schlegel et al. 2012) are nowadays able to solve problems in hilly terrain and forests, which were addressed above as deficits of footprint models. Therefore we can finally state that the future of footprint models will not be a further improvement of the classical footprint models (as it was the issue in this book), but the development of adequate

LES models, which need now no high sophisticated computer centre because of much faster processors, will be the future.

Summarizing these remarks, we believe that the main thrust of the development in footprint technology is now done and made this technology to an in most cases easily applicable method. This gives us the opportunity that the book will give the basics for footprints in micrometeorology and ecology also for the future. Some improvements are possible but without significant changes. The challenge over the next years will either be to make footprint technology more applicable to applied technologies of atmospheric gas exchange and to develop the next generation of models used in the interpretation of experimental data with LES and higher-order closure technology.

References

Belcher SE, Harman IN, Finnigan JJ (2012) The wind in the willows: flows in forest canopies in complex terrain. Ann Rev Fluid Mech 44:479–504

Foken T, Leclerc MY (2004) Methods and limitations in validation of footprint models. Agric Forest Meteorol 127:223–234

Gloor M, Bakwin P, Hurst D, Lock L, Draxler R and Tans P (2001) What is the concentration footprint of a tall tower? J Geophys Res 106(D16):17,831–17,840

Göckede M, Markkaken T, Mauder M, Arnold K, Leps JP, Foken T (2005) Validation of footprint models using natural tracer measurements from a field experiment. Agric Forest Meteorol 135:314–325

Göckede M, Thomas C, Markkanen T, Mauder M, Ruppert J, Foken T (2007) Sensitivity of Lagrangian stochastic footprints to turbulence statistics. Tellus 59B:577–586

Klaassen W, Sogatchev A (2006) Flux footprint simulation downwind of a forest edge. Bound-Layer Meteorol 121:459–473

Kljun N, Rotach MW, Schmid HP (2002) A three-dimensional backward Lagrangian footprint model for a wide range of boundary layer stratification. Bound-Layer Meteorol 103:205–226

Kljun N, Kormann R, Rotach M, Meixner FX (2003) Comparison of the Lagrangian footprint model LPDM-B with an analytical footprint model. Bound-Layer Meteorol 106:349–355

Kljun N, Calanca P, Rotach M, Schmid HP (2004) A simple parameterization for flux footprint predictions. Bound-Layer Meteorol 112:503–523

Kormann R, Meixner FX (2001) An analytical footprint model for non-neutral stratification. Bound-Layer Meteorol 99:207–224

Markkanen T, Steinfeld G, Kljun N, Raasch S, Foken T (2009) Comparison of conventional Lagrangian stochastic footprint models against LES driven footprint estimates. Atmos Chem Phys 9:5575–5586

Markkanen T, Steinfeld G, Kljun N, Raasch S, Foken T (2010) A numerical case study on footprint model performance under inhomogeneous flow conditions. Meteorol Z 19:539–547

Mauder M, Desjardins R, MacPherson I (2008) Creating surface flux maps from airborne measurements: application to the Mackenzie Area GEWEX Study MAGS 1999. Bound-Layer Meteorol 129:431–450

Metzger S, Junkermann W, Mauder M, Beyrich F, Butterbach-Bahl K, Schmid HP, Foken T (2012) Eddy-covariance flux measurements with a weight-shift microlight aircraft. Atmos Meas Tech 5:1699–1717

Neftel A, Spirig C, Ammann C (2008) Application and test of a simple tool for operational footprint evaluations. Environ Pollut 152:644–652

Raupach MR, Thom AS (1981) Turbulence in and above plant canopies. Ann Rev Fluid Mech 13:97–129

Raupach MR, Coppin PA, Legg BJ (1986) Experiments on scalar dispersion within a model plant canopy. Part I: the turbulence structure. Bound-Layer Meteorol 35:21–52

Rinne J, Taipale R, Markkanen T, Ruuskanen TM, Hellén H, Kajos MK, Vesala T, Kulmala M (2007) Hydrocarbon fluxes above a Scots pine forest canopy: measurements and modeling. Atmos Chem Phys 7:3361–3372

Rotach M et al (2005) BUBBLE—an urban boundary layer meteorology project. Theory Appl Climat 81:231–261

Schlegel F, Stiller J, Bienert A, Maas H-G, Queck R, Bernhofer C (2012) Large-eddy simulation of inhomogeneous canopy flows using high resolution terrestrial laser scanning data. Bound-Layer Meteorol 142:223–243

Schuepp PH, Leclerc MY, MacPherson JI, Desjardins RL (1990) Footprint prediction of scalar fluxes from analytical solutions of the diffusion equation. Bound-Layer Meteorol 50:355–373

Sogachev A, Lloyd J (2004) Using a one-and-a-half order closure model of atmospheric boundary layer for surface flux footprint estimation. Bound-Layer Meteorol 112:467–502

Sogachev A, Leclerc MY (2011) On concentration footprints for a tall tower in the presence of a nocturnal low-level jet. Agric Forest Meteorol 151:755–764

Steinfeld G, Raasch S, Markkanen T (2008) Footprints in homogeneously and heterogeneously driven boundary layers derived from a Langrangian stochastic particle model embedded into large-eddy simulation. Bound-Layer Meteorol 129:225–248

Strong C, Fuentes JD, Baldocchi D (2004) Reactive hydrocarbon flux footprints during canopy senescence. Agric Forest Meteorol 127:159–173

Su H-B, Shaw RH, Paw UKT, Moeng C-H, Sullivan PP (1998) Turbulent statistics of neutrally stratified flow within and above sparse forest from large-eddy simulation and field observations. Bound-Layer Meteorol 88:363–397

Vesala T, Kljun N, Rannik U, Rinne J, Sogachev A, Markkanen T, Sabelfeld K, Foken T, Leclerc MY (2008a) Flux and concentration footprint modelling: state of the art. Environ Pollut 152:653–666

Vesala T et al (2008b) Surface–atmosphere interactions over complex urban terrain in Helsinki. Finland Tellus B 60:188–199

Vesala T, Kljun N, Rannik Ü, Rinne J, Sogatchev A, Markkanen T, Sabelfeld K, Foken T and Leclerc MY (2010) Flux and concentration footprint modelling. In: Hanrahan G (ed.) Modelling of pollutants in complex environmental systems, vol 2, ILM Publications, St. Albans, Glendale, pp 339–355

Zhang G, Leclerc MY, Karipot A (2010) Local flux-profile relationships of wind speed and temperature in a canopy layer in atmospheric stable conditions. Biogeosciences 7:3625–3636

Glossary

The glossary was partly used from Foken (2008) or adapted from Glickman (2000).

Advection Transport of properties of the air (momentum, temperature, water vapor, etc.) by the wind. As a rule, horizontal transport is understood. Because of advection, air properties change in both horizontal coordinates, and the conditions are no longer homogeneous. Vertical advection is the vertical movement of air due to mass continuity rather than buoyancy, i.e. convection.

Biome Communities of plants and animals living in similar climate conditions.

Boundary layer atmospheric Lowest part of the troposphere near the ground (approx. 500 m–2 km high) where the friction stress decrease with height.

Coherence Coherence expresses the degree of the in-phase relationship between two signals. Coherent structures in atmospheric turbulence research are examined in time series of velocity, temperature and other scalars. They are significantly larger and of a larger persistence than the small, more local eddies (e.g. squall lines, convective cells).

Coriolis force Fictitious force in a rotating coordinate system, and named after the mathematician Coriolis (1792–1843). It is a force normal to the velocity vector causing a deflection to the right in the Northern hemisphere and to the left in the Southern hemisphere.

Coriolis parametert is twice the value of the angular velocity of the Earth for a certain location of the latitude φ: $f=2\ \Omega \sin \varphi$. At the equator $f = 0$ on the Northern hemisphere positive and on the Southern hemisphere negative.

Counter gradient Gradient directed against the flux, typically for coherent structures or strong buoyancy fluxes.

Dissipation Conversion of kinetic energy by work against the viscous stresses. Under turbulent conditions, it is the conversion of the kinetic energy of the smallest eddies into heat.

M. Y. Leclerc and T. Foken, *Footprints in Micrometeorology and Ecology*,
DOI: 10.1007/978-3-642-54545-0,

Eulerian coordinates Coordinate system (e.g. Cartesian) which describes the properties of a fluid in space at a given time. Eulerian coordinates are fixed and are in contrast with Lagrangian coordinates in which the frame of reference moves with the fluid particle.

Fetch Windward distance from a measuring point to a change of the surface properties or an obstacle; extent of a measuring field for micrometeorological research.

Flux, buoyancy Buoyancy flux is the vertical flux of the virtual potential temperature.

Flux gradient similarity Proportionality of the flux and the gradient: Log-lin profile in the case of neutral stratification, for atmospheric stability given by the Monin-Obukhov similarity theory.

Flux variance similarity is the proportionality of the flux and the variance of the relevant scalar.

Gravity wave A wave generated between layers of the atmosphere with restoring forces that include buoyancy and gravity.

Inversion An air layer where the temperature increases with the altitude instead of the usual decrease. Inversions are of two types; surface inversion due to longwave radiation from the ground, and elevated or free inversions e.g. at the top of the atmospheric boundary layer.

Lagrangian coordinates Coordinate system that moves with the fluid element by assigning them coordinates that do not very in time.

Leaf area index Ratio of the leaf area (upper side) within a vertical cylinder to the bottom area of the cylinder.

Low-level jet Vertical band of strong winds in the lower part of the atmospheric boundary layer. For stable stratification, the low-level jet develops at the upper border of the nocturnal surface inversion. Typical heights are 100–300 m, and sometimes lower.

Lidar (Light Detection And Ranging) Optical remote sensing instrument using laser light. The backscatter signals gives information about the properties of the atmosphere.

Mixed layer A layer of strong vertical mixing due to convection resulting in vertically-uniform values of potential temperature and wind speed but decreasing values of moisture. It is often capped by an inversion layer (see above).

Mixing layer Layer of mixing of initially two co-flowing streams with different velocities, e.g. above large roughness elements like forests.

Parameterization Representation of complicated relations in models by more simple combinations of parameters, which are often only valid under certain circumstances.

Planar fit method Method used to minimize in an ensemble sense the mean vertical velocity for a place parallel to streamlines; the method results are valid as long as surface properties are constant.

Prandtl number Ratio of the molecular temperature conductance and the kinematic viscosity. The turbulent Prandtl number is defined for the ratio of the turbulent diffusion coefficient for heat and momentum.

Rossby wave Planetary wave on a uniform current in a two-dimensional nondivergent fluid system, rotating with varying angular speed about the local vertical.

Scale, sub-grid Length-scales below the adequately resolved scale of a model.

Schmidt number Ratio of the molecular diffusion and the kinematic viscosity. The turbulent Schmidt number is defined for the ratio of the turbulent diffusion coefficient of a water vapour or another scalar and momentum.

Scintillometer Instrument which measures the scintillation of the air (a measure of the fluctuation of the refraction index) using optical or radio waves. When similarity laws are used, the signal can be transformed into turbulent fluxes.

Sodar (SOund/SOnic Detecting And Ranging) Remote sensing instrument which uses sound. The backscatter signals gives information about the properties of the atmosphere.

Stability of the stratification The static stability separates turbulent and laminar flow conditions depending on the gradient of the potential temperature (see below). If the potential temperature decreases with height, then the atmospheric stability is unstable, but if it increases with height, then the stratification is stable. Due to the effects of vertical wind shear, the statically-stable range is turbulent up to the critical Richardson number.

Temperature, potential The temperature of a dry air parcel that is moved adiabatically to a pressure of 1000 hPa, see Eq. (2.15).

Temperature, virtual The temperature of a dry air parcel if it had the same density as a moist air parcel. The virtual temperature is slightly higher than the temperature of moist air, see Eq. (2.12).

Wind geostrophic Wind above the atmospheric boundary layer where pressure gradient force and Coriolis force (see above) are in equilibrium.

References

Foken T (2008) Micrometeorology. Springer, Berlin, p 308
Glickman TS (ed) (2000) Glossary of Meteorology. American Meteorology Society, Boston, pp 855

About the Authors

Monique Y. Leclerc was born in Montreal, Canada in 1955. She graduated from McGill University in 1979 and went on to the University of Guelph where she obtained her doctorate degree in 1987. Following her Ph.D., she joined Utah State University when she spent some time in a dual appointment with the National Center for Atmospheric Research (NCAR). In 1990, she joined the ranks of the Physics Department at the University of Quebec at Montreal until 1995 when she joined the University of Georgia. Professor Leclerc received the American Meteorological Society Award for 'Outstanding Contributions to the Field of Biometeorology'. She became D. W. Brooks Distinguished Research Professor in 2008, and then Regents Professor, the highest honor at the University of Georgia. Professor Leclerc has published hundreds of publications, advised governments and foreign ministries on matters of climate change, and given lectures worldwide across all continents. She has given invited lectures at prestigious institutions and has served on dozens of national panels assessing surface-atmosphere exchange and climate change research around the globe.

Thomas Foken was born in 1949 in Zwickau, Germany. He studied meteorology in Berlin, and received his Ph.D. in 1978 in Leipzig. His research topic was marine micrometeorology. He received his second doctor's degree in 1990 in Berlin. From 1981 to 1997 he was the Head of Laboratories of the Meteorological Service of the GDR (after 1990 the German Meteorological Service) at the Meteorological Observatories in Potsdam and Lindenberg. In 1997, Dr. Foken became Professor of Micrometeorology at the University of Bayreuth. He has taught courses on micrometeorology and geoecology in Berlin, Potsdam and Bayreuth; he has published numerous scientific papers, and also participated in national and international experiments and research projects. He received the American Meteorological Society Award for 'Outstanding Contributions to the Field of Biometeorology'.

M. Y. Leclerc and T. Foken, *Footprints in Micrometeorology and Ecology*,
DOI: 10.1007/978-3-642-54545-0,

Index

A
Aggregation
  flux, 58
  parameter, 58
Airborne measurements, 190
Air pollution, 214
Anemometer
  sonic, 2
Approach
  sub-grid scale, 60
  tile, 59
Area of interest, 207
Atmospheric boundary layer, 21, 103
  convective, 112
Averaging, 58

B
Blending height, 49
Bowen-ratio method, 174
Buoyancy flux, 29

C
Clearcut, 106, 122
Coherent structure, 38
Constant flux layer, 23
Convection
  free, 49
Convective velocity, 41
Coriolis force, 110

D
Deardorff-velocity. *See* Convective velocity
Density, 27
Diffusion coefficient, 26
Displacement height, 34

E
Eddy-covariance method, 181
  1-dimensional, 183
  data analysis, 205
  data quality, 186, 209
  generalized, 184
  used footprint models, 181, 188
Ekman layer, 22
Emission
  odor, 135
  trace gases, 135
Enhanced Vegetation Index, 166
Enhancement factor, 38
Entrainment zone, 21
European wind atlas, 163
Experiment
  footprint validation, 13
  micrometeorological, 3

F
Fetch, 45, 202
Fick's diffusion law, 50
Flux-profile similarity, 25
Footprint
  airborne measurements, 190
  analysis, 201
  analytical model, 72, 202
  application, 199
  backward model, 86, 90, 105
  climatology, 8, 206
  definition, 6
  dimension, 8
  easily applicable model, 217
  eddy-covariance method, 205, 209
  flux, 7
  forward model, 85, 88
  function, 76

M. Y. Leclerc and T. Foken, *Footprints in Micrometeorology and Ecology*,
DOI: 10.1007/978-3-642-54545-0, © Springer-Verlag Berlin Heidelberg 2014

Footprint (*cont.*)
further developments, 225
higher-order closure model, 91, 109, 127
history, 4
hybrid model, 93
in-canopy, 115
Lagrangian model, 84, 105, 115, 203
large-Eddy simulation model, 92, 112, 116
limits, 218
model, 7, 12, 71
model validation, 12, 145
profile method, 179
scalar, 7
scintillometer, 189
tall tower, 104
wind turbine, 215
Footprint approach
Cai and Leclerc (2007), 94
Haenel and Grünhage (1999), 78
Horst and Weil (1992, 1994), 75
Hsieh et al. (2009), 96
Kaharabata et al. (1997), 78
Kljun et al. (2002), 89
Kljun et al. (2004), 82
Kormann and Meixner (2001), 79
Leclerc and Thurtell (1990), 88
Luhar and Rao (1994), 96
Prabha et al. (2008), 94
Rannik-Göckede, 140
Sabelfeld (1998)–Rannik (2000), 89
Schmid (1994, 1997), 77
Schmid and Oke (1990), 74
Schuepp et al. (1990), 73
Sogachev and Leclerc (2011), 98
Steinfeld et al. (2008), 95
Forest edge, 126
Friction velocity, 26

G
Gaussian distribution functions, 50
Grid schema, 159

H
Height
aerodynamic, 34
canopy, 34
geometric, 34
Hill, 120, 132
History, 1

I
Inertial sublayer, 23
Integral turbulence characteristics, 40
Internal boundary layer, 45, 201
mechanical, 46
thermal, 45, 48, 201
Inversion, 22

K
K-approach, 25, 56
Kolmogorov's micro-scale, 52

L
Land cover, 161
Land use
classification, 168
Large Eddy Simulation, 11, 56
Latent heat flux, 26
Length scale
Eulerian, 53
integral turbulent, 53
Lagrangian, 53

M
Mass flux, 26
Measuring site
evaluation, 204
selection, 199
Mixed layer, 21
Mixing-layer, 38
Model
analytical, 8
diffusion, 50
higher order closure, 55
Lagrangian, 9, 52
second-order closure, 131
trajectory, 104
Momentum flux, 26
Monin-Obukhov similarity theory, 29, 39
Morphometric method, 35

N
Normalized difference vegetation index, 166

O
Oasis effect, 48
Obukhov length, 29
Obukhov parameter, 29

P
Parameter
  effective, 58
Power law, 39
Prandtl layer, 22
Prandtl number
  turbulent, 26
Pressure, 27
Probability density functions, 50
Profile method, 171
  accuracy, 176
  three and more levels, 172
  two levels, 174

R
Remote-sensing, 164
  technical characteristics, 166
Residual layer, 22
Reynolds averaging, 26
Richardson number
  bulk, 33
  critical, 33
  flux, 33
  gradient, 33
Roughness length, 28, 34, 163
Roughness sublayer, 37, 114

S
Scale, 24
Schmidt number
  turbulent, 27
Scintillometer, 188
Sensible heat flux, 26
Shape parameter, 12, 51
Source area
  history, 6
Surface
  characteristics, 159, 162
  complex, 2
  thermal heterogeneous, 121
Surface layer, 22

T
Temperature
  potential, 28
  virtual, 27
  virtual potential, 29
Theory
  localized near-field, 10, 114
Time scale
  Lagrangian, 53
Topography
  complex, 131
Tower
  tall, 103
Turbulence
  isotropic, 52
  spectra, 53

U
Universal function, 30
Upscaling, 213
Urban area, 137

V
Validation, 145
  model against model, 147
  model against natural tracer, 153
  model against tracer, 151
  quality, 153
Viscous sublayer, 23
von-Kármán constant, 27

W
Water vapour flux, 26
Wind energy, 215
Wind profile
  in the canopy, 35
  logarithmical, 34

Z
Zero-plane displacement, 34

MIX
Papier aus verantwortungsvollen Quellen
Paper from responsible sources
FSC® C105338

If you have any concerns about our products,
you can contact us on
ProductSafety@springernature.com

In case Publisher is established outside the EU,
the EU authorized representative is:
Springer Nature Customer Service Center GmbH
Europaplatz 3, 69115 Heidelberg, Germany

Printed by Libri Plureos GmbH
in Hamburg, Germany